SUBHARMONIC FUNCTIONS

VOLUME 1

L.M.S. MONOGRAPHS

Editors: P. M. Cohn *and* G. E. H. Reuter

Published for the London Mathematical Society

by Academic Press Inc. (London) Ltd.

SUBHARMONIC FUNCTIONS

VOLUME I

W. K. HAYMAN, F.R.S.
Imperial College of Science and Technology
University of London

and

the late

P. B. KENNEDY
University of York

1976

ACADEMIC PRESS
LONDON NEW YORK SAN FRANCISCO

A Subsidiary of Harcourt Brace Jovanovich, Publishers

ACADEMIC PRESS INC. (LONDON) LTD.
24/28 Oval Road
London NW1

United States Edition published by
ACADEMIC PRESS INC.
111 Fifth Avenue
New York, New York 10003

Library of Congress Catalog Card Number: 76-1084
ISBN: 0-12-334801-3

PRINTED IN GREAT BRITAIN BY
PAGE BROS (NORWICH) LTD, NORWICH

0. Preface

Let $f(z)$ be regular in a domain D in the open z plane. Then

$$u(z) = \log |f(z)| \tag{0.1}$$

is subharmonic (s.h.) in D. In other words $u(z)$ is upper semi-continuous (u.s.c.) in D, $-\infty \leqslant u(z) < +\infty$ and if $|z - z_0| \leqslant r$ lies in D, then $u(z_0)$ is not greater than the average of $u(z)$ on the circumference $|z - z_0| = r$.

In this book we study the problem of extending the properties of the particular s.h. functions of the type (0.1) to general s.h. functions in domains of m-dimensional Euclidean space R^m, where $m \geqslant 2$. The general study of s.h. and related classes of functions is frequently called potential theory. It is now a vast subject, to which it would be hard to do justice in a single work. For a good history including over 200 references the reader is referred to the valuable article by Brelot [1972]. Among many books the following may be mentioned: Kellogg [1929], Rado [1937], Tsuji [1959], Brelot [1960], Landkof [1972], Bauer [1966], Carleson [1967], Martensen [1968], Helms [1969], Du Plessis [1970], Fuglede [1972]. Of these Tsuji's book is closest to the present volume. However, Tsuji is solely concerned with functions in the plane and his is more a book on function theory than is mine. The same is true of Carleson's book. Most of the other works quoted above are barely related to function theory at all but concentrate on other aspects of potential theory. I gladly acknowledge my indebtedness in particular to the books by Brelot and Carleson, particularly for Chapter 5. The present volume thus steers an intermediate course between books on various aspects of abstract potential theory and books on functions theory using some potential theoretic techniques.

An outline of the present book is as follows. In Chapter 1 we develop some introductory material on set theory, various classes of functions such as u.s.c., convex, real analytic functions and the class C^n. We continue by proving a suitable form of Green's theorem which allows us to develop the basic properties of harmonic functions in R^m, including Poisson's formula which with its generalisation to s.h. functions plays a fundamental role in the theory. In the equation (0.1) harmonic functions $u(z)$ arise from functions $f(z)$ without zeros. They will always form an important subclass in our general study. We introduce the problem of Dirichlet, which asks for a harmonic function in a domain with preassigned boundary values, and prove the uniqueness of the solution if it exists. The existence of a solution is proved only in the case of the hyperball. The more general cases are deferred to Chapters 2 and 5.

In Chapter 2 we introduce s.h. functions, give various examples of such functions and develop their elementary properties, such as the maximum principle. We also show that $u(x)$ is harmonic if and only if u and $-u$ are both subharmonic. We use Perron's [1923] method to solve the problem of Dirichlet for continuous boundary values and regular domains, and in particular domains having the property of Poincaré–Zaremba† that every boundary point is the vertex of a cone lying entirely outside the domain. Having done this we can develop convexity theorems for averages or means of s.h. functions on hyperspheres. We also discuss the question of harmonic extensions of semi-continuous functions on the boundary of domains to harmonic functions in the interior. This subject will be further developed in Chapter 5. We conclude the chapter by introducing Littlewood's concept of subordination for regular functions $f(z)$ in the unit disk. Here the techniques of the chapter are used to prove Littlewood's theorem that various averages of $|f(z)|$ are decreased by subordination. Various consequences of this result are also proved.

Suppose that $f(z)$ is regular in a domain D. Let e be a subset of D and let $\mu(e)$ be the number of zeros of f on e, counting multiplicity. Thus μ is a positive measure, finite on compact subsets E of D. Given such a set E, we write

$$p(z) = \int_E \log|z - \zeta|\, d\mu(e_\zeta). \tag{0.2}$$

Evidently $p(z) = \log|P(z)|$, where

$$P(z) = \prod_{n=1}^{N} (z - \zeta_n),$$

and the ζ_n are the zeros of $f(z)$ on E. We define $u(z)$ by (0.1) and deduce that

$$u(z) = p(z) + h(z), \tag{0.3}$$

where

$$h(z) = \log|f(z)/P(z)|$$

is harmonic in the interior of E. Thus (0.3) represents $u(z)$ locally as the sum of a potential plus a harmonic function.

† Poincaré [1890], Zaremba [1909]. The latter paper contains the cone condition for the first time.

In Chapter 3 we develop the fundamental result of F. Riesz [1926, 1930], which states that a corresponding result is valid for general s.h. functions. The only difference is that the measure is now an arbitrary positive measure in D, finite on compact subsets of D and vanishing on open sets where $u(z)$ is harmonic. For functions $u(x)$ s.h. in a domain D of R^m, where $m > 2$, (0.2) is replaced by

$$p(x) = \int_E -|x - \xi|^{2-m} \, \mathrm{d}\mu(e_\xi). \tag{0.2$'$}$$

The measure μ is uniquely determined by u. Conversely the potentials (0.2), respectively (0.2)′, corresponding to any such measure μ, are always s.h. in the whole of space and harmonic outside E. Thus the study of s.h. functions is equivalent to the study of potentials $p(x)$, which explains the name potential theory. The case (0.1) corresponds to plane measures μ assuming only integer values. On the other hand if $u \in C^2(D)$, i.e. if u has continuous second partial derivatives in D, then

$$\mathrm{d}\mu = \frac{1}{e_m} \nabla^2(u) \, \mathrm{d}x, \tag{0.4}$$

where $e_2 = 2\pi$, $e_m = 4\pi^{\frac{1}{2}m}/\Gamma(\frac{1}{2}m - 1)$, $m \geqslant 3$, $\mathrm{d}x = \mathrm{d}x_1 \, \mathrm{d}x_2 \ldots \mathrm{d}x_m$ denotes volume in R^m and

$$\nabla^2 u = \sum_{\nu=1}^{m} \frac{\partial^2 u}{\partial x_\nu^2}$$

is the Laplacian. For general functions the equation (0.4) remains valid in the sense of distributions.

There are many proofs of Riesz's theorem but I know of none that is really simple. A fair amount of measure theory is required at this stage and we develop the relation between positive measures and positive linear functionals which is another pregnant contribution of F. Riesz [1909]. This idea is also useful in other connections. The measure μ then occurs as a positive linear functional L_u on the class $C_0^\infty(D)$ of functions having continuous partial derivatives of all orders and vanishing outside a compact subset of D. For such functions v we can define $L_u(v)$ by (0.4), i.e.

$$L_u(v) = \frac{1}{e_m} \int v \nabla^2 u \, \mathrm{d}x = \frac{1}{e_m} \int u \nabla^2 v \, \mathrm{d}x.$$

Then $L_u(v)$ extends uniquely to the class $C_0(D)$ of functions v continuous in D and vanishing outside a compact subset of D and μ is the measure associated with this functional.

The first half of Chapter 3 is devoted to the development of the above ideas culminating in Riesz's theorem. In the remainder of the chapter various consequences are deduced. Let e be a suitable set on the frontier of a bounded regular domain D. The harmonic measure $\omega(x, e)$ is the harmonic extension of the characteristic function of e into D. If $u(x)$ is s.h. in the closure of D we obtain the Poisson–Jensen formula

$$u(x) = \int_F u(\xi)\, \mathrm{d}\omega(x, e_\xi) - \int_D g(x, \xi, D)\, \mathrm{d}\mu e_\xi. \tag{0.5}$$

Here F is the frontier of D and $g(x, \xi, D)$ is the Green's function of D. We have to assume that F has zero m-dimensional measure and that D is regular. These assumptions are removed in Chapter 5.

The formula (0.5) was developed by R. Nevanlinna [see e.g. 1929] for the case (0.1) and was used by him as a starting point for the theory of meromorphic functions which bears his name. We use it to extend Nevanlinna's first fundamental theory to s.h. functions in a hyperball in R^m. As a final application we obtain a representation for functions which are s.h. and bounded above in R^m, where $m \geqslant 3$. If $m = 2$ such functions are necessarily constant. It turns out that such functions are given by the analogue of (0.5) with F the point at ∞, $u(\xi) \equiv C$, on F, where C is the least upper bound of $u(x)$, and $g(x, \xi) = -|x - \xi|^{2-m}$. We deduce that

$$u(x) \to C \quad \text{as} \quad x \to \infty$$

outside a rather small exceptional set and in particular as $x \to \infty$ along almost all fixed lines.

This leads us naturally to the question of the general behaviour of functions s.h. in the whole of R^m, a question which is considered in Chapter 4. The order of such functions is defined in terms of the Nevanlinna characteristic and the representations in terms of the Riesz mass μ for general functions and functions of finite order are proved which reduce to the theorems of Weierstrass [1876] and Hadamard [1893] respectively in the case (0.1), when $f(z)$ is entire.

Next we obtain some inequalities relating

$$B(r) = \sup_{|x|=r} u(x),$$

$$N(r) = d_m \int_0^r \frac{n(t)\, \mathrm{d}t}{t^{m-2}},$$

where $d_m = \max(1, m-2)$, $n(t)$ is the Riesz mass in $|x| \leqslant t$, and the Nevanlinna characteristic

$$T(r) = \frac{1}{c_m r^{m-1}} \int u^+(x)\, \mathrm{d}\sigma(x).$$

Here $u^+(x) = \max(u(x), 0)$, c_m is the surface area of the unit sphere $|x| = 1$ in R^m and $d\sigma(x)$ denotes surface area on $|x| = r$. We also define

$$m(r, u) = \frac{1}{c_m r^{m-1}} \int u^-(x)\, d\sigma(x),$$

where $u^-(x) = \sup(-u(x), 0)$. Then the first fundamental theorem referred to above states that

$$T(r) = m(r) + N(r) + u(0).$$

Thus we can define the deficiency of u as

$$\delta(u) = \varliminf_{r\to\infty} \frac{m(r)}{T(r)} = 1 - \varlimsup_{r\to\infty} \frac{N(r)}{T(r)}.$$

For functions of order $\lambda < 1$ in R^m we obtain the sharp lower bound for $\delta(u)$ in terms of λ. In strict analogy to the case of entire functions, the extremals are functions having all their Riesz mass on a ray through the origin. A later result of Dahlberg [1972] who obtained the exact lower bound for

$$\varliminf_{r\to\infty} \frac{B(r)}{T(r)}$$

as a function of λ is only referred to and a somewhat weaker but much simpler inequality is proved.

In the second half of the chapter we develop various analogues of theorems related to the asymptotic values of entire functions. Suppose that $f(z)$ is such a function. We say that $f(z)$ has the asymptotic value a along the path Γ if

$$f(z) \to a \quad \text{as} \quad z \to \infty \text{ along } \Gamma.$$

If $f(z)$ has N distinct finite asymptotic values a_ν then it was noted by Heins [1959] that for large K the set $u(z) > K$ has at least N components in the plane, where $u(z)$ is given by (0.1). Using only this fact Heins showed that if $u(z)$ is s.h. in the plane and $u(z) > K$ has at least $N \geqslant 2$ components then $u(z)$ has order at least $\frac{1}{2}N$. This result generalizes a famous theorem of Ahlfors [1930].

We defer a full proof of the theorem of Heins to Chapter 7 in Volume 2. Here we prove quite simply a less precise but more general result. This shows that if $u(x)$ is s.h. in R^m and the set $u(x) > 0$ has $N \geqslant 2$ components then the order of $u(x)$ is at least

$$A(m)\, N^{1/(m-1)},$$

where the positive constant $A(m)$ depends only on m. Examples show that this lower bound has the right order of magnitude as $N \to \infty$ for fixed m.

Finally we prove some generalisations of a classical theorem of Iversen [1915–16]. His original result states that if $f(z)$ is a non-constant entire function, then $f(z)$ has ∞ as an asymptotic value, i.e.

$$f(z) \to \infty \text{ as } \quad z \to \infty \text{ along a path } \Gamma.$$

If we define $u(z)$ by (0.1) we deduce that

$$u(z) \to +\infty \text{ as } z \to \infty \text{ along } \Gamma. \tag{0.6}$$

We prove that in this form (0.6) extends to general s.h. functions $u(x)$ in R^m, except for the trivial case when $u(x)$ is bounded above. This case was considered at the end of Chapter 3. In this complete form (0.6) was first proved by Fuglede [1975]. For $m = 2$ the result is due to Talpur [1975] who also proved [1976] a somewhat weaker result for $m > 2$, replacing the path Γ by a chain of continua. We also ask how quickly convergence to ∞ must take place in (0.6). It turns out that if the order λ of $u(x)$ is finite, there always exists an asymptotic path Γ, on which $u(x)$ has order λ. If λ is infinite there exist paths on which $u(x)$ has arbitrarily large finite order, but not necessarily a path on which $u(x)$ has infinite order.

In the final Chapter 5 we develop some deeper results of classical potential theory. Let E be a compact set in R^m. Let

$$K_0(x) = \log |x|$$
$$K_\alpha(x) = -|x|^{-\alpha}, \alpha > 0.$$

Finally let μ be a unit mass distribution on E and set

$$V(x) = \int_E K_\alpha(x - \xi)\, d\mu e_\xi.$$

We choose μ in such a way that

$$I_\alpha(\mu) = \int_E V(x) \cdot d\mu(x) = \iint_{E \times E} K_\alpha(x - y)\, d\mu(x)\, d\mu(y)$$

has its largest value V_α. If $V_\alpha = -\infty$, we define the α-capacity $C_\alpha(E)$ to be zero and say that E is α-polar. Otherwise we define

$$C_\alpha(E) = e^{V_0}, \qquad \alpha = 0$$
$$C_\alpha(E) = (-V_\alpha)^{1/\alpha}, \qquad \alpha > 0.$$

We shall be mainly concerned with the case $\alpha = m - 2$; the corresponding α-polar sets will simply be called polar [Brelot, 1941].

According to recent results of Choquet [1955 and 1958] the notion of capacity thus defined and in particular the notion of polar sets can be extended to arbitrary Borel-measurable sets in R^m. Choquet's beautiful theory will be developed in the last part of the chapter.

It turns out that polar sets play the role of negligible or null-sets in many problems of potentials theory. The main aim of this chapter is to prove such results and their consequences. After obtaining some introductory material, we prove in Section 5.3 that if E is a compact set or the countable union of compact sets, then there exists a s.h. function which is $-\infty$ on E without being identically $-\infty$, if and only if E is polar. This result is extended to general Borel sets E at the end of the chapter. Next we investigate the metric size of α-polar sets in Section 5.4 in terms of Hausdorff measure. If $h_\alpha(E)$ denotes α-dimensional Hausdorff measure, then if $C_\alpha(E) = 0$, $h_\beta(E) = 0$ for any $\beta > \alpha$, while if $h_\alpha(E) < \infty$, we deduce $C_\alpha(E) = 0$. Thus for instance polar sets in R^2 necessarily have zero length, and hence are totally disconnected, but if $m \geqslant 3$ sets of finite length are polar in R^m.

In the next two sections we turn to some further properties of polar sets. We first show that they can be neglected, from the point of view of the maximum principle, for s.h. functions which are bounded above. In other words if $u(x)$ is s.h. and bounded above in a domain D and

$$\overline{\lim_{x \to \zeta}}\, u(x) \leqslant 0$$

for a set e of boundary points ζ of D which is not empty and whose complement in the boundary of D is polar, then $u(x) < 0$ or $u(x) \equiv 0$ in D. We deduce that if $u(x)$ is s.h. and bounded above in a domain D outside a polar set E, then $u(x)$ can be uniquely extended to the whole of D as a s.h. function. If $u(x)$ is bounded and harmonic in D outside E, then u becomes harmonic in the whole of D [Brelot, 1934]. If E has positive capacity the above results fail.

Next we prove the result of Evans [1933] that the set of irregular boundary points of an arbitrary bounded domain is polar. Thus this set can be ignored in the solution of the problem of Dirichlet. Following Wiener [1924] we deduce that the problem of Dirichlet possesses a unique solution for an arbitrary bounded domain and arbitrary continuous boundary values f if we demand that the solution be bounded and assumes the required boundary values at the regular boundary points of D. Since the solution at a fixed point x in D is clearly a positive linear functional on f we obtain Brelot's [1939a] extended solution for a function f which is integrable with respect to harmonic measure. This leads also to a generalisation of the Poisson–Jensen formula

(0.5) to bounded domains which need no longer be regular. We also obtain a generalized Green's function for an arbitrary domain D in R^m provided only that $m > 2$ or $m = 2$ and the complement of D is not polar. The symmetry property of the Green's function is also deduced.

The last section of Chapter 5 is devoted to Choquet's theory of capacitability. The definition of capacity extends in an obvious way from compact sets to open sets. If E is an arbitrary bounded set the inner and outer capacities of E are defined respectively as the upper bound of capacities of compact sets contained in E and the lower bound of compacities of open sets containing E. So far the only essential properties of our set function $C(E)$ defined on compact sets are that (i) $C(E) \geqslant 0$, (ii) $C(E)$ increases with E $(C(E_1) \leqslant C(E_2)$ if $E_1 \subset E_2)$ and (iii) that $C(E)$ is u.s.c., i.e. given $\varepsilon > 0$ and E we can find E' containing E in its interior such that $C(E') < C(E) + \varepsilon$.

Following Choquet we prove that if in addition

$$\text{(iv) } C(E_n) \to C(E)$$

when E_n is an expanding sequence of sets such that $E_n \to E$, then all Borel sets are capacitable, i.e. their inner and outer capacities are the same. To prove this we follow the exposition of Carleson [1967]. We also show that (iv) holds for $C_{m-2}(E)$ by proving for $m > 2$ the property of strong subadditivity

$$C(E_1 \cup E_2) + C(E_1 \cap E_2) \leqslant C(E_1) + C(E_2)$$

with $C(E) = C_{m-2}(E)^{m-2}$. For $m = 2$ a modified proof of (iv) has to be given.

In a second volume we hope to prove some further results concerning s.h. functions largely in the plane. There will be a chapter on the minimum and maximum of subharmonic functions in the plane, one on asymptotic values and related matters and one on examples arising from techniques for approximating s.h. functions in the plane by functions given by (0·1) where $f(z)$ is entire. It is also hoped to cover some other recent developments and applications such as the theory of Baernstein [1975].

Acknowledgements

I have already acknowledged gratefully my debt to previous works on the subject. In addition I owe a particular debt to Professor Brelot for much helpful guidance concerning historical matters. It is often easier to find one's own proof of a classical theorem than to track down the original proof. I apologize in advance for any errors of attribution. Without Professor Brelot's help there would have been many more.

Professor Drasin read several of the chapters and I am grateful for his valuable comments. Mr G. Cámera read the whole book very carefully and corrected many inaccuracies. He, Mr P. Rippon and Mr G. Higginson read all the proofs and prepared the index. Prof. G. E. H. Reuter, the editor of the series, also read the proofs and gave me much helpful advice. Miss P. Edge, Mrs J. Grubb and Miss A. Mills combined to create order out of some very ugly pencilled scrawls and showed remarkable intuition in distinguishing between virtually indistinguishable symbols. It is a pleasure to thank them now.

I must end on a rather sadder note. This project was started by my friend and former student Professor P. B. Kennedy. His tragic death in 1967 made it impossible for him to finish the work. I hope it may serve as a memorial to a deeply conscientious and extremely charming person whom I still miss greatly.

W. K. HAYMAN

September 1975

Contents

Chapter 3. Representation Theorems

Chapter 4. Functions Subharmonic in Space

Chapter 5. Capacity and Null Sets

CHAPTER 1

Preliminary Results

1.0. INTRODUCTION

In this chapter we develop the theory of the calculus in Euclidian space R^m of m dimensions together with the theory of harmonic functions as far as will be necessary for our later applications. In the first 3 sections we discuss sets in R^m and various classes of real functions such as convex and semicontinuous functions. In the next section we give a sketch of integration theory ending with Green's Theorem and in Section 1.5 we apply the previous analysis to harmonic functions in R^m.

1.1. BASIC RESULTS FROM SET THEORY

We shall be concerned with functions $f(x)$ where x denotes a point in the Euclidian space R^m of m dimensions ($m \geqslant 1$). The applications to classical function theory occur in the plane case, i.e. $m = 2$, but there are a number of significant extensions and questions of interest in the case $m > 2$. Let x be the point with coordinates $(x_1, \ldots, x_m)$ in m dimensions. We write

$$|x| = \sqrt{(x_1^2 + x_2^2 + \ldots + x_m^2)}$$

for the distance of x from the origin 0, i.e. $(0, 0, \ldots, 0)$. If $x(x_1, x_2, \ldots, x_m)$ and $y(y_1, \ldots, y_m)$ are two points, then $x \mp y$ denotes the points whose jth coordinates are $x_j \mp y_j$ respectively and $x \cdot y = \sum_1^m x_\mu y_\mu$. If λ is real we write λx for the point with coordinates λx_ν.

If $m = 2$ we frequently write z instead of x and we identify a point with coordinates x, y with the complex number $z = x + iy$. In general it will be clear from the context, which notation is being used.

If x_0 is a point in space and r a positive number we write

$$D(x_0, r) = \{x \mid |x - x_0| < r\}$$

$$C(x_0, r) = \{x \mid |x - x_0| \leqslant r\}$$

and

$$S(x_0, r) = \{x \mid |x - x_0| = r\}.$$

A set $D(x_0, r)$ will be called a *ball* (*disk* if $m = 2$). A set $C(x_0, r)$ will be called a *closed ball* (*closed disk*, if $m = 2$). A set $S(r_0, r)$ will be called a *sphere*, (*circle* if $m = 2$). An arbitrary set of points E containing a ball $D(x_0, r)$ will be called a *neighbourhood* of x_0. An *open* set is one which is a neighbourhood of all its points. A set E is *closed* if its complement, $\mathscr{C}E$, consisting of all points of R^m not in E, is open. A set which is contained in some ball is called *bounded*. A set which is both bounded and closed is called *compact*.

We write $\varnothing$ for the empty set containing no point and $\bigcup E_\alpha$ and $\bigcap E_\alpha$ for the *union* and *intersection* of a collection of sets E_α.

A point x is called a *limit point* of a set E if every neighbourhood of x contains more than one point of E. It is evident that in this case every such neighbourhood contains infinitely many points of E. Clearly a set is closed if and only if it contains all its limit points. If E is a set we denote by $\bar{E}$ the closure of E, i.e. the set consisting of all points and limit points of E. The closure of a set is closed, i.e. $\bar{\bar{E}} = \bar{E}$. A point which belongs to both $\bar{E}$ and $\overline{\mathscr{C}E}$ is called a *frontier point* of E. The *frontier* or *boundary* of E is the union of all frontier points of E.

If E is a set, a pair of sets E_1, E_2 is called a *partition* of E, if

$$E_1 \cup E_2 = E, \quad \bar{E}_1 \cap E_2 = E_1 \cap \bar{E}_2 = \varnothing \quad \text{and} \quad E_1 \neq \varnothing, \ E_2 \neq \varnothing.$$

In other words E_1, E_2 are disjoint subsets of E, neither of which contains a limit point of the other, whose union is the whole of E, and neither of which is empty. If E does not permit a partition then E is said to be *connected*. An open connected set is called a *domain*. A compact connected set containing at least two points is called a *continuum*.

It is frequently important to consider sets which are subsets of a fixed set or space S. For such sets all definitions are made as before except that we now confine ourselves to points of S. Thus a *neighbourhood in S* of a point x_0 in S is any subset of S which contains all the points of S which lie in some ball $D(x_0, r)$. A subset of S which is a neighbourhood in S of all its points is called *open in S* or *relatively open*, a subset E of S is called *closed in S* or *relatively closed* if every limit point of E which lies in S also lies in E. The following result plays a fundamental role in set theory and is frequently used as a definition of compact sets.

THEOREM 1.1. *If E is a set in R^m the following three properties are equivalent*

(i) *E is compact*, i.e. *bounded and closed.*

(ii) *(The Weierstrass property). If x_n is a sequence of points in E, there exists a subsequence x_{n_p} converging to a limit x in E as $p \to \infty$.*

(iii) *(The Heine–Borel property). If E is contained in a union of open sets G_α, then there exists a finite subsystem $G_{\alpha_1}, G_{\alpha_2}, \ldots, G_{\alpha_n}$ of the sets G_α, whose union covers E.*

(i) ⇒ (iii). Suppose that (i) holds. Since E is bounded E will lie in some hypercube H_0, i.e. a set of the form $a_j \leqslant x_j \leqslant b_j$, $j = 1$ to m, where x^j is the jth coordinate of the point x. Also H_0 is the union of 2^m hypercubes of the form

$$a_{j,1} \leqslant x_j \leqslant b_{j,1},$$

where either $a_{j,1} = a_j, b_{j,1} = \frac{1}{2}(a_j + b_j)$ or $a_{j,1} = \frac{1}{2}(a_j + b_j), b_{j,1} = b_j$.

Suppose now that G_α is a system of open sets covering E and that no finite subsystem of the G_α covers E. Then the same must be true for the points of E in at least one of these hypercubes H_1 say, and the sides of H_1 have half the lengths of the corresponding sides of H_0. Continuing in this way we obtain a sequence H_N of hypercubes such that

$$H_{N+1} \subset H_N$$

and each H_N has sides whose lengths are 2^{-N} times the lengths of the corresponding sides of H_0, and the part of E in H_N cannot be covered by finitely many G_α. If H_N is given by

$$a_{j,N} \leqslant x_j \leqslant b_{j,N}$$

then $a_{j,N}$ is a bounded increasing sequence of N and so tends to a limit a_j as $N \to \infty$. Also

$$b_{j,N} - a_{j,N} \to 0, \quad \text{so that} \quad b_{j,N} \to a_j$$

as $N \to \infty$. Let a be the point $(a_1, a_2, \ldots, a_m)$. Then if a is not a point of E then a is not a limit point of E, since E is closed, and so there is a neighbourhood M of a which contains no point of E. Since H_N lies in M for all large N this contradicts our assumption that the part of H_N in E cannot be covered by a finite number of the G_α. Again if a is in E then a lies in G_α for some α and thus G_α contains H_N for all large N, since G_α is open. This again gives a contradiction. Thus (iii) must hold.

(iii) ⇒ (ii). Suppose that (ii) is false. Then there exists a sequence $x_n \in E$ having no limit point in E and so if x is any point of E, x is not a limit point of the sequence x_n. Thus there exists a ball $D(x, r)$ containing x_n for only finitely many n. The union of these balls $D(x, r)$ forms an open covering of E. But any finite subcovering contains only finitely many of the points x_n and so cannot cover E. Thus (iii) must be false. Hence if (iii) is true so is (ii).

(ii) $\Rightarrow$ (i). Suppose that E is unbounded. Then there exists a sequence x_n of points in E, such that $|x_n| > n$. Clearly x_n has no limit point, so that (ii) is false. Suppose next that E is not closed. Then if x is a limit point of E, which is not in E, we can find $x_n \in E$, such that $|x - x_n| < 1/n$. Then the sequence x_n converges to x and so has no limit point other than x, so that x_n has no limit point in E. Thus again (ii) is false. Thus if (ii) is true so is (i).

1.2. VARIOUS CLASSES OF FUNCTIONS

1.2.1. Semicontinuous functions

We shall be concerned mainly with real and sometimes with complex-valued functions $f(x)$ on sets in R^m. A real function $f(x)$ defined in a set E in R^m is said to be upper semi-continuous (u.s.c) on E if

(i) $-\infty \leqslant f(x) < \infty, \qquad x \in E.$

(ii) The sets $\{x | x \in E, f(x) < a\}$ are open in E for $-\infty < a < +\infty$. A function $f(x)$ is said to be lower semi-continuous (l.s.c) in E if $-f(x)$ is u.s.c. in E. If $f(x)$ is both u.s.c. and l.s.c. then $f(x)$ is said to be continuous.

We need

THEOREM 1.2. *If $f(x)$ is* u.s.c. *on a (non-empty) compact set E, then $f(x)$ attains its maximum on E, i.e. $\exists\, x_0 \in E$, such that*

$$f(x_0) \geqslant f(x), \qquad x \text{ on } E.$$

Suppose that

$$M = \sup_{x \in E} f(x).$$

If $f(x)$ is not bounded above on E, we set $M = +\infty$. If $f(x) \equiv -\infty$, then we can take for x_0 any point on E and there is nothing to prove. Otherwise there exists a sequence x_n on E, such that

$$f(x_n) \to M.$$

For if M is finite we can take for x_n any point such that

$$f(x_n) > M - \frac{1}{n}.$$

If $M = +\infty$, we take x_n such that

$$f(x_n) > n.$$

By taking a subsequence if necessary we may assume, in view of Theorem 1.1, that

$$x_n \to \xi \text{ in } E, \text{ as } n \to \infty.$$

Suppose now that

$$f(\xi) = \lambda < M.$$

Choose μ such that $\lambda < \mu < M$. Then the set

$$G(\mu) = \{x \mid x \in E \quad \text{and} \quad f(x) < \mu\}$$

is open in E since f is u.s.c. in E and also $G(\mu)$ contains ξ. Thus $G(\mu)$ contains all points of E in some ball $D(\xi, r)$ and hence $G(\mu)$ contains x_n for all sufficiently large n. Thus

$$f(x_n) \leqslant \mu$$

for all sufficiently large n and this gives a contradiction. Thus $f(\xi) \geqslant M$, so that M is finite. It follows from the definition of M that $f(\xi) \leqslant M$ and so $f(\xi) = M$.

THEOREM 1.3. *If $f_n(x)$ is a decreasing sequence of* u.s.c. *functions defined on a set E, then $f(x) = \lim_{n\to\infty} f_n(x)$ is* u.s.c. *on E.*

Since $f_n(x)$ is decreasing for each fixed x on E and $f_1(x) < +\infty$, the limit $f(x)$ exists for each x on E and $f(x) \leqslant f_1(x) < +\infty$. Let $E(a)$ be the set of all points on E for which $f(x) < a$. We have to show that $E(a)$ is open. Let ξ be a point of $E(a)$. Since $f_n(\xi) \to f(\xi)$ and $f(\xi) < a$, we can find n such that

$$f_n(\xi) < a.$$

Since $f_n(\xi)$ is u.s.c., the set $E_n(a)$ of all points x in E where $f_n(x) < a$ is open in E and so is a neighbourhood N of ξ in E. Clearly $f(x) \leqslant f_n(x) < a$ in N, so that $E(a)$ contains the neighbourhood N of ξ in E. Since ξ is any point of $E(a)$, $E(a)$ is open. In the opposite direction we can prove

THEOREM 1.4. *If $f(x)$ is* u.s.c. *on a set E then there exists a decreasing sequence $f_n(x)$ of functions continuous on E, such that*

$$f_n(x) \to f(x) \text{ as } n \to \infty.$$

Suppose first that $f(x) > 0$ on E. Since f is u.s.c. on E, given any point ξ in E, there exists $\delta = \delta(\xi)$, such that

$$\sup f(x) < +\infty, \quad \text{for} \quad x \in D(\xi, \delta) \cap E.$$

If this inequality is valid for some value of ξ and all positive δ then it is clearly true for all ξ and all positive δ and in this case we define $\delta(\xi) = 1$ for all ξ. Otherwise we define $\delta(\xi)$ to be the least upper bound of all such δ.

We note that $\delta(\xi)$ is continuous on E and in fact if ξ_1, ξ_2 are in E we have

$$|\delta(\xi_2) - \delta(\xi_1)| \leqslant |\xi_2 - \xi_1|. \tag{1.2.1}$$

For suppose that $\varepsilon > 0$. Then we have, for some finite M,

$$f(x) < M \quad \text{for} \quad |x - \xi_1| < \delta(\xi_1) - \varepsilon.$$

Thus this inequality holds in particular for

$$|x - \xi_2| < \delta(\xi_1) - \varepsilon - |\xi_2 - \xi_1|$$

and hence

$$\delta(\xi_2) \geqslant \delta(\xi_1) - \varepsilon - |\xi_2 - \xi_1|.$$

Since ε is arbitrary, we deduce that

$$\delta(\xi_2) \geqslant \delta(\xi_1) - |\xi_2 - \xi_1|.$$

Similarly

$$\delta(\xi_1) \geqslant \delta(\xi_2) - |\xi_2 - \xi_1|$$

and (1.2.1) follows.

We next set for $h < \delta(\xi)$

$$M(\xi, h) = \sup_{x \in E,\, |x - \xi| \leqslant h} f(x),$$

and

$$f_n(x) = \frac{2n}{\delta(x)} \int_0^{[\delta(x)]/2n} M(x, t)\, dt. \tag{1.2.2}$$

Then we prove that the sequence $f_n(x)$ has the required properties. In the first instance it is clear that $M(\xi, h)$ is a finite non-decreasing function of h for $0 < h < \delta(\xi)$, and since $f(x)$ is u.s.c.

$$M(\xi, h) \to f(\xi) \quad \text{as} \quad h \to 0.$$

Thus the integral mean

$$I(\xi, h) = \frac{1}{h} \int_0^h M(\xi, t)\, dt$$

exists for $h < \delta$ and also increases with h. In fact if $0 < h_1 < h_2 < \delta$

$$\begin{aligned} h_1 h_2[I(\xi, h_2) - I(\xi, h_1)] &= h_1 \int_0^{h_2} M(\xi, t)\, dt - h_2 \int_0^{h_1} M(\xi, t)\, dt \\ &= (h_1 - h_2) \int_0^{h_1} M(\xi, t)\, dt + h_1 \int_{h_1}^{h_2} M(\xi, t)\, dt \\ &\geqslant h_1(h_2 - h_1)[M(\xi, h_1) - M(\xi, h_1)] = 0. \end{aligned}$$

Thus $I(\xi, h)$ increases with h and clearly $I(\xi, h) \to f(\xi)$ as $h \to 0$. Hence the sequence $f_n(x)$ decreases with increasing n and $f_n(x) \to f(x)$ as $n \to \infty$. It remains to show that $f_n(x)$ is continuous for a fixed n. In view of (1.2.1) $\delta(\xi)$ is continuous, so that it is enough to show that

$$\int_0^{\delta(\xi)/2n} M(\xi, t)\,dt$$

is a continuous function of ξ for a fixed n.

Suppose that $0 < f(x) < M$ for $|x - \xi_1| < \frac{5}{6}\delta(\xi_1)$, and that $|\xi_2 - \xi_1| \leqslant \eta < \frac{1}{4}\delta(\xi_1)$, $0 < h_1 < \frac{1}{3}\delta(\xi_1)$, $0 < h_2 < \frac{1}{3}\delta(\xi_1)$ and $|h_2 - h_1| < \eta$. The ball $D(\xi_2, t)$ is contained in the ball $D(\xi_1, t + \eta)$ and so

$$M(\xi_2, t) \leqslant M(\xi_1, t + \eta).$$

Thus

$$\begin{aligned}\int_0^{h_2} M(\xi_2, t)\,dt &\leqslant \int_\eta^{h_2+\eta} M(\xi_1, t)\,dt \\ &\leqslant \int_0^{h_1} M(\xi_1, t)\,dt + \int_{h_1}^{h_2+\eta} M(\xi_1, t)\,dt \\ &\leqslant \int_0^{h_1} M(\xi_1, t)\,dt + 2M\eta.\end{aligned}$$

Similarly

$$\int_0^{h_1} M(\xi_1, t)\,dt \leqslant \int_0^{h_2} M(\xi_2, t)\,dt + 2M\eta,$$

so that

$$\left|\int_0^{h_1} M(\xi_1, t)\,dt - \int_0^{h_2} M(\xi_2, t)\,dt\right| \leqslant 2M\eta.$$

Choosing

$$h_1 = \frac{\delta(\xi_1)}{4n}, \qquad h_2 = \frac{\delta(\xi_2)}{4n},$$

we deduce from (1.2.1) that

$$|h_2 - h_1| \leqslant \frac{|\xi_2 - \xi_1|}{4n} \leqslant \frac{\eta}{4},$$

so that if

$$|\xi_2 - \xi_1| < \eta < \tfrac{1}{2}\delta(\xi_1)$$

we conclude that

$$\left| \int_0^{[\delta(\xi_1)]/2n} M(\xi_1, t)\,dt - \int_0^{[\delta(\xi_2)]/2n} M(\xi_2, t)\,dt \right| < 2M\eta.$$

Thus

$$\int_0^{[\delta(\xi)]/2n} M(\xi, t)\,dt$$

is a continuous function of ξ, and hence so are the functions $f_n(\xi)$. This proves Theorem 1.4, when $f(\xi) > 0$ on E.

If $f(x) > -K$ on E we can obtain the required conclusion as before by considering $f(x) + K$ instead of $f(x)$. We deduce that the functions $f_n(x)$ are still continuous.

In the general case we construct $f_n(x)$ as above but with the functions $g_n(x) = \max(f(x), -n)$ replacing $f(x)$. Then $f_n(x)$ is still continuous and since $g_n(x)$ decreases with increasing n, we deduce the same for $f_n(x)$, since evidently $\delta(\xi)$ is the same for all the functions $f_n(x)$. To prove that $f_n(x) \to f(x)$, suppose first that $f(\xi) = -\infty$. Then given any constant K we have if n is large

$$f(x) < K, \qquad |x - \xi| < \frac{\delta(\xi)}{2n}.$$

Thus if n is large so that $-n < K$, $g_n(x) < K$ in this range and so $f_n(\xi) < K$. Thus $f_n(\xi) \to -\infty$. If $f(\xi) > -\infty$, then if $-n < f(\xi)$, we have

$$g_n(\xi) = f(\xi),$$

and so $M(\xi, t)$ is the same for $f(x)$ as $g_n(x)$. Thus $f_n(x) \to f(x)$ by the same argument as before. This completes the proof of Theorem 1.4.

Examples

1. If $u_1, u_2, \ldots, u_n$ are u.s.c. on E prove that $u = \max\limits_{1 \leqslant k \leqslant n} u_k$, and $v = \sum\limits_{k=1}^{n} \lambda_k u_k$ are u.s.c. on E, where λ_1 to λ_n are positive or zero.

2. Prove that a function $f(x)$ defined on a set E is u.s.c. on E if and only if the following condition holds. Given $x_0 \in E$ and $K > f(x_0)$, $\exists\, \delta$, such that if $|x - x_0| < \delta$ and $x \in E$, then $f(x) < K$.

3. If $u(x)$ is u.s.c. in E and takes there only values lying in the interval (a, b) and if $f(t)$ is continuous and increasing in (a, b), show that $f[u(x)]$ is u.s.c. on E.

4. If $u(x)$ is u.s.c. on a compact set E and $U = \sup\limits_{x \in E} u(x)$ show that $u(x) = U$ on a compact subset of E.

5. If $v(x)$ is defined and bounded above on a set E and $u(x) = \limsup v(y)$ as $y \to x$ from E, show that $u(x)$ is u.s.c. on the closure $\bar{E}$ of E. (If $M(x, t) = \sup_{y \in D(x,t) \cap E} v(y)$, then $u(x)$ is defined formally as $u(x) = \lim_{t \to 0+} M(x, t)$.)

1.2.2. The classes C^n and A

A function $f(x)$ defined in a domain D in R^m is said to belong to C if $f(x)$ is continuous in D. If further $n \geqslant 0$ and all the partial derivatives of f of orders up to n belong to C, then $f(x)$ is said to belong to C^n. If $f(x) \in C^n$ for every positive integer n, we say that $f(x) \in C^\infty$. We note that if $f(x) \in C^n$ then the values of the various partial derivatives of $f(x)$ of all orders up to n are independent of the order of differentiation. To see this it is enough to prove that

$$\frac{\partial^2 f}{\partial x_1 \partial x_2} = \frac{\partial^2 f}{\partial x_2 \partial x_1},$$

for $m = 2$ and $f \in C^2$. For then the general result for C^n can be proved by induction on n.

Now we have

$$\Delta(h) = f(x_1 + h, x_2 + h) + f(x_1, x_2) - f(x_1, x_2 + h) - f(x_1 + h, x_2)$$
$$= \phi(x_1 + h) - \phi(x_1).$$

say, where

$$\phi(x) = f(x, x_2 + h) - f(x, x_2).$$

Thus we deduce from the mean-value theorem that

$$\Delta(h) = h\phi'(x_1 + \theta_1 h) = h\left\{\frac{\partial f}{\partial x_1}(x_1 + \theta_1 h, x_2 + h) - \frac{\partial f}{\partial x_1}(x_1 + \theta_1 h, x_2)\right\}$$

$$= h^2 \frac{\partial}{\partial x_2}\frac{\partial}{\partial x_1} f(x_1 + \theta_1 h, x_2 + \theta_2 h),$$

where

$$0 < \theta_1 < 1, \qquad 0 < \theta_2 < 1.$$

Since

$$\frac{\partial^2 f}{\partial x_2 \partial x_1} \in C$$

by hypothesis we deduce that

$$\frac{\partial^2 f}{\partial x_2 \partial x_1} = \lim_{h \to 0} \frac{\Delta(h)}{h^2} = \frac{\partial^2 f}{\partial x_1 \partial x_2},$$

similarly.

A function $f(x)$ in a domain D is said to be analytic or to belong to A, if for each $\xi \in D$, there exists $\varepsilon > 0$, such that for $h = (h_1, h_2 \ldots h_m)$, $|h| < \varepsilon$, we have

$$f(\xi + h) = \sum_{(m)} a_{(m)} h^{(m)},$$

where (m) ranges over all sets of m non-negative integers $(n_1, n_2, \ldots, n_m)$, $h^{(m)} = h_1^{n_1} h_2^{n_2} \ldots h_m^{n_m}$, $a_{(m)}$ is independent of h, and the series is absolutely convergent.

It is easy to see that, just as in the case $m = 1$, a multiple power series convergent in $|h| < \varepsilon$ can be differentiated term by term partially with respect to the h_ν any number of times inside the ball $|h| < \varepsilon$, and that the series remains continuous, being uniformly convergent for $|h| < \lambda\varepsilon$, $0 < \lambda < 1$. By setting $h = 0$ after differentiation we deduce that

$$n_1! n_2! \ldots n_m! a_{(m)} = \frac{\partial^n f(\xi)}{\partial \xi_1^{n_1} \partial \xi_2^{n_2} \ldots \partial \xi_m^{n_m}},$$

where $n = n_1 + n_2 + \ldots n_m$. In particular $A \subset C^\infty$. To see that A is a proper subclass of C^∞ we prove the following

THEOREM 1.5. *Identity Theorem. If $f \in A$ in D and $f \equiv 0$ in a hyperball $D(x_0, r)$ lying in D, then $f \equiv 0$ in D.*

Let E_1 be the set of all points ξ of D such $f \equiv 0$ in some neighbourhood $D(\xi, h)$ of D. Clearly E_1 is open. Next let x be a limit point of D. Then there exists a sequence $x_n \in E_1$, such that $x_n \to x$ in D. Since $f \equiv 0$ in a neighbourhood of each point x_n, all the partial derivatives of f vanish at x_n and so by continuity at x. Thus the Taylor series of f at x reduces identically to zero and so $f \equiv 0$ near x, since $f \in A$. Thus E_1 is closed in D. Since D is a domain, and E_1 is not empty, $E_1 = D$. This proves Theorem 1.5.

Example. The function

$$f(x) = \exp\{-|x|^{-2}\}, \qquad |x| > 0, \qquad f(0) = 0,$$

belongs to C^∞ but not to A in R^m.

We can easily prove by induction that any partial derivative $\phi(x)$ of $f(x)$ can be written in the form

$$\phi(x) = \frac{P(x)}{|x|^k} f(x), \qquad x \neq 0$$

where $P(x)$ is a polynomial in $x_1, x_2, \ldots, x_m$, and k is an even integer. Thus

$$\phi(x) \to 0 \quad \text{as} \quad x \to 0,$$

and we deduce that $\phi(x)$ exists as a partial derivative also at $x = 0$ and $\phi(0) = 0$. Thus $f(x) \in C^\infty$. Further the Taylor series of $f(x)$ vanishes identically at $x = 0$, but $f(x)$ does not vanish for $x \neq 0$ and so $f(x)$ is not given by its Taylor series in any neighbourhood of $x = 0$, so that $f(x) \notin A$.

To see that Theorem 1.5 fails for C^∞, we can define

$$f(x) = 0, x_1 \leqslant 0, \qquad f(x) = \exp(-x_1^{-2}), \qquad x_1 > 0.$$

We easily see that $f(x) \in C^\infty$, and that $f(x)$ vanishes in the half-space $x_1 < 0$ without vanishing identically.

1.3. CONVEX FUNCTIONS

A real function $f(x)$, of one real variable x defined in an interval I, is called *convex* in I if, for every $x_1, x_2 \in I$ with $x_1 < x_2$, and every linear function $l(x)$ satisfying

$$l(x_1) \geqslant f(x_1), \qquad l(x_2) \geqslant f(x_2), \tag{1.3.1}$$

we have also

$$l(x) \geqslant f(x) \qquad (x_1 < x < x_2). \tag{1.3.2}$$

If we choose, in particular, that linear function $l(x)$ which *coincides* with $f(x)$ at x_1 and x_2, we get

$$f(x) \leqslant \frac{x_2 - x}{x_2 - x_1} f(x_1) + \frac{x - x_1}{x_2 - x_1} f(x_2) \qquad (x_1 < x < x_2). \tag{1.3.3}$$

Geometrically, (1.3.3) means that if we graph $f(x)$ we find that each chord of the curve lies above the curve.

A subharmonic function is an extension from one to two dimensions of the notion of a convex function. Moreover we shall find that certain integral means, and other functions associated with subharmonic functions, are themselves convex. We therefore need a brief discussion of convex functions.

By a simple manipulation we get from (1.3.3)

$$\frac{f(x) - f(x_1)}{x - x_1} \leqslant \frac{f(x_2) - f(x_1)}{x_2 - x_1} \leqslant \frac{f(x_2) - f(x)}{x_2 - x}. \tag{1.3.4}$$

We prove now

THEOREM 1.6. *If f is convex in an open interval I, then $f(x)$ possesses at each point of I left and right derivatives, which are increasing functions of x and coincide outside a countable set. The left derivative is continuous to the right and the left derivative is not greater than the right derivative. In particular $f(x)$ is continuous in I.*

We assume that x_1, x, x_2 are 3 points in I, such that $x_1 < x < x_2$, and hence that (1.3.3) and (1.3.4) holds, i.e.

$$\frac{f(x) - f(x_1)}{x - x_1} \leqslant \frac{f(x_2) - f(x_1)}{x_2 - x_1}.$$

Thus for $h > 0$

$$f(x_1, h) = \frac{f(x_1 + h) - f(x_1)}{h} \tag{1.3.5}$$

increases with h and so this ratio tends to a limit or right derivative $f_2(x)$ as $h \to 0+$. Similarly $f(x, h)$ approaches the left derivative $f_1(x)$ as $h \to 0-$. By letting x_1 and x_2 tend to x simultaneously we deduce from (1.3.4) that

$$f_1(x) \leqslant f_2(x).$$

Also we have proved that $f(x, h)$ is an increasing function of h, provided that $h \neq 0$ and $x + h$ lies in I. Hence if δ is a small positive number

$$f(x, -\delta) \leqslant f_1(x) \leqslant f_2(x) \leqslant f(x, \delta),$$

so that $f_1(x)$, $f_2(x)$ are both finite at every point of I. In other words the left and right derivatives exist and are finite. Hence $f(x)$ is clearly continuous. We note next that if x_1, x_2 lie in I and $x_2 - x_1 = \delta > 0$, then

$$f_1(x_1) \leqslant f_2(x_1) \leqslant f(x_1, \delta) = f(x_2, -\delta) \leqslant f_1(x_2) \leqslant f_2(x_2). \tag{1.3.6}$$

Thus $f_1(x)$ and $f_2(x)$ are both increasing functions of x.

In order to complete our proof we need

LEMMA 1.1. *If $F(x)$ is an increasing function of x in an open interval I, then the left and right limits $F(x - 0)$, $F(x + 0)$ exist everywhere in I and coincide with $F(x)$ outside a finite or countable set E. Thus $F(x)$ is continuous outside E.*

If x is in I, then $F(x + h)$ is an increasing function of h provided that $x + h$ lies in I. It is easy to see that $F(x + h) \to F(x + 0)$ as $h \to 0$ from above, where $F(x + 0)$ is the greatest lower bound of $F(x + h)$ for $x + h$ in I and $h > 0$. Similarly $F(x + h) \to F(x - 0)$, as $h \to 0$ from below, where $F(x - 0)$ is the least upper bound of $F(x + h)$ for $h < 0$. Since F is monotonic we have for $h_1 < 0 < h_2$

$$F(x + h_1) \leqslant F(x) \leqslant F(x + h_2)$$

and on letting h_1, h_2 tend to zero we deduce that

$$F(x - 0) \leqslant F(x) \leqslant F(x + 0).$$

We define $\delta(x) = F(x + 0) - F(x - 0)$ to be the jump of $F(x)$ at the point x

and deduce that $F(x)$ is continuous at the point $x = \xi$, if and only if $\delta(\xi) = 0$.

Suppose now that $x_1 = \xi'_1 < \xi_1 < \xi'_2 < \ldots < \xi'_{n+1} = x_2, x_1 \in I, x_2 \in I$. Then from the monotonicity of $F(x)$ we deduce that for $1 \leqslant j \leqslant h$

$$\delta(\xi_j) = F(\xi_j + 0) - F(\xi_j - 0) \leqslant F(\xi'_{j+1}) - F(\xi'_j).$$

Thus

$$\sum_{j=1}^{n} \delta(\xi_j) \leqslant \sum_{j=1}^{n} (F(\xi'_{j+1}) - F(\xi'_j)) = F(\xi'_{n+1}) - F(\xi'_1) = F(x_2) - F(x_1). \tag{1.3.7}$$

This inequality is valid for any finite number of distinct points ξ_j in the interval (x_1, x_2). Hence there can be at most 2^N such points ξ for which

$$2^{-N}[F(x_2) - F(x_1)] \leqslant \delta(\xi) \leqslant 2^{1-N}[F(x_2) - F(x_1)].$$

By enumerating these points in order of increasing ξ for $N = 1, 2, 3, \ldots$ in turn we obtain an enumeration of all points in (x_1, x_2) for which $\delta(\xi) > 0$, and so the set of these points is countable. Since the whole interval I is the union of a countable number of open intervals whose closures lie in I, and a countable union of countable sets is countable, we deduce Lemma 1.1.

We can now complete the proof of Theorem 1.6. Let E be the set where at least one of $f_1(x)$, $f_2(x)$ is discontinuous. Then E is finite or countable by Lemma 1.1. Hence the complement of E is dense in I. Let x_1 be a point in this complement. Then by (1.3.6) $f_1(x_1) \leqslant f_2(x_1)$. On the other hand if $x_2 > x_1$ it follows from (1.3.6) that

$$f_2(x_1) \leqslant f_1(x_2).$$

Since $f_1(x)$ is continuous at $x = x_1$ we may let x_2 tend to x_1 in this inequality and obtain $f_2(x_1) \leqslant f_1(x_1)$. Thus $f_1(x_1) = f_2(x_1)$ so that f_1 and f_2 coincide outside E. Thus for x_1 outside the countable set E the left and right derivatives of $f(x)$ exist and are equal at $x = x_1$ so that $f'(x_1)$ exists.

It remains to prove that $f_1(x)$ is continuous on the left and that $f_2(x)$ is continuous on the right. We confine ourselves to proving the second statement, since the proof of the first is similar. It follows from (1.3.6) that if $x_1 < x_2$, then

$$f_2(x_1) \leqslant f_2(x_2),$$

so that on letting x_2 tend to x_1 we deduce that

$$f_2(x_1 + 0) \geqslant f_2(x_1).$$

On the other hand it follows from the definition of $f_2(x_1)$ that, given $\varepsilon > 0$, we may choose $x_2 > x_1$, so that

$$f_2(x_1) \geqslant \frac{f(x_2) - f(x_1)}{x_2 - x_1} - \varepsilon.$$

It then follows from the continuity of $f(x)$ that we can choose x_3, so that $x_1 < x_3 < x_2$ and

$$f_2(x_1) \geqslant \frac{f(x_2) - f(x_3)}{x_2 - x_3} - 2\varepsilon.$$

Using the definition of $f_2(x_3)$ we deduce

$$f_2(x_1) \geqslant f_2(x_3) - 2\varepsilon,$$

and hence

$$f_2(x_1) \geqslant f_2(x_1 + 0) - 2\varepsilon.$$

Since ε is arbitrary we deduce that

$$f_2(x_1) \geqslant f_2(x_1 + 0), \text{ so that } f_2(x_1) = f_2(x_1 + 0).$$

This completes the proof of Theorem 1.6.

Examples

1. Prove that, with the above notation, $f_1(x)$ and $f_2(x)$ have the same points of discontinuity.

2. Prove that $f_2(x_1 - 0) = f_1(x - 0) = f_1(x)$, $f_1(x + 0) = f_2(x + 0)$
$= f_2(x)$

for each point x of I.

3. Prove that if $x_1 < x_2$ and x_1, x_2 lie in I, then $f_1(x)$ and $f_2(x)$ are Riemann integrable over $[x_1, x_2]$ and

$$\int_{x_1}^{x_2} f_1(x)\,dx = \int_{x_1}^{x_2} f_2(x)\,dx = f(x_2) - f(x_1).$$

4. It follows from a Theorem of Lebesgue† that an increasing function has a derivative almost everywhere. Prove that if $f_1(x)$ is differentiable at $x = x_1$ then so is $f_2(x)$ and the derivatives are equal.

† See e.g. Titchmarsh, [1939, p. 358].

5. Prove that a function $f(x)$ is convex in the open interval I if and only if $f(x)$ has the form

$$f(x) = \int_{x_0}^{x} \phi(t)\, dt,$$

where x_0 is any point of I and $\phi(t)$ is increasing in I.

6. If $f''(x)$ exists for x in I, show that $f(x)$ is convex in I if and only if $f''(x) \geqslant 0$ at every point of I. (Hint. Show that $f'(x)$ increases if and only if $f''(x) \geqslant 0$).

7. If $f(x)$ is convex and $f(x) \leqslant 0$ for $a \leqslant x \leqslant b$ and further $f(a) \leqslant 0$, $f(b) \leqslant 0$ and $f(x_0) \geqslant 0$ for some x_0 such that $a < x_0 < b$, prove that $f(x) \equiv 0$, $a < x < b$. (By (1.3.3) $f(x) \leqslant 0$ in $[a, b]$. If $f(x_1) < 0$, where for instance $x_0 < x_1 < b$, then (1.3.3) applied with a, x_0, x_1 instead of x_1, x, x_2 leads to a contradiction).

8. If $f(x)$ is convex in (a, b) show that the limits $f(a + 0)$ and $f(b - 0)$ exist as limits which are finite or $+\infty$. If the limits are finite and $f(a)$, $f(b)$ are defined arbitrarily, show that $f(x)$ remains convex in $[a, b]$ if and only if $f(a) \geqslant f(a + 0)$, $f(b) \geqslant f(b - 0)$.

9. If $f(x)$ is convex in I show that

$$\frac{f(x_2) - f(x_1)}{x_2 - x_1}$$

is an increasing function of either x_1 or x_2 when the other variable is kept fixed, $x_1, x_2 \in I$, and $x_1 \neq x_2$.

10. If $f(x)$ is convex in the interval (a, ∞), show that

$$\alpha = \lim_{x \to +\infty} \frac{f(x)}{x}$$

exists and $-\infty < \alpha \leqslant +\infty$.

11. If $f(x)$ is convex in $(-\infty, +\infty)$ show that the limits

$$\alpha_1 = \lim_{x \to +\infty} \frac{f(x)}{x}, \qquad \alpha_2 = \lim_{x \to -\infty} \frac{f(x)}{-x} \quad \text{exist,}$$

and that $\alpha_1 \leqslant 0$ and $\alpha_2 \leqslant 0$ only if $f(x) \equiv$ constant.

1.4. INTEGRATION THEORY AND GREEN'S THEOREM

1.4.1. The Lebesgue integral

We shall be concerned with volume integrals in R^m. Let x denote a general point in R^m, and let $f(x)$ denote in the first instance a bounded positive Borel measurable† function of compact support, i.e. vanishing outside a sufficiently large ball. We recall the definition and basic properties of the Lebesgue integral. This integral

$$I(f) = \int f(x)\,\mathrm{d}x = \int f(x_1, x_2, \ldots x_m)\,\mathrm{d}x_1\,\mathrm{d}x_2 \ldots \mathrm{d}x_m$$

can be defined as follows. Suppose that $f(x) < M$, and let

$$0 = M_1 < M_2 < \ldots M_k = M$$

be a partition of the interval $[0, M]$. Let E_ν be the set in space such that $M_\nu < f(x) \leqslant M_{\nu+1}$. Then E_ν is a bounded Borel measurable set and so has a Lebesgue measure $m(E_\nu)$. We now write

$$s(\Delta) = \sum_{\nu=1}^{k-1} M_\nu m(E_\nu), \qquad S(\Delta) = \sum_{\nu=1}^{k-1} M_{\nu+1} m(E_\nu),$$

and can show easily that for any two distinct partitions Δ, Δ'

$$s(\Delta) \leqslant S(\Delta').$$

Thus if I_1 is the lower bound of all the sums $S(\Delta)$, and I_2 the upper bound of all the sums $s(\Delta)$ for varying Δ, we deduce that

$$I_1 \leqslant I_2.$$

Also if Δ is such that $M_{\nu+1} - M_\nu < \varepsilon$ for each ν, we deduce that

$$S(\Delta) - s(\Delta) \leqslant \varepsilon \sum_{\nu=1}^{k-1} m(E_\nu) = \varepsilon m(E),$$

where E is a bounded set outside which f vanishes. Since ε can be as small as we please, it follows that $I_1 = I_2$. We now define

$$I(f) = I_1 = I_2.$$

If f is now a general Borel measurable function, which is nonnegative but not necessarily bounded, let $E_N = C(O, N)$. We define

$$f_N(x) = \min\{f(x), N\}, \ x \text{ in } E_N$$
$$= 0 \qquad\qquad , x \text{ outside } E_N.$$

It is evident that $f_N(x)$ increases with N and hence so does $I(f_N)$. We now define

$$I(f) = \lim_{N\to\infty} I(f_N),$$

and note that $I(f)$ always exists as a finite or infinite limit. If $I(f)$ is finite, f is said to be (Lebesgue) integrable.

† This concept is defined and developed in detail in section 3.1.

Next if E is a general Borel measurable set, and f is nonnegative and Borel measurable on E we define the function $\phi(x)$ by setting

$$\begin{aligned}\phi(x) &= f(x),\ x \text{ on } E\\ &= 0 \quad ,\ x \text{ outside } E.\end{aligned}$$

and define

$$\int_E f(x)\,\mathrm{d}x = I(\phi).$$

Finally if f is measurable on a Borel set E but not necessarily of constant sign, we define

$$f_1 = \max(f, 0), \qquad f_2 = \min(f, 0),$$

so that

$$f = f_1 + f_2,$$

and

$$\int_E f(x)\,\mathrm{d}x = \int_E f_1(x)\,\mathrm{d}x - \int_E - f_2(x)\,\mathrm{d}x,$$

provided at least one of f_1, f_2 is integrable over E. If both f_1 and f_2 are integrable, then f is said to be integrable.

We take for granted the usual properties of the Lebesgue integral but recall specifically Fubini's Theorem, which enables us to reduce the n-dimensional integral to repeated 1-dimensional integrals. The result may be stated as follows.†

THEOREM 1.7. (Fubini's Theorem). *Suppose that f is integrable over a set E, and for each point $(x_2, \ldots, x_n)$ let $E(x_2, \ldots, x_n)$ be the set of all x_1 for which $(x_1, x_2, \ldots, x_n)$ lies in E. Then*

$$\int_E f(x_1, x_2, \ldots, x_n)\,\mathrm{d}x_1\,\mathrm{d}x_2 \ldots \mathrm{d}x_n$$

$$= \int \mathrm{d}x_2\,\mathrm{d}x_3 \ldots \mathrm{d}x_n \int_{E(x_2, \ldots, x_n)} f(x_1, x_2, \ldots, x_n)\,\mathrm{d}x_1.$$

We note that since E and f are Borel measurable, so are the sets $E(x_2, \ldots x_n)$ and $f(x_1, x_2 \ldots x_n)$ for each fixed point $(x_2 \ldots x_n)$ in $(n-1)$-space. Thus the

† A more general form of this result will be proved in Theorem 3.5. For the original result, see Fubini [1907].

right-hand side makes sense. If f takes both signs, it is possible that

$$\int_{E(x_2, \cdots, x_n)} f(x_1, \ldots, x_n)\, dx_1$$

does not exist, because f is not integrable for certain values of $(x_2, \ldots, x_n)$ but the set of such points $(x_2, \ldots, x_n)$ necessarily has $(n-1)$ dimensional measure zero and so may be omitted from the range of integration.

In future all sets E and functions f will be assumed to be Borel measurable. We shall also need one further result from general integration theory.

THEOREM 1.8. *Let* $y = T(x)$ *be a* $(1, 1)$ *transformation from a domain* D *of points* x, $(x_1, \ldots, x_m)$ *in* R^m *to a domain* D' *of points* y, $(y_1, \ldots, y_m)$ *in* R^m. *Suppose further that* $T(x)$ *has continuous partial derivatives*

$$y_{\mu,\nu} = \frac{\partial y_\mu}{\partial x_\nu}, \qquad \mu, \nu = 1 \text{ to } m$$

such that the Jacobian determinant

$$J(x) = \frac{\partial(y_1, \ldots y_m)}{\partial(x_1, \ldots x_m)} = |y_{\mu,\nu}|$$

does not vanish in D. *Then if* $f(y)$ *is integrable over* D', *we have*

$$\int_{D'} f(y)\, dy = \int_D f\{T(x)\}\, |J(x)|\, dx,$$

in the sense that the integral on the right-hand side exists and is equal to that on the left-hand side.

1.4.2. Surface integrals

We are now in a position to define parametric surfaces and integrals over them. A parametric hypersurface in R^m is defined as the image S under a $(1, 1)$ map $y = T(x)$ of a domain D in $(m-1)$-dimensional space $(x_1, x_2, \ldots, x_{m-1})$ into points $(y_1, \ldots, y_m)$ of R^m. In addition we shall assume that the map belongs to C^1, i.e. that all the partial derivatives

$$y_{\mu\nu} = \frac{\partial y_\mu}{\partial x_\nu}, \qquad \mu = 1 \text{ to } m, \qquad \nu = 1 \text{ to } m-1,$$

exist and are continuous in D, and that at each point of D at least one of the

Jacobians

$$J_\nu(x) = \frac{\partial(y_1, y_2, y_{\nu-1}, y_{\nu+1}, \ldots, y_m)}{\partial(x_1, \ldots, x_{m-1})} = (-1)^{\nu-1} \det y_{\mu\nu},$$

$$\mu = 1, 2, \ldots, \nu - 1, \nu + 1, m,$$

$$\nu = 1 \text{ to } m.$$

does not vanish.

With each point y of S we associate the two unit normal vectors

$$\mathbf{t} = \left\{\frac{J_\nu}{\sqrt{(\Sigma_{\nu=1}^{m} J_\nu^2)}}\right\} \nu = 1 \text{ to } m \text{ and } -\mathbf{t} = \left\{\frac{-\mathrm{J}_\nu}{\sqrt{\Sigma J_\nu^2}}\right\}, \ \nu = 1 \text{ to } m,$$

and note that the pair $(\mathbf{t}, -\mathbf{t})$ varies continuously, as x varies over D and so as y varies over S. The tangent plane at a point $\eta = (\eta_1, \ldots, \eta_n)$ is defined as the hyperplane through η orthogonal to $\mathbf{t}$, i.e.

$$\sum_{\nu=1}^{m} J_\nu(y_\nu - \eta_\nu) = 0.$$

We note that these definitions are independent of the particular parametrizations chosen.

For suppose, e.g. that $J_m(x) \neq 0$, at

$$x = (\xi_1, \ldots \xi_{m-1})$$

and that

$$T(\xi) = (\eta_1, \eta_2, \ldots, \eta_m).$$

Then the map from x to $(y_1, y_2 \ldots y_{m-1})$ is locally reversible so that x becomes a continuously differentiable function of $(y_1, \ldots, y_{m-1})$ near $(\eta_1, \ldots, \eta_{m-1})$ and hence so does y_m. Also we have

$$\mathrm{d}y_\mu = \sum_{\nu=1}^{m-1} y_{\mu\nu}\, \mathrm{d}x_\nu, \qquad \mu = 1 \text{ to } m$$

so that

$$\sum_{\mu=1}^{m} J_\mu \, \mathrm{d}y_\mu = 0,$$

since the left-hand side can be expressed as a linear function in $\mathrm{d}x_\nu$, in which the coefficients are determinants with two equal columns. Thus

$$\frac{\partial y_m}{\partial y_\mu} = \frac{-J_\mu}{J_m}, \qquad \mu = 1 \text{ to } m - 1,$$

so that the ratios $J_1 : J_2 : \ldots : J_m$ depend only on S as a set of points and not on the particular parametrization. Thus the pair of vectors (**t**, $-$**t**) also depends only on S.

Given now any function f on S we define the surface integral

$$\int_S f\,d\sigma = \int_D f[T(x)]\sqrt{\left\{\sum_{\nu=1}^{m-1} J_\nu(x)^2\right\}}\,dx \tag{1.4.1}$$

Similarly we define for $\nu = 1$ to $m-1$

$$\int_S f\,d\sigma_\nu = \int_D f[T(x)]\varepsilon J_\nu(x)\,dx, \tag{1.4.2}$$

where $\varepsilon = \mp 1$ and the sign of εJ_ν is determined by assigning at each point of S one of the two normal vectors **t** and $-$**t** and choosing the same sign for εJ_ν as that occurring in the νth coordinate of t. Using Theorem 1.8 we see again that these definitions do not depend on the parametrization. For we can divide our surface into patches S' where one of the coordinates, e.g. y_m can be expressed as a function of the others, and if D' is the corresponding, domain of points $(y_1, y_2, \ldots, y_{m-1})$, then

$$\int_{S'} f\,d\sigma = \int_{D'} f[y_1 \ldots y_{m-1}, y_m(y_1 \ldots y_{m-1})] \sqrt{\left\{1 + \sum_{\nu=1}^{m-1}\left(\frac{\partial y_m}{\partial y_\nu}\right)^2\right\}}\,dy_1 \ldots dy_{m-1} \tag{1.4.3}$$

$$\int_{S'} f\,d\sigma_m = \int_{D'} \varepsilon f[y_1, y_2 \ldots, y_m(y_1, y_2, \ldots, y_{m-1})]\,dy_1\,dy_2 \ldots dy_{m-1}$$

The part of S where $J_\nu \neq 0$ can be split up into a countable union of disjoint Borel sets, in each of which y_ν is locally a function of the other coordinates and the part of S' where $J_\nu = 0$ contributes zero to the integral (1.4.2). Thus in all cases the integrals (1.4.1) and (1.4.2) depend only on S and in the case of (1.4.2) the local definition of the unit normal.

It is frequently not possible to parametrize a whole surface in the way indicated above. Thus we finally come to the notion of hypersurface. A hypersurface is the union S of a finite or countable number of parametric surfaces S_μ defined as above, such that

(i) Each S_μ meets at most a finite number of different S_ν.

(ii) If $\Gamma_{\mu\nu}$ is the part of the boundary of S_ν which lies in S_μ, then $\Gamma_{\mu\nu}$ has a zero surface area, in the sense that if $f = 1$ in $\Gamma_{\mu\nu}$, $f = 0$ elsewhere then

$$\int_{S_\mu} f\,d\sigma = 0.$$

Then our integrals can be extended in an evident way to integrals over S. In fact we define

$$\int_S = \sum_\nu \int_{S_\nu^0},$$

where S_ν^0 denotes the interior (in the topology of the $(m-1)$ dimensional domain D_ν of which S_ν is the image) of the part of S_ν which is not in $S_1, S_2, \ldots, S_{\nu-1}$.

For future reference, and as our most important application, we note that the hypersphere S

$$\Sigma\, x_\nu^2 = 1,$$

is a hypersurface in our sense. For near any point of S, x_ν can be locally expressed in terms of the other coordinates, provided that $x_\nu \neq 0$ at the point. We choose the patches $1/(2m) < x_\nu \leqslant 1$, $-1/(2m) \geqslant x_\nu \geqslant -1$, $\nu = 1$ to m, and note that they cover the sphere and that their boundaries $x_\nu = \mp 1/(2m)$ have zero area.

1.4.3. Domains and their frontier surfaces

Let D be a domain in R^m, suppose that P is a boundary point of D and that in some neighbourhood V of P all boundary points of D lie on a surface S passing through P. It is evident that if V is small enough the two unit normals $\mathbf{t}_1, \mathbf{t}_2$ to S at P do not meet S again in V. Thus $\mathbf{t}_1, \mathbf{t}_2$ lie either wholly inside or wholly outside D.

Suppose for definiteness that P is the origin and that the tangent plane to S at P is not parallel to the x_1 axis. Then S can be written

$$x_1 = f(x_2, \ldots x_m)$$

near P, where f is a differentiable function and $f(0, 0, \ldots, 0) = 0$. Let V^+, V^- be the two sets

$$x_1 > f(x_2 \ldots x_m), \quad \sum_2^m x_\nu^2 < \varepsilon; x_1 < f(x_2, \ldots, x_m), \quad \sum_2^m x_\nu^2 < \varepsilon$$

respectively. Clearly any two points in V^+ can be joined in V^+. For let

$$\delta = \sup_{\Sigma_2^m x_\nu^2 \leqslant \varepsilon} |f(x_2, \ldots, x_m)|.$$

We can join a point $(x_1, x_2, \ldots, x_m)$ to $(2\delta, x_2, \ldots, x_m)$ in V^+ and then join $(2\delta, x_2, \ldots, x_m)$ to $(2\delta, x_2', \ldots, x_m')$ and then finally to $(x_1', x_2', \ldots, x_m')$ by 3 straight line segments. These lie in V^+ and so, if ε and hence δ are small enough, in V. Hence either none or both these points must lie in D, if x_1 and

ε are small enough. Similarly all or no points of V^- which are close enough to the origin lie in D. If D contains all points in V^+, but no points in V^- near the origin, we say D lies above S at the origin. In this case we define the *inward* and *outward* normal to be those of $\mathbf{t}_1, \mathbf{t}_2$ respectively having positive and negative x_1-components.

If D contains points of V^- but not of V^+, we define the inward normal as having a negative x_1-component and the outward normal as having a positive x_1-component. In either case we say that D has S as one-sided boundary.

If D contains near P points of both V^+ and V^-, we say that D has $\mathscr{S}$ as two-sided boundary locally. Since D is a domain and so consists entirely of interior points, D cannot lie entirely on S and so must have S as one-sided or two-sided boundary near every point.

Clearly all points of the inward normal, which are sufficiently near P, lie in D and all points of the outward normal near P lie outside D, if S is the one-sided boundary of D.

We now make the following

DEFINITION. *A domain D in R^m is said to be admissible if*

(i) *D is bounded*
(ii) *The boundary S of D is the union of a finite number of disjoint hypersurfaces.*
(iii) *at each point of S, S is the one-sided boundary of D.*
(iv) *The set of points of S, where the tangent plane is parallel to one of the axes has zero area.*
(v) *No line parallel to the coordinate axes meets S in more than a finite number of points.*

1.4.4. Green's theorem†

We end this section by proving Green's Theorem.

THEOREM 1.9. *Suppose that D is an admissible domain with boundary S in R^m and that $u \in C^1$ and $v \in C^2$ in $\bar{D}$. Then*

$$\int_S u \frac{\partial v}{\partial n} \, d\sigma = -\int_D \left\{ \sum_\nu \frac{\partial u}{\partial x_\nu} \frac{\partial v}{\partial x_\nu} + u \nabla^2 v \right\} dx \tag{1.4.4}$$

where

$$\nabla^2 = \sum_{\nu=1}^{m} \frac{\partial^2}{\partial x_\nu^2}$$

† The integral identities and the function associated with Green appear in Green [1828].

is Laplace's operator. Hence if $u, v \in C^2$ in $\bar{D}$ we have

(1.4.5) $$\int_S \left(u\frac{\partial v}{\partial n} - v\frac{\partial u}{\partial n}\right) d\sigma = \int_D (v\nabla^2 u - u\nabla^2 v)\, dx.$$

Here $\partial/\partial n$ denotes differentiation along the inward normal into D.

Let $P(x) \in C^1$ in $\bar{D}$ and consider

(1.4.6) $$I_1 = \int_D \frac{\partial P}{\partial x_1} dx.$$

Let D_1 be the projection of $\bar{D}$ onto the hyperplane $x_1 = 0$ and let E_1 be the projection of the part of S where the tangent plane is orthogonal to $x_1 = 0$. By hypothesis E_1 has $(m-1)$-dimensional measure zero. If $\xi = (0, \xi_2, \ldots, \xi_m)$ is a point of $D_1 - E_1$, then the line $x_\nu = \xi_\nu$, $\nu = 2$ to m meets D in a finite number of straight line segments

$$\xi_{1,2\mu} < x_1 < \xi_{1,2\mu+1}, \mu = 0 \text{ to } k = k(\xi).$$

Hence by Fubini's theorem, and since E_1 has zero $(m-1)$-dimensional measure

$$I_1 = \int_{D_1 - E_1} dx_2\, dx_3 \ldots dx_m \sum_{\mu=0}^{k} \int_{\xi_{1,2\mu}}^{\xi_{1,2\mu+1}} \frac{\partial P}{\partial x_1} dx_1.$$

Also since $P \in C^1$, $\partial P/\partial x_1$ is continuous in x_1 for fixed $(x_2, \ldots, x_m)$ and $\xi_{1,2\mu} \leqslant x_1 \leqslant \xi_{1,2\mu+1}$. Thus

$$\int_{\xi_{1,2\mu}}^{\xi_{1,2\mu+1}} \frac{\partial P}{\partial x_1} dx_1 = P(\xi_{1,2\mu+1}, x_2, \ldots, x_m) - P(\xi_{1,2\mu}, x_2, \ldots, x_m).$$

Thus

(1.4.7) $$I_1 = \int_{D_1 - E_1} \sum_{\mu=0}^{2k} (-1)^{\mu-1} P(\xi_{1,\mu}, x_2 \ldots, x_m)\, dx_2\, dx_3\, dx_m.$$

Consider now

(1.4.8) $$J_1 = \int_S P(x)\, d\sigma_1,$$

where the integral is defined in accordance with (1.4.2). We need to assign at each point of S one of the fixed normal vectors and for this purpose we choose the inward normal from S into D. We divide S into those points S_1 where the normal is parallel to the plane $x_1 = 0$ and the remainder S_2. On S_1 we have $J_1 = 0$ for any local parametrization and so

$$\int_{S_1} P\, d\sigma_1 = 0.$$

Near any point of S_2 we can locally express x_1 in terms of the remaining coordinates $(x_2, \ldots, x_m)$ and so

$$\int_{S_2} P\, d\sigma_1 = \int_{S_2} \mp P(x_1, x_2 \ldots x_m)\, dx_2\, dx_3 \ldots dx_m. \tag{1.4.9}$$

Here the sign + is chosen if the inward normal makes an acute angle with the vector pointing in the direction of the positive x_1 axis, i.e. if x_1 increases along the inward normal, and the sign—is chosen otherwise.

We can write the right-hand side of (1.4.9) as

$$\int_{D_2} \Sigma \mp P(x_1, x_2 \ldots x_m)\, dx_2\, dx_3 \ldots dx_m.$$

where D_2 denotes the subset of D_1 onto which S_2 projects and the sum is taken for fixed $(x_2, \ldots, x_m)$ over all points $(x_1, x_2, \ldots, x_m)$ which lie on S and are such that the tangent plane is not parallel to the x_1 axis. Thus D_2 includes all points of D_1 together possibly with some points of E_1 namely those for which the line $x_\nu =$ constant, $\nu = 2$ to m meets S in some points where the tangent plane is parallel to the x_1 axis and others, where it is not. Since E_1 has measure zero, it follows that

$$\begin{aligned} J_1 &= \int_{D_2} \Sigma \mp P(x_1, \ldots, x_m)\, dx_2\, dx_3 \ldots dx_m \\ &= \int_{D_1} \Sigma \mp P(x_1, \ldots, x_m)\, dx_2\, dx_3 \ldots dx_m. \end{aligned} \tag{1.4.10}$$

Now we note that for $(\xi_2, \ldots, \xi_m)$ in D_1 the relevant points $(x_1, \xi_2, \ldots, \xi_m)$ are precisely the points $x_1 = \xi_{1,\mu}$, $\mu = 0$ to $2k$. Also since the segments $\xi_{1,2\mu} < x_1 < \xi_{2,2\mu+1}$ lie in D, these segments described in the direction of increasing x_1 make an acute angle with the inward normal at $x_1 = \xi_{1,2\mu}$ and an obtuse angle at $x_1 = \xi_{1,2\mu+1}$. Thus the sign at $\xi_{1,\mu}$ is + when μ is even and − when μ is odd. Using (1.4.7), (1.4.8) and (1.4.10), we deduce that

$$I_1 = -J_1,$$

i.e.

$$\int_D \frac{\partial P}{\partial x_1}\, dx = -\int_S P\, d\sigma_1.$$

Similarly we have for $\nu = 1$ to m

$$\int_D \frac{\partial P}{\partial x_\nu}\, dx = -\int_S P\, d\sigma_\nu. \tag{1.4.11}$$

We now note that if $(t_1, \ldots, t_{m-1})$ gives a suitable local parametrization of S, near ξ $(\xi_1, \xi_2, \ldots, \xi_m)$, then we have at ξ

$$\frac{\partial v}{\partial n} = \varepsilon \lim_{h \to 0} \frac{v\left(\xi_1 + \frac{J_1}{J}h, \xi_2 + \frac{J_2}{J}h, \ldots, \xi_m + \frac{J_m}{J}h\right) - v(\xi_1, \ldots, \xi_m)}{h}$$

where $\varepsilon = \mp 1$,

$$J_\nu = (-1)^{\nu-1} \frac{\partial(x_1, x_2, \ldots, x_{\nu-1}, x_{\nu+1}, \ldots, x_m)}{\partial(t_1, t_2, \ldots, t_{m-1})}$$

and

$$J = \sqrt{\left(\sum_{\nu=1}^{m} J_\nu^2\right)}.$$

Thus

$$\frac{\partial v}{\partial n} = \varepsilon \sum_{\nu=1}^{m} \frac{J_\nu}{J} \frac{\partial v}{\partial x_\nu},$$

$$\frac{\partial v}{\partial n} \mathrm{d}\sigma = \varepsilon \sum_{\nu=1}^{m} \frac{\partial v}{\partial x_\nu} \frac{J_\nu}{J} \mathrm{d}\sigma.$$

The sign ε ascribed is that associated with the inward normal into D and thus εJ_ν has positive sign if x_ν increases in the direction of the inward normal and εJ_ν has negative sign otherwise. Thus (1.4.2) and (1.4.11) show that

$$\int_S u \frac{\partial v}{\partial n} \mathrm{d}\sigma = \sum_{\nu=1}^{m} \int_S u \frac{\partial v}{\partial x_\nu} \mathrm{d}\sigma_\nu = -\sum_{\nu=1}^{m} \int_D \frac{\partial}{\partial x_\nu}\left(u \frac{\partial v}{\partial x_\nu}\right) \mathrm{d}x$$
$$= -\int_D \left\{u \nabla^2 v + \sum_{\nu=1}^{m} \frac{\partial u}{\partial x_\nu} \frac{\partial v}{\partial x_\nu}\right\} \mathrm{d}x.$$

This proves (1.4.4). Also (1.4.5) follows by subtraction and the proof of Theorem 1.9 is complete.

1.5. HARMONIC FUNCTIONS

1.5.1. Green's function and Poisson's integral†

If $u \in C^2$ in a domain D, and $\nabla^2 u = 0$ there, then u is said to be harmonic in D. We proceed to investigate the properties of harmonic functions.

† Poisson [1827].

If $u = f(R)$ is harmonic in a domain in R^m, where

$$R = \left(\sum_{\nu=1}^{m} x_\nu^2\right)^{\frac{1}{2}}$$

is the distance of the point $x(x_1, x_2, \ldots, x_m)$ from the origin, then

$$\sum_{\nu=1}^{m} \frac{\partial^2 u}{\partial x_\nu^2} = \frac{(m-1)}{R} f'(R) + f''(R) = 0,$$

so that

$$\begin{aligned} u &= A \log R + B, \qquad m = 2 \\ &= AR^{2-m} + B, \qquad m > 2. \end{aligned}$$

This leads to the following

DEFINITION. *The function $g(x, \xi, D)$ is said to be a (classical) Green's function of x with respect to the bounded domain D in R^m and the point ξ of D, if*

(i) *g is a harmonic function of x in D except at the point $x = \xi$;*
(ii) *g is continuous in $\bar{D}$ except at $x = \xi$ and $g = 0$ on the boundary of D;*
(iii) *$g + \log|x - \xi|$ remains harmonic at $x = \xi$ if $m = 2$,*
$g - |x - \xi|^{2-m}$ remains harmonic at $x = \xi$ if $m > 2$.

It will be shown in Theorem 1.14 that $g(x, \xi, D)$ is unique (if it exists). For the time being we do not need this result and so will speak about any function g having the properties (i) to (iii) as being a Green's function.

We note

THEOREM 1.10. *If $D = D(0, R)$ and ξ a point of D, $\xi' = \xi R^2 |\xi|^{-2}$, and if for $m = 2$*

$$g(x, \xi, D) = \log \frac{|x - \xi'||\xi|}{|x - \xi| R}, \qquad \xi \neq 0; \qquad g(x, 0, D) = \log \frac{R}{|x|};$$

while for $m > 2$

$$g(x, \xi, D) = |x - \xi|^{2-m} - \{|\xi||x - \xi'|/R\}^{2-m}, \qquad \xi \neq 0;$$

$$g(x, 0, D) = |x|^{2-m} - R^{2-m};$$

then $g(x, \xi, D)$ is a (classical) Green's function of D.

It is evident from the above analysis that the function $g(x, \xi, D)$ as defined above has the properties (i) and (iii). Further, if $|x| = R$,

$$|x - \xi'|^2 = |x|^2 + |\xi'|^2 - 2(\Sigma x_\nu \xi_\nu) \frac{R^2}{|\xi|^2}$$

$$= R^2 + \frac{R^4}{|\xi|^2} - 2(\Sigma x_\nu \xi_\nu) \frac{R^2}{|\xi|^2}$$

$$= \frac{R^2 |x - \xi|^2}{|\xi|^2},$$

so that

$$|\xi| |x - \xi'| = R |x - \xi|.$$

Thus $g(x, \xi, D) = 0$ when $|x| = R$, and the proof of Theorem 1.10 is complete.

We deduce

THEOREM 1.11. (Poisson's Integral). *If u is harmonic in $D(x_0, R)$ and continuous in $C(x_0, R)$ then for $\xi \in D(x_0, R)$ we have*

$$u(\xi) = \frac{1}{c_m} \int_{S(x_0, R)} \frac{R^2 - |\xi - x_0|^2}{R |x - \xi|^m} u(x) \, d\sigma_x \tag{1.5.1}$$

where $d\sigma_x$ denotes an element of surface area of $S(x_0, R)$ and $c_m = 2\pi^{m/2}/\Gamma(m/2)$.

We suppose, as we may do without loss of generality, that $x_0 = 0$, since otherwise we may consider $u(x_0 + x)$ instead of $u(x)$.

We assume first that u remains harmonic and so belongs to C^2 in $D(0, R')$ for some $R' > R$. Thus we may apply Theorem 1.9 with D_ε instead of D, where ε is a small positive number and

$$D_\varepsilon = (|x| < R) \cap (|x - \xi| > \varepsilon).$$

We use the function $g = g(x, \xi, D)$ of Theorem 1.10 instead of u. Also $S = S(0, R) \cup S(\xi, \varepsilon)$. Since u and g are harmonic in D_ε, we see from (1.4.5) that

$$\int_{D_\varepsilon} (u \nabla^2 g - g \nabla^2 u) \, dx = 0$$

Also on $S(0, R)$ we have $g = 0$. We next calculate $\partial g / \partial n$ on $S(0, R)$. We denote ξ by A, ξ' by A' and a general point x in space by Q. Let O be the origin, set $OQ = r$, $OA = \rho$ and let θ be the angle QOA. Then

$$AQ^2 = r^2 + \rho^2 - 2r\rho \cos \theta,$$

$$A'Q^2 = r^2 + \frac{R^4}{\rho^2} - \frac{2rR^2}{\rho} \cos \theta.$$

Thus

$$AQ \frac{\partial}{\partial r} AQ = r - \rho \cos \theta, \qquad A'Q \frac{\partial}{\partial r} A'Q = r - \frac{R^2}{\rho} \cos \theta.$$

If $m = 2$, we have for Q on $S(0, R)$, since $r = R$, $A'Q = (AQ)(R/\rho)$ in this case,

$$\frac{\partial g}{\partial n} = -\frac{\partial}{\partial r}\log\frac{A'Q}{AQ} = \frac{r - \rho\cos\theta}{AQ^2} - \frac{r - (R^2/\rho)\cos\theta}{A'Q^2}$$

$$= \frac{1}{AQ^2}\left[R - \rho\cos\theta - \frac{\rho^2}{R^2}\left(R - \frac{R^2}{\rho}\cos\theta\right)\right] = \frac{(R^2 - \rho^2)}{R.AQ^2}.$$

Also if $m > 2$ we have similarly

$$\frac{\partial g}{\partial n} = -\frac{\partial}{\partial r}\left[\frac{1}{AQ^{m-2}} - \left(\frac{R}{\rho}\frac{1}{A'Q}\right)^{m-2}\right]$$

$$= (m-2)\left[\frac{r - \rho\cos\theta}{(AQ)^m} - \left(\frac{R}{\rho}\right)^{m-2}\frac{r - (R^2/\rho)\cos\theta}{(A'Q)^m}\right]$$

$$= \frac{(m-2)}{(AQ)^m}\left[R - \rho\cos\theta - \frac{\rho^2}{R^2}\left(R - \frac{R^2}{\rho}\cos\theta\right)\right] = \frac{(m-2)(R^2 - \rho^2)}{R(AQ)^m}.$$

Thus we have on $S(0, R)$

$$\frac{\partial g}{\partial n} = \frac{(R^2 - \rho^2)}{R[R^2 + \rho^2 - 2R\rho\cos\theta]}, \qquad m = 2;$$

$$\frac{\partial g}{\partial n} = \frac{(m-2)(R^2 - \rho^2)}{R(R^2 + \rho^2 - 2R\rho\cos\theta)^{m/2}}, \qquad m > 2.$$

Next we note that on $S(\xi, \varepsilon)$ we have as $\varepsilon \to 0$

$$u(x) = u(\xi) + O(\varepsilon), \qquad \frac{\partial u}{\partial n} = O(1),$$

$$g(x) = o(\varepsilon^{1-m}); \quad \frac{\partial g}{\partial n} = -\varepsilon^{-1} + O(1), \qquad m = 2;$$

$$\frac{\partial g}{\partial n} = -(m-2)\varepsilon^{1-m} + O(1), \qquad m > 2.$$

and

$$\int_{S(\xi,\varepsilon)} d\sigma = \varepsilon^{m-1}\int_{S(0,1)} d\sigma = c_m\varepsilon^{m-1},$$

where c_m is a constant. Thus

$$\int_{S(\xi,\varepsilon)}\left(u\frac{\partial g}{\partial n} - g\frac{\partial u}{\partial n}\right)d\sigma = \begin{cases} -c_2 u(\xi) + o(1), & m = 2; \\ -(m-2)c_m u(\xi) + o(1), & m > 2. \end{cases}$$

Hence Theorem 1.8 gives finally

$$u(\xi) = \frac{1}{c_m} \int_{S(0,R)} \frac{(R^2 - \rho^2) u(x) \, d\sigma_x}{R(R^2 + \rho^2 - 2R\rho \cos \theta)^{m/2}},$$

where the constant c_m is the surface area of the hypersphere of radius one in m-dimensional space. It is an easy exercise to show that

$$c_m = \frac{2\pi^{m/2}}{\Gamma(m/2)}.$$

We have assumed that u remains harmonic in some $D(0, R')$ for $R' > R$. If this is not the case we apply the result of Theorem 1.11 with R_1 instead of R, where $R_1 < R$ and let R_1 tend to R from below. Since $u(x)$ is continuous on $C(0, R)$ it is evident that the right-hand side of (1.5.1) remains continuous as R_1 tends to R and Theorem 1.11 follows in the general case.

1.5.2. The maximum principle for harmonic functions

The Poisson integral is a powerful tool for deducing properties of harmonic functions. We start with

THEOREM 1.12. *The maximum principle. If u is harmonic in a bounded domain D in R^m and continuous in $\bar{D}$ and if $u \leqslant M$ on the boundary of D, then $u < M$ in D or else $u \equiv M$ in D.*

Let M' be the maximum of u in $\bar{D}$. The maximum is attained since $\bar{D}$ is compact. If $M' < M$ there is nothing to prove. If $M' = M$ and $u < M$ in D, there is again nothing to prove. Thus we may assume that $M' \geqslant M$ and that $u(\xi) = M'$ for some ξ in D.

Let E_1, E_2 be the points in D where $u < M'$ and $u = M'$ respectively. Clearly every point of D belongs to E_1 or E_2. Also, since u is continuous, E_1 is open. We show next that E_2 is open.

In fact let ξ be a point of E_2 and suppose that $D(\xi, r) \subset D$. Then we show that $D(\xi, r) \subset E_2$. In fact if $\rho < r$ it follows from Theorem (1.10) that

$$u(\xi) = \frac{\rho^{1-m}}{c_m} \int_{S(\xi,\rho)} u(x) \, d\sigma_x$$

and hence that

$$0 = u(\xi) - M' = \frac{\rho^{1-m}}{c_m} \int_{S(\xi,\rho)} (u(x) - M') \, d\sigma_x.$$

Here the integrand is non-positive and since $u(x)$ is continuous, $u(x) \equiv M'$

on $S(\xi, \rho)$ for $\rho < r$, i.e.

$$u(x) = M' \text{ in } D(\xi, r).$$

Thus E_2 is open.

Hence E_1 and E_2 are both open and since D is connected either E_1 or E_2 is empty. Thus $u < M'$ in D, in which case $M' \leqslant M$ and our result is proved, or $u \equiv M'$ in D, and since $u \leqslant M$ on the (non-empty) frontier of D, and u is continuous, we again deduce that $M' \leqslant M$. This proves Theorem 1.12.

Let D be a bounded domain in space with frontier S. *The problem of Dirichlet*† consists in finding a function u, harmonic in D continuous in $\bar{D}$ and assuming preassigned continuous values in S. We now deduce

THEOREM 1.13. *The solution to the problem of Dirichlet is unique, if it exists.*

In fact if u_1, u_2 solve the problem of Dirichlet with given boundary values for a domain D, then $u = u_1 - u_2$ is harmonic in D continuous in $\bar{D}$ and $u = 0$ on the boundary of D. Thus by Theorem 1.12, $u \leqslant 0$ in D and similarly $-u \leqslant 0$, i.e. $u \equiv 0$ in D. We also have

THEOREM 1.14. *The* (*classical*) *Green's function* $g(x, \xi, D)$ *is uniquely defined by the properties* (i) to (iii) *of section* 1.4.2 (*if it exists*). *Further* $g(x, \xi, D) > 0$ *in* D.

In fact if $g_1(x), g_2(x)$ have the properties (i), (ii) and (iii) for a given domain D and point ξ, then $g_1 - g_2$ remains harmonic at every point of D including ξ, since near ξ we can write

$$g_1 - g_2 = g_1 + \log|x - \xi| - \{g_2 + \log|x - \xi|\}$$

if $m = 2$,

$$g_1 - g_2 = (g_1 - |x - \xi|^{2-m}) - (g_2 - |x - \xi|^{2-m})$$

if $m > 2$. Also $g_1 - g_2 = 0$ on the boundary of D and so by the maximum principle $g_1 = g_2$ in D.

Next if Δ is the part of D outside a small neighbourhood of ξ, then g is continuous in $\bar{\Delta}$, harmonic in Δ and by (ii) and (iii) $g \geqslant 0$ on the boundary of Δ. Thus by the maximum principle applied to g we see that $g > 0$ in Δ or $g \equiv 0$ in Δ. The second possibility is excluded by (ii) if we choose our neighbourhood of ξ sufficiently small.

† Riemann [1857] refers to Dirichlet in discussing this problem.

1.5.3. Analyticity

We prove

THEOREM 1.15. *If $u(x)$ is harmonic in a domain D, then $u(x) \in A$ in D.*

To see this we note that if $|x - x_0| = R, |\xi - x_0| = \rho, (x - x_0).(\xi - x_0) = t$, and $\rho < \frac{1}{3}R$, then

$$|x - \xi|^{-m} = (R^2 - 2t + \rho^2)^{-m/2}$$
$$= R^{-m} \sum_0^{\infty} b_\nu \left(\frac{2t - \rho^2}{R^2}\right)^\nu$$

where the b_ν are binomial coefficients. Also if x_μ, ξ_μ are the coordinates of $x - x_0$, $\xi - x_0$ respectively, we have

$$|2t - \rho^2| = |2\Sigma x_\mu \xi_\mu - \Sigma \xi_\mu^2| \leqslant 2\Sigma |x_\mu| |\xi_\mu| + \Sigma \xi_\mu^2 \leqslant \rho^2 + 2R\rho \leqslant \tfrac{7}{9}R^2.$$

Thus for $|x - x_0| = R$, $|\xi - x_0| < \frac{1}{3}R$ and $\xi - x_0 = (\xi_1, \xi_2, \ldots, \xi_m)$, we can write

$$\frac{R^2 - |\xi - x_0|^2}{R} |x - \xi|^{-m} = \Sigma a_\mu(x) \xi^{(\mu)}$$

where $\xi^{(\mu)} = \xi_1^{\mu_1} \xi_2^{\mu_2} \ldots \xi_m^{\mu_m}$ and the sum is taken over all sets of m non-negative integers $(\mu) = (\mu_1, \ldots, \mu_m)$ and is absolutely and uniformly convergent. Hence we can substitute this series in (1.5.1) if $u(x)$ is harmonic in $C(x_0, R)$ and integrate term by term with respect to x. We obtain an expansion

$$u(\xi) = \sum_{(\mu)} a_{(\mu)} \xi^{(\mu)}$$

which is uniformly and absolutely convergent for $|\xi - x_0| < \frac{1}{3}R$. Since such an expansion is valid near any point x_0 of D, $u(\xi) \in A$. We recall that by the analysis in Section 1.2.2 it follows in particular that $u(x) \in C^\infty$ and also that in view of Theorem 1.5, $u(x)$ is determined throughout D by its values in any open set lying in D.

A somewhat deeper analysis shows that the series for $u(\xi)$ converges absolutely for $\rho < (R/\sqrt{2})$, and this constant is sharp.†

1.5.4. The problem of Dirichlet for a hyperball

We proceed to show that the Poisson integral can be used to solve the

† Kiselman [1969], Hayman [1970].

problem of Dirichlet for a hyperball. We have more precisely

THEOREM 1.16. *Suppose that the function $f(x)$ is continuous on $S(x_0, R)$ and for $\xi \in D(x_0, R)$ let*

$$u(\xi) = \frac{1}{c_m}\int_{S(x_0, R)} f(x)\frac{(R^2 - |\xi - x_0|^2)}{R|\xi - x|^m}\, d\sigma_x, \tag{1.5.2}$$

where $d\sigma_x$ denotes an element of surface area of $S(x_0, R)$. Then $u(\xi)$ solves the problem of Dirichlet for $D(x_0, R)$, with boundary values $f(x)$.

We prove our result in a number of stages.

(i) *The function $u(\xi)$ is harmonic in $D(x_0, R)$.* In fact it is evident that we may differentiate under the integral sign any number of times with respect to the coordinates of ξ, since $f(x)$ is continuous, and

$$K(x, \xi) = \frac{1}{c_m}\frac{R^2 - |\xi - x_0|^2}{R|\xi - x|^m} \tag{1.5.3}$$

has continuous partial derivatives of all orders in the coordinates of (x, ξ) jointly for $\xi \neq x$. Thus it is enough to show that $K(x, \xi)$ is harmonic in ξ for fixed x. We may suppose without loss in generality that $x = 0$, since we can write $\xi - x$ for x, without changing the harmonicity. Let $\xi = (\xi_1, \ldots, \xi_m)$, $x_0 = (x_1, \ldots, x_m)$, $x = 0$, then

$$K(x, \xi) = \frac{\Sigma x_\nu^2 - \Sigma(\xi_\nu - x_\nu)^2}{(\Sigma\xi_\nu^2)^{m/2}} = \frac{2\Sigma x_\nu \xi_\nu}{(\Sigma\xi_\nu^2)^{m/2}} - (\Sigma\xi_\nu^2)^{1-(m/2)}.$$

The second term is harmonic as we saw in Section 1.5.1. Also if u is harmonic, then so is $\partial u/\partial\xi_\nu$, since u is real analytic and so

$$\nabla^2 \frac{\partial u}{\partial \xi_\nu} = \frac{\partial}{\partial \xi_\nu}\nabla^2 u = 0.$$

If $R = (\Sigma\xi_\nu^2)^{\frac{1}{2}}$ then, for $m = 2$, $\log R$ is harmonic and hence so is

$$\frac{x_1\xi_1 + x_2\xi_2}{R^2} = \left(x_1\frac{\partial}{\partial\xi_1} + x_2\frac{\partial}{\partial\xi_2}\right)\log R,$$

and for $m > 2$ we see similarly that

$$\frac{\Sigma x_\nu \xi_\nu}{R^m} = \frac{-1}{m-2}\left(\Sigma x_\nu \frac{\partial}{\partial\xi_\nu}\right)R^{2-m}$$

is harmonic. Thus $K(x, \xi)$ is harmonic and so is $u(\xi)$.

(ii) *If* $m \leqslant f(x) \leqslant M$ *on* $S(x_0, R)$, *then* $m \leqslant f(x) \leqslant M$ *in* $D(x_0, R)$. If $f(x) = C =$ constant then by Theorem 1.11 we see that $u(\xi) = C$. Hence, for any constant C,

$$u(\xi) - C = \int_{S(x_0, R)} K(x, \xi)(f(x) - C)\, d\sigma_x.$$

Since $f(x) \leqslant M$, and $K(x, \xi) > 0$, we see that $u(\xi) \leqslant M$. Similarly $u(\xi) \geqslant m$.

(iii) *If* N_1 *is a neighbourhood of* x_1 *on* $S(x_0, R)$, *and* $m \leqslant f(x) \leqslant M$ *in* N_1, *then*

$$m \leqslant \underline{\lim}\, u(\xi) \leqslant \overline{\lim}\, u(\xi) \leqslant M,$$

where the limits are taken as $\xi \to x_1$ *from* $D(x_0, R)$.

We write $S(x_0, R) = N_1 \cup N_2$,

$$u(\xi) - M = \int_{N_1} K(x, \xi)(f(x) - M)\, d\sigma_x + \int_{N_2} K(x, \xi)(f(x) - M)\, d\sigma_x$$
$$= I_1 + I_2$$

Also by hypothesis $I_1 \leqslant 0$. For $x \in N_2$

$$K(x, \xi) = \frac{R^2 - |\xi - x_0|^2}{c_m R |\xi - x|^m} \to 0$$

uniformly as $\xi \to x_1$, since then $|\xi - x_0|^2 \to |x_1 - x_0|^2 = R^2$, but $|\xi - x| \to |x_1 - x|$ which is bounded below for x outside N_1. Thus $I_2 \to 0$ as $\xi \to x_1$. Hence

$$\overline{\lim_{\xi \to x_1}}\, u(\xi) \leqslant M,$$

and similarly we have

$$\underline{\lim_{\xi \to x_1}}\, u(\xi) \geqslant m.$$

(iv) *If* $x_1 \in S(x_0, R)$ *then*

$$u(\xi) \to f(x_1) \quad \text{as} \quad \xi \to x_1.$$

In fact since $f(x)$ is continuous at x_1 we can apply (iii) with $m = f(x_1) - \varepsilon$, $M = f(x_1) + \varepsilon$. On letting ε tend to zero, we deduce (iv) from (iii).

Now Theorem 1.16 follows from (i) and (iv).

1.5.5. The mean-value property

It follows from Theorem 1.11 that if $u(x)$ is harmonic in a neighbourhood of $C(x_0, R)$, then

$$u(x_0) = \frac{1}{c_m R^{m-1}} \int_{S(x_0, R)} u(x)\, d\sigma_x \tag{1.5.4}$$

where $d\sigma_x$ denotes an element of surface area on $S(x_0, R)$. We can integrate the right-hand side with respect to R from $R = 0$ to ρ, if $u(x)$ is harmonic in $C(x_0, \rho)$ and deduce that

$$u(x_0) = \frac{1}{d_m \rho^m} \int_{C(x_0, \rho)} u(x)\, dx \tag{1.5.5}$$

where dx denotes m-dimensional volume and

$$d_m \rho^m = \int_{C(0, \rho)} dx = \frac{c_m \rho^m}{m} = \frac{\pi^{m/2}}{\Gamma(\frac{1}{2}m + 1)} \rho^m$$

is the volume of the m-dimensional ball of radius ρ. Either of the properties (1.5.4) and (1.5.5) can be taken as defining harmonic functions. It is clear that (1.5.5) implies (1.5.4) so we prove a result using only (1.5.4).

THEOREM 1.17. *If $u(x)$ is continuous in a domain D of R^m and for each $x_0 \in D$ the equation* (1.5.5) *holds for some arbitrarily small ρ then $u(x)$ is harmonic in D.*

Let $C(x_0, r)$' be a hyperball lying in D and let $v(x)$ solve the problem of Dirichlet in $C(x_0, r)$ for the boundary values $u(x)$ in $S(x_0, r)$. By Theorem 1.16 $v(x)$ exists uniquely and is given by the Poisson integral.

Set

$$h(x) = v(x) - u(x)$$

for x in $D(x_0, r)$. It is enough to prove that $h(x) \equiv 0$ in $D(x_0, r)$, since $v(x)$ is harmonic in $D(x_0, r)$ by Theorem 1.16. We note that $h(x)$ is continuous on the closed ball $C = C(x_0, r)$ and vanishes on the boundary $S = S(x_0, r)$. Thus if $h(x)$ is not identically zero $h(x)$ must have a positive maximum or a negative minimum in $C(x_0, r)$ and so in $D = D(x_0, r)$. Suppose, e.g. that

$$m = \sup_{x \in C} h(x) > 0.$$

The set of points E in D such that $h(\xi) = m$ is compact and not empty, since $h(x)$ is continuous, and so we can find a point $\xi_0 \in E$ such that $|\xi_0 - x_0|$ is maximal. Since $u(x)$ satisfies (1.5.5) for some arbitrarily small ρ and $v(x)$ is harmonic, $h(x)$ satisfies (1.5.5) for the same values of ρ and so we can find ρ as small as we please such that

$$\int_{C(\xi_0, \rho)} [h(x) - h(\xi_0)]\, dx = 0.$$

Here the integrand is continuous and non-positive in $C(\xi_0, \rho)$ and so this integrand must be identically zero. In particular

$$h(\xi_1) = m, \quad \text{where} \quad \xi_1 = \xi_0 + \frac{\rho}{2}(\xi_0 - x_0).$$

But

$$|\xi_1 - x_0| = \left|\left(1 + \frac{\rho}{2}\right)(\xi_0 - x_0)\right|$$

which contradicts our assumption that $|\xi_0 - x_0|$ was maximal subject to $h(\xi_0) = m$. Thus $h(x)$ must be identically zero and Theorem 1.17 is proved.

Examples

1. *Schwarz's reflection principle.* If u is harmonic in a domain D in $x_1 > 0$ whose frontier contains an open subset Δ of $x_1 = 0$ and if u remains continuous on Δ and $u = 0$ there, prove that u can be continued as a harmonic function into the domain $D_1 = D \cup \Delta \cup D'$, where D' consists of the reflection of D in the hyperplane $x_1 = 0$.

(Set $u(x_1, x_2, \ldots, x_m) = -u(-x_1, x_2, \ldots, x_m)$, $x \in D'$, and use Theorem 1.17).

2. If $u \in C$ in a domain D in R^m and if the partial derivatives $(\partial^2 u/\partial x_\nu^2)$ exist at each point of D and satisfy $\nabla^2 u = 0$, prove that u is harmonic in D.

(Let D' be a bounded domain whose closure lies in D. Show that $u + \varepsilon|x|^2$ satisfies a maximum principle in D' for every $\varepsilon > 0$, and hence that u satisfies a maximum principle in D'. Deduce that u coincides in every hyperball in D with the Poisson integral of its boundary values.)

1.5.6. Harnack's Inequality and Harnack's Theorem

We proceed to prove the m-dimensional form of an inequality due to Harnack [1886].

THEOREM 1.18. *Suppose that $u(x)$ is harmonic and positive in $D(x_0, r)$. Then for $|\xi - x_0| = \rho < r$ we have*

$$\frac{(r-\rho)r^{m-2}}{(r+\rho)^{m-1}}u(x_0) \leqslant u(\xi) \leqslant \frac{(r+\rho)r^{m-2}}{(r-\rho)^{m-1}}u(x_0).$$

We have, by Theorem 1.16 with $\rho < R < r$,

$$u(\xi) = \int_{S(x_0, R)} u(x)\, K(x, \xi)\, d\sigma_x,$$

where $K(x, \xi)$ is given by (1.5.3) so that

$$\frac{R^2 - \rho^2}{R(R + \rho)^m} \leqslant c_m K(x, \xi) \leqslant \frac{R^2 - \rho^2}{R(R - \rho)^m}.$$

Also setting $\rho = 0$ we obtain

$$u(x_0) = \frac{1}{c_m R^{m-1}} \int_{S(x_0, R)} u(x)\, d\sigma_x.$$

This gives at once the desired inequality with R instead of r. We obtain our result by allowing R to tend to r from below.

We note that the inequalities of Theorem 1.18 are sharp by considering the functions

$$u(x) = \frac{r^2 - |x - x_0|^2}{r|x - x_1|^m},$$

where $|x_1 - x_0| = r$, which were shown to be harmonic in Section 1.5.4.

We also deduce

THEOREM 1.19. *If $u(x)$ is harmonic and not constant in R^m and*

$$A_1(r) = \sup_{|x|=r} u(x), \qquad A_2(r) = \inf_{|x|=r} u(x),$$

then

$$\overline{\lim_{r\to\infty}} \frac{A_2(r)}{r} < 0 < \underline{\lim_{r\to\infty}} \frac{A_1(r)}{r}.$$

In particular $u(x)$ cannot be bounded above or below.

We apply Theorem 1.18 to $A_1(r) - u(x)$ in $|x| < r$, and obtain for $|x| = \rho < r$.

$$A_1(r) - u(x) \leqslant \frac{(r + \rho)r^{m-2}}{(r - \rho)^{m-1}}[A_1(r) - u(0)],$$

i.e.

$$u(x) \geqslant \frac{(r + \rho)r^{m-2}}{(r - \rho)^{m-1}} u(0) - \frac{(r + \rho)r^{m-2} - (r - \rho)^{m-1}}{(r - \rho)^{m-1}} A_1(r).$$

We suppose that there exists a sequence $r = r_n$ such that

$$\frac{A_1(r_n)}{r_n} \to \alpha \leqslant 0.$$

Then we deduce by allowing r to tend to infinity through the values of this sequence that for any x in space

$$u(x) \geqslant u(0).$$

This contradicts the maximum principle unless $u(x)$ is constant. Thus

$$\underline{\lim}_{r \to \infty} \frac{A_1(r)}{r} > 0$$

and similarly

$$\overline{\lim}_{r \to \infty} \frac{A_2(r)}{r} < 0.$$

Examples

1. If $u(x)$ is positive harmonic in $D(0, r)$, prove that at $x = 0$

$$\left|\frac{\partial u}{\partial x_1}\right| \leqslant \frac{m}{r} u(0).$$

2. If u is harmonic and satisfies $u < M$ in $D(x_0, r)$, and $u(x_0) = 0$ show that

$$\left|\frac{\partial u}{\partial x_1}\right| < \frac{A(\lambda)M}{r}$$

in $D(x_0, \lambda r)$ if $0 < \lambda < 1$, where $A(\lambda)$ depends on λ only. Hence or otherwise show that if $p_1 + p_2 + \cdots p_m = p$

$$\left|\frac{\partial^p u}{\partial x_1^{p_1} \partial x_2^{p_2} \ldots \partial x_m^{p_m}}\right| < \frac{A(\lambda, p)M}{r^p} \text{ in } D(x_0, \lambda r),$$

where $A(\lambda, p)$ depends on λ and p only.

3. If $u(x)$ is harmonic in space and does not reduce to a polynomial prove that in the notation of Theorem 1.19

$$\lim_{r \to \infty} \frac{\log|A_j(r)|}{\log r} \to \infty.$$

(If the lower limit is $p < \infty$, show that all the partial derivatives of order greater than p vanish identically).

We have next *Harnack's Theorem.*†

THEOREM 1.20. *Suppose that $u_n(x)$ is a monotonic increasing sequence of harmonic functions in a domain D. Then either $u_n(x)$ diverges to $+\infty$ everywhere in D or*

$$u_n(x) \to u(x)$$

uniformly in every compact subset of D and $u(x)$ is harmonic in D.

† Harnack [1886].

Clearly $u(x)$ exists everywhere in D as a finite or infinite limit. Suppose that $u(x_0) < +\infty$ for at least one x_0 in D. Then we have for $n > m > N_0(\varepsilon)$

$$u_n(x_0) - u_m(x_0) < \varepsilon.$$

Then if $D(x_0, r)$ lies in D we have for $|x - x_0| = \rho < r$ by Theorem 1.18

$$0 < u_n(x) - u_m(x) < \varepsilon \frac{(r+\rho)r^{m-2}}{(r-\rho)^{m-1}}, \qquad n > m > N_0(\varepsilon).$$

Thus $u_n(x)$ converges uniformly in $C(x_0, \rho)$ for $\rho < r$ to a limit $u(x)$. Thus $u(x)$ is finite and continuous in $C(x_0, \rho)$. Similarly from the left hand inequality of Theorem 1.18 we see that if $u(x_0) = \infty$ then $u(x) = \infty$ in $D(x_0, r)$. Thus the sets where $u(x) = \infty$ and $u(x) < \infty$ are both open in D and so one of these sets must be empty. If e.g. $u(x) < \infty$ in D, then the convergence is locally uniform in D. We now note that if $C(x_0, r)$ lies in D and $|\xi - x_0| = \rho < r$

$$u_n(\xi) = \int_{S(x_0, r)} u_n(x)\, K(x, \xi)\, d\sigma_x,$$

where $K(x, \xi)$ is the Poisson–Kernel given by (1.5.3). Letting n tend to infinity we deduce that

$$u(\xi) = \int_{S(x_0, r)} u(x)\, K(x, \xi)\, d\sigma_x$$

and in view of Theorem 1.16 we deduce that $u(\xi)$ is harmonic. This proves Theorem 1.20.

1.5.7. Conclusion

Harmonic functions in the plane have a particular importance in view of their connection with regular functions. Suppose that $u(x, y)$ is harmonic in a plane domain D and set $z = x + iy$

$$f(z) = \frac{\partial u}{\partial x} - i\frac{\partial u}{\partial y} = U + iV.$$

Then, since $u \in C^2$,

$$\frac{\partial^2 u}{\partial x\, \partial y} = \frac{\partial U}{\partial y} = -\frac{\partial V}{\partial x},$$

and since u is harmonic

$$\frac{\partial^2 u}{\partial x^2} + \frac{\partial^2 u}{\partial y^2} = \frac{\partial U}{\partial x} - \frac{\partial V}{\partial y} = 0.$$

Thus $f(z) = U + iV$ satisfies the Cauchy–Riemann equations everywhere and since $U, V \in C^1$, $f(z)$ is a regular function. If we define near any fixed point z_0 of D

$$F(z) = \int_{z_0}^{z} f(\xi)\,d\xi = u_1 + iv_1,$$

we easily see that u_1, v_1 are harmonic functions and that $u(z) = u_1(z) + u(z_0)$. Thus u is locally the real part of the regular function $F(z) + u(z_0)$. This result together with the use of the conjugate harmonic function $v_1(z)$ enables many of the properties of harmonic functions in the plane to be derived very simply. However once Theorems 1.12 and 1.16 have been obtained these results can be used, as we have seen, to provide for many of the standard properties of harmonic functions proofs which are valid also in space of higher dimensions.

Example

If $u(z)$ is harmonic in a plane domain D and $z = t(w)$ is regular in D' and maps D' into D, prove that $u[t(w)]$ is harmonic in D'.

Chapter 2

Subharmonic Functions

2.0. INTRODUCTION

In this chapter we develop the definition and simple properties of subharmonic (s.h.) functions. They are related to harmonic functions just as convex functions are related to linear functions in one dimension. After giving the definition and a few examples we develop the maximum principle which is one of the key properties of s.h. functions. From this we deduce that s.h. functions lie in any disk below the Poisson integral of their function values on the circumference of the disk. This leads us to the central part of the chapter, which is Perron's method of solving the problem of Dirichlet by means of s.h. functions, whose lack of analyticity makes them a very flexible tool. Convexity theorems for mean-values on hyperspheres follow and these will play a fundamental role in the following chapter. We finish the chapter with a short discussion of subordination for regular functions in a disk, where subharmonic functions have an attractive application.

2.1. DEFINITION AND SIMPLE EXAMPLES

We have seen in Theorem 1.17 that harmonic functions can be defined in terms of a mean value property. If we replace equality by inequality in this relation we obtain the subharmonic functions. We may consider such functions as lying below harmonic functions, just as convex functions lie below linear functions. In fact convex functions are one-dimensional subharmonic functions. These ideas suggest the following

DEFINITION†. *A function $u(x)$ defined in a domain D of R^m is said to be subharmonic (s.h.) in D if*

(i) $-\infty \leqslant u(x) < +\infty$ *in D.*

† This elegant formulation appears to be due to F. Riesz [1926, 1930].

(ii) *$u(x)$ is* u.s.c. *in D.*

(iii) *If x_0 is any point of D then there exist arbitrarily small positive values of r such that*†

$$u(x_0) \leqslant \frac{1}{c_m r^{m-1}} \int_{S(x_0, r)} u(x)\, d\sigma(x),$$

where $d\sigma(x)$ denotes surface area on $S(x_0, r)$.

It follows from (1.5.4) that real multiples of harmonic functions are s.h. Conversely we shall see that u is harmonic if and only if u and $-u$ are s.h. We note some properties of s.h. functions. The first two are very simple and their proof is left to the reader.

Examples

1. *If $u_1, \ldots, u_k$ are* s.h. *in D and $t_1, \ldots, t_k$ are nonnegative real numbers, then*

$$u = \sum_{\nu=1}^{k} t_\nu u_\nu$$

is s.h.

2. *If $u_1, \ldots, u_k$ are* s.h. *in D then so is $u(x) = \sup_{\nu = 1 \text{ to } k} u_\nu(x)$.*

3. *If $u \in C^2$ in D, then u is* s.h. *in D if and only if $\nabla^2 u \geqslant 0$ in D.* If $0 < \rho < r$ it follows from Theorem 1.9 in this case that if $C(x_0, r)$ lies in D, then

$$\int_{S(x_0, r)} - \int_{S(x_0, \rho)} \frac{\partial u}{\partial r}\, d\sigma = \int_{\Delta(x_0, \rho)} \nabla^2 u\, dx,$$

where $\Delta(x_0, \rho) = D(x_0, \rho) - C(x_0, r)$ and $\dfrac{\partial}{\partial r}$ denotes the differentiation in the direction of increasing r. Suppose first that $\nabla^2 u \geqslant 0$ in D. Then given x_0 in D and r sufficiently small, we deduce that $\mu(r) = r^{m-1} J'(r)$ is an increasing function of r, where

$$J(r) = \frac{1}{c_m r^{m-1}} \int_{S(x_0, r)} u(x)\, d\sigma(x).$$

Evidently $\mu(0) = 0$. Thus $J'(r) \geqslant 0$ for $r > 0$, and so

$$J(r) \geqslant \lim_{r \to 0+} J(r) = u(x_0).$$

Since $u(x)$ is continuous it follows that $u(x)$ is s.h. in D.

† Here and subsequently $c_m = 2\pi^{m/2}/\Gamma(m/2)$ as in Theorem 1.1. Some authors assume that $u(x)$ is not identically $-\infty$, but we regard the function which is identically $-\infty$ as s.h.

Suppose next that $\nabla^2 u < 0$ for some point x_0 of D. Then we deduce from the continuity of $\nabla^2 u$ that $\nabla^2 u < 0$ near x_0 and hence that $\mu(r)$ decreases and so is negative for small r. Thus

$$J(r) < J(0)$$

for small r and so $u(x)$ is not s.h. near x_0.

4. *If $f(z)$ is a regular function of the complex variable z in a plane domain D, then $u(z) = \log|f(z)|$ is s.h. in D.*

Clearly properties (i) and (ii) are satisfied. Also (iii) is obvious if z_0 is a zero of $f(z)$ since then $u(z_0) = -\infty$. If $f(z_0) \neq 0$, then $\log f(z)$ is regular and so $u(z) = \log|f(z)|$ is harmonic and so subharmonic near z_0.

2.2 JENSEN'S INEQUALITY†

In order to obtain some other classes of s.h. functions we need an inequality for integrals. We state this in a form which is not the most general possible, but which covers all the applications we have in mind.

THEOREM 2.1. *Suppose that $u(x)$ is a function of x defined on a set E of R^m, and that $\phi(u)$ is a convex function of u on an interval containing the range of values assumed by $u(x)$ on E. Then we have*

$$\phi\left\{\frac{1}{\mu(E)}\int_E u(x)\,d\mu(x)\right\} \leqslant \frac{1}{\mu(E)}\int_E \phi[u(x)]\,d\mu(x), \tag{2.2.1}$$

provided that E is a measurable set in R^m and $d\mu$ denotes Lebesgue measure, or E is on a hypersurface S on R^m and $d\mu$ denotes surface area on S, and further that $0 < \mu(E) < \infty$ and that $u(x)$ is integrable on E.

The inequality is to be understood in the following sense. By hypothesis

$$I = \frac{1}{\mu(E)}\int_E u(x)\,d\mu(x)$$

is finite. If the range of $u(x)$ is contained in a finite closed interval $[a, b]$ then I also lies in $[a, b]$. If the interval containing the range of $u(x)$ is open or semi-open, we define

$$\phi(a) = \lim_{x\to a+} \phi(x), \qquad \phi(b) = \lim_{x\to b-} \phi(x).$$

In view of the convexity of $\phi(x)$ these limits exist, but may be infinite. Thus the left hand side is well defined in (2.2.1). The inequality is to be interpreted in

† Jensen [1905].

the sense that if the left-hand side is $-\infty$ no assertion is made. If the left-hand side is finite then the right-hand side exists as a finite integral satisfying (2.2.1) or else the right-hand side is $+\infty$. If the left-hand side is $+\infty$, then so is the right-hand side.

2.2.1. We first need the discrete analogue of Theorem 2.1. This is

LEMMA 2.1. *Suppose that $\phi(u)$ is convex in $[a, b]$, that $t_i \geqslant 0, i = 1$ to k and that $\sum_{i=1}^{k} t_i = 1$. Then if $a \leqslant u_i \leqslant b$, $i = 1$ to k, we have*

$$\phi\left(\sum_{i=1}^{k} t_i u_i\right) \leqslant \sum_{i=1}^{k} t_i \phi(u_i).$$

We suppose without loss in generality that $a \leqslant u_1 \leqslant u_2 \ldots \leqslant u_k \leqslant b$ and proceed by induction on k. Lemma 2.1 is trivial if $k = 1$. For $k = 2$ the inequality is equivalent to the definition (1.3.3.) of convexity. Suppose that the result has already been proved for $k - 1 \geqslant 2$. We suppose that $0 < t_k < 1$, since otherwise the inequality follows either from the result for $k - 1$ or for 2.

Write

$$t_1 u_1 + t_2 u_2 + \ldots + t_{k-1} u_{k-1} = (1 - t_k) v_k.$$

Since $\sum_{i=1}^{k-1} t_i = 1 - t_k$, we deduce that $u_1 \leqslant v_k \leqslant u_{k-1}$, so that v_k lies in $[a, b]$. Hence in view of (1.3.3) we deduce that

$$\begin{aligned}\phi[(1 - t_k)v_k + t_k u_k] &\leqslant (1 - t_k)\phi(v_k) + t_k \phi(u_k)\\ &\leqslant \sum_{i=1}^{k-1} t_i \phi(u_i) + t_k \phi(u_k) = \sum_{i=1}^{k} t_i \phi(u_i)\end{aligned}$$

by the inductive hypothesis. This proves the Lemma.

We proceed to prove Theorem 2.1 and suppose first of all that $u(x)$ is a simple function on the set E, i.e. that $u(x)$ assumes only a finite number of different values $u_1, u_2, \ldots, u_k$. Let E_i be the subset of E where $u = u_i$ and set $\mu(E_i) = \mu_i$, $\mu(E) = \mu$, $\mu_i/\mu = t_i$. Then the numbers t_i, u_i satisfy the hypotheses of Lemma 2.1. Thus we deduce from Lemma 2.1 that

$$\phi\left\{\frac{1}{\mu(E)}\int_E u(x)\,\mathrm{d}\mu(x)\right\} = \phi\left\{\sum_{i=1}^{k} t_i u_i\right\} \leqslant \sum_{i=1}^{k} t_i \phi(u_i) = \frac{1}{\mu(E)}\int_E \phi[u(x)]\,\mathrm{d}\mu(x).$$

Thus the result is proved in this case.

We suppose next that $u(x)$ is bounded on E, $a \leqslant u \leqslant b$ and that $\phi(u)$ is convex on $[a, b]$. Let N be a large positive integer. We write

$$u_i = a + \frac{i(b-a)}{N}, \qquad 0 \leqslant i \leqslant N,$$

and set

$$u_N(x) = u_i, \qquad \text{if} \quad u_{i-1} < u(x) \leqslant u_i,\ i > 1;$$
$$u_N(x) = u_1, \qquad \text{if} \quad u_0 \leqslant u(x) \leqslant u_1.$$

Then $u_N(x)$ is a simple function, so that

$$\phi\left\{\frac{1}{\mu(E)}\int_E u_N(x)\,\mathrm{d}\mu(x)\right\} \leqslant \frac{1}{\mu(E)}\int_E \phi[u_N(x)]\,\mathrm{d}\mu(x). \tag{2.2.2}$$

We note that $u_N(x) \to u(x)$ as $N \to \infty$, uniformly on E, and since $\phi(u)$ is convex and so continuous and uniformly continuous on $[a, b]$ by Theorem 1.6 $\phi[u_N(x)]$ tends to $\phi[u(x)]$ uniformly on E. Thus

$$\int_E u_N(x)\,\mathrm{d}\mu(x) \to \int_E u(x)\,\mathrm{d}\mu(x)$$

$$\int_E \phi[u_N(x)]\,\mathrm{d}\mu(x) \to \int_E \phi[u(x)]\,\mathrm{d}\mu(x)$$

and so in view of (2.2.2) we deduce (2.2.1).

We next suppose that the smallest interval I containing the range of values of $u(x)$ for x on E is not a closed interval, so that I may be open or semi-open stretching to infinity. Let I_n be an expanding sequence of closed intervals $a_n \leqslant x \leqslant b_n$, whose limit is I. Thus if a, b are the lower and upper bounds of x in I, we choose $a_n = a$ for all n if a lies in I and otherwise take a_n to be strictly decreasing and tending to a as $n \to \infty$. Similarly if b lies in I, we choose $b_n = b$ for all n and otherwise choose b_n to be strictly increasing and tending to b as $n \to \infty$.

Let E_n be the subset of E, where $a_n \leqslant u(x) \leqslant b_n$. Then $u(x)$ is bounded in E and $\phi[u]$ is convex in $[a_n, b_n]$. Also E_n tends to E and so $\mu(E_n)$ tends to $\mu(E)$ as $n \to \infty$. Thus $\mu(E_n)$ is positive for all sufficiently large n and so we can apply (2.2.1) with E_n instead of E. Thus

$$\phi\left\{\frac{1}{\mu(E_n)}\int_{E_n} u(x)\,\mathrm{d}\mu(x)\right\} \leqslant \frac{1}{\mu(E_n)}\int_{E_n} \phi[u(x)]\,\mathrm{d}\mu(x). \tag{2.2.3}$$

Let E_n' be the set where $u(x) < a_n$ and E_n'' the set where $u(x) > b_n$. Since $u(x)$ is integrable over E, it follows that

$$\int_{E_n'} u(x)\,\mathrm{d}\mu(x) \to 0, \qquad \int_{E_n''} u(x)\,\mathrm{d}\mu(x) \to 0, \tag{2.2.4}$$

as $n \to \infty$.

Since $\phi'(x)$ increases, we have for $x_0 \in (a, b)$, $x_0 < x < b$

$$\phi'(x) \geqslant \phi'(x_0),$$

so that

$$\phi(x) \geqslant \phi(x_0) + \phi'(x_0)(x - x_0).$$

If $\phi'(x) < 0$, for $a < x < b$ we see that $\phi(x)$ is decreasing in (a, b) and so $\phi(x)$ is bounded above as $x \to b$. Thus in this case $\phi(u)$ is bounded as $u \to b$, unless $b = +\infty$, in which case

$$\phi(u) = O(|u|), \quad \text{as} \quad u \to +\infty.$$

Thus in this case we have by (2.2.4)

$$\int_{E_n''} \phi(u(x))\,\mathrm{d}\mu(x) = O\left\{\int_{E_n''} u(x)\,\mathrm{d}\mu(x)\right\} + O(\mu(E_n'')) \to 0 \quad \text{as} \quad n \to \infty.$$

On the other hand if $\phi'(u)$ is finally positive, then either $\phi(u)$ is bounded as $u \to b$ or $\phi(u) \to +\infty$ as $u \to b$. In the former case

$$\int_{E_n''} \phi(u(x))\mathrm{d}\mu(x) \to 0. \tag{2.2.5}$$

In the latter either

$$\int_{E_n''} \phi(u(x))\,\mathrm{d}\mu(x) = +\infty \quad \text{for each } n \tag{2.2.6}$$

or (2.2.5) holds. Summing up we see that in each case either (2.2.5) or (2.2.6) holds. A similar conclusion holds for E_n'. In view of (2.2.4) we see that

$$\frac{1}{\mu(E_n)}\int_{E_n} u(x)\,\mathrm{d}\mu(x) \to \frac{1}{\mu(E)}\int_E u(x)\,\mathrm{d}\mu(x) = I,$$

say, where $a \leqslant I \leqslant b$. Since $\phi(u)$ is convex and so continuous it follows that

$$\phi\left\{\frac{1}{\mu(E_n)}\int_{E_n} u(x)\,\mathrm{d}\mu(x)\right\} \to \phi(I),$$

where $\phi(I)$ is defined by continuity in the limiting cases $I = a, b$. Thus in view of (2.2.3), (2.2.5) and (2.2.6) we deduce that

$$\phi(I) \leqslant \lim \frac{1}{\mu(E_n)}\int_{E_n} \phi[u(x)]\,\mathrm{d}\mu(x) \leqslant \frac{1}{\mu(E)}\int_E \phi[u(x)]\,\mathrm{d}\mu(x).$$

This completes the proof of Theorem 2.1.

2.3. SOME FURTHER CLASSES OF SUBHARMONIC FUNCTIONS

We can use Theorem 2.1 to prove the following

THEOREM 2.2. *If $u(x)$ is s.h. in a domain D and $\phi(u)$ is convex and increasing on the range R of values assumed by $u(x)$ in D or if $u(x)$ is harmonic in D and $\phi(u)$ is convex on R, then $\phi[u(x)]$ is s.h. in D.*

Suppose first that $u(x)$ is harmonic in D. Then $\phi(u)$ is convex and so continuous on R and hence $\phi[u(x)]$ is continuous and finite. Thus properties (i) and (ii) of the definition are satisfied. Also

$$u(x_0) = \frac{1}{c_m r^{m-1}} \int_{S(x_0, r)} u(x)\, d\sigma(x),$$

so that

$$\begin{aligned}\phi[u(x_0)] &= \phi\left(\frac{1}{c_m r^{m-1}} \int_{S(x_0, r)} u(x) d\sigma(x)\right)\\ &\leqslant \frac{1}{c_m r^{m-1}} \int_{S(x_0, r)} \phi[u(x)]\, d\sigma(x)\end{aligned}$$

by Theorem 2.1. This completes the proof of the subharmonicity in this case.

Next if $u(x)$ is s.h. in D, $\phi(u)$ is continuous and increasing on the range of values R assumed by $u(x)$ in D (including possibly $u = -\infty$) and so $\phi[u(x)]$ is u.s.c. and not equal to $+\infty$ on R. Thus it remains to prove (iii). We have, since $\phi(u)$ is increasing and $u(x)$ is s.h.,

$$\begin{aligned}\phi[u(x_0)] &\leqslant \phi\left(\frac{1}{c_m r^{m-1}} \int_{S(x_0, r)} u(x)\, d\sigma(x)\right)\\ &\leqslant \frac{1}{c_m r^{m-1}} \int_{S(x_0, r)} \phi[u(x)]\, d\sigma(x)\end{aligned}$$

by Theorem 2.1. This completes the proof of Theorem 2.2.

COROLLARY 1. *If $u(x)$ is s.h. in a domain D then so are $e^{\lambda u}$ for $\lambda > 0$ and $[u^+(x)]^k$ for $k \geqslant 1$, where $u^+(x) = \max(u(x), 0)$.*

COROLLARY 2. *If $u(x)$ is harmonic in D and $k \geqslant 1$, then $|u(x)|^k$ is s.h. in D.*

COROLLARY 3. *If $f(z)$ is regular in a plane domain D then the functions $(\log^+|f|)^k$ for $k \geqslant 1$ and $|f|^\lambda$ for $\lambda > 0$ are s.h. in D, where*

$$\log^+|f| = \max(\log|f|, 0).$$

It follows from Example 2 of Section 2.1. that, in Corollary 1, $u^+(x)$ is s.h. in D. Since u^k is convex and increasing in $(0, \infty)$ for $k \geqslant 1$, and $e^{\lambda u}$ is convex and increasing in $(-\infty, +\infty)$ for $\lambda > 0$, Corollary 1 follows.

If $u(x)$ is harmonic then so is $-u(x)$ and hence $|u(x)| = \max[u(x), -u(x)]$ is s.h. Hence $|u(x)|^k$ is s.h. by Corollary 1.

If $f(z)$ is regular in a plane domain D then $u = \log|f|$ is s.h. by example 4 of Section 2.1. Hence so are $(\log^+|f|)^k$ and $|f|^\lambda = e^{\lambda u}$ by Corollary 1. This completes the proof of the corollaries.

The properties of Corollary 3 and Example 4 of section 2.1. are the key to many of the applications of s.h. functions in regular function theory. Many properties of $u(z) = \log|f(z)|$ where f is regular extend to general plane subharmonic functions and frequently this approach leads to simpler proofs.

2.4. THE MAXIMUM PRINCIPLE

We prove

THEOREM 2.3. *Suppose that $u(x)$ is* s.h. *in a domain D of R^m and that, if ξ is any boundary point of D and $\varepsilon > 0$, we can find a neighbourhood N of ξ such that*

$$u(x) < \varepsilon \quad \text{in} \quad N \cap D. \tag{2.4.1}$$

Then $u(x) < 0$ in D or $u(x) \equiv 0$. If D is unbounded we consider $\xi = \infty$ to be a boundary point of D and assume that (2.4.1) holds when N is the exterior of some hyperball $|x| > R$.

We need

LEMMA 2.2. *If $u(x)$ is* s.h. *and $u(x) \leqslant 0$ in $D(x_0, r)$ and $u(x_0) = 0$, then $u(x) \equiv 0$ in $S(x_0, \rho)$ for some arbitrarily small ρ.*

It follows from property (iii) that we can find ρ as small as we please such that

$$u(x_0) = 0 \leqslant \frac{1}{c_m \rho^{m-1}} \int_{S(x_0, \rho)} u(x)\, d\sigma(x).$$

Since $u(x) \leqslant 0$, we deduce that

$$\int_{S(x_0, \rho)} u(x)\, d\sigma(x) = 0.$$

Suppose that there exists x_1 in $S(x_0, \rho)$ such that $u(x_1) < 0$. Then by (ii) of (2.1) we can find a neighbourhood N_1 of x_1, such that $u(x) < -\eta$ in N_1, where $\eta > 0$. If N_2 is the intersection of N_1 and $S(x_0, \rho)$ and E_2 is the complement of N_2 on $S(x_0, \rho)$ then

$$\int_{S(x_0,\rho)} u(x)\, d\sigma(x) = \int_{N_2} + \int_{E_2} \leqslant \int_{N_2} u(x)\, d\sigma(x) \leqslant -\eta \int_{N_2} d\sigma(x) < 0,$$

giving a contradiction. Thus $u(x) \equiv 0$ in $S(x_0, \rho)$ and Lemma 2.2. is proved.

We proceed to prove Theorem 2.2. Let

$$M = \sup_{x \in D} u(x).$$

If $M < 0$ there is nothing to prove. Suppose next that $M > 0$, and let x_n be a sequence of points in D such that

$$u(x_n) \to M.$$

By taking a subsequence if necessary, we may assume that $x_n \to \xi$. Since $M > 0$ this contradicts our basic hypothesis with $\varepsilon = M/2$, if ξ is a frontier point of D. Thus ξ is a point of D. Also since $u(x)$ is u.s.c. we obtain a contradiction of $u(\xi) < M$. Since $u(\xi) \leqslant M$ by hypothesis we must have $u(\xi) = M$.

Thus if E is the set of all points of D for which $u(x) = M$, we see that E is not empty. If $M = 0$ and $u < 0$ in D there is again nothing to prove. So we may assume in all cases that $M \geqslant 0$ and that the set E where $u(x) = M$ is not empty. Since u is u.s.c., E is closed. We proceed to prove that E contains the whole of D.

For suppose contrary to this that x_1, x_2 are points of D such that $u(x_1) < M = u(x_2)$. Then since D is a domain, we can join x_1, x_2 by a polygonal path $x_1 = \xi_1, \xi_2, \ldots, \xi_n = x_2$ in D, so that each straight line segment $\xi_j \xi_{j+1} \in D$ for $j = 1$ to $n - 1$. Let j be the last integer so that $u(\xi_j) < M$. Then $u(\xi_{j+1}) = M$. Let

$$x(t) = (1 - t)\, \xi_j + t\xi_{j+1},$$

and let t_0 be the lower bound of all t in $0 < t < 1$, such that $x(t) \in E$. Since E is closed $x_0 = x(t_0) \in E$. We now apply Lemma 2.2 to $u(x) - M$ and deduce that there exists ρ, such that $0 < \rho < |x_0 - \xi_j|$ and $S(x_0, \rho) \subset E$. Also $S(x_0, \rho)$ meets the segment $[\xi_j, x_0]$, which contradicts the definition of t_0. Thus E contains the whole of D and $u(x) \equiv M$ in D.

If D is bounded, D contains at least one finite frontier point ξ and if $M > 0$ we obtain a contradiction. Thus $M \leqslant 0$, and $u \equiv M$ in D or $u < M$ in D. This proves Theorem 2.3. If D is unbounded, D contains the frontier point ∞, and we again obtain a contradiction.

We deduce immediately.

THEOREM 2.4. *Suppose that* $u(x)$ *is* s.h. *and* $v(x)$ *is harmonic in a bounded domain D and that*

$$\overline{\lim_{x \to \xi}}\, \{u(x) - v(x)\} \leqslant 0$$

as x approaches any boundary point ξ of D from inside D. Then $u(x) < v(x)$ in D or $u(x) \equiv v(x)$ in D.

We apply Theorem 2.3. to $h(x) = u(x) - v(x)$ which is s.h. in D and satisfies the hypotheses of Theorem 2.3 with $M = 0$. Thus $h(x) < 0$ in D or $h(x) \equiv 0$ in D and this is Theorem 2.4.

Under the hypotheses of Theorem 2.4 we say that $v(x)$ is a harmonic majorant of $u(x)$. Such harmonic majorants play a fundamental role in the theory.

2.5. S.H. FUNCTIONS AND THE POISSON INTEGRAL

We shall say that $u(x)$ is s.h. on a set E if u is s.h. on an open set containing E. With this definition we have

THEOREM 2.5. *Suppose that $u(x)$ is* s.h. *in $C(x_0, R)$. Then for $\xi \in D(x_0, R)$ we have*

$$u(\xi) \leqslant \int_{S(x_0, R)} u(x)\, K(x, \xi)\, d\sigma_x \tag{2.5.1}$$

where $K(x, \xi)$ is the Poisson kernel given by (1.5.3) *and* $d\sigma_x$ *denotes an element of surface area of* $S(x_0, R)$.

The integral is to be interpreted as a Lebesgue integral. Since $u(x)$ is u.s.c. $u(x)$ is bounded above on $S(x_0, R)$ and so the integral is finite or $-\infty$. In the latter case Theorem 2.5 is to be interpreted as stating that $u(\xi) = -\infty$.

We proceed to prove Theorem 2.5. Since $u(x)$ is u.s.c. on $S(x_0, R)$ we can by Theorem 1.4 find a sequence $u_n(x)$ of functions continuous on $S(x_0, R)$, such that

$$u_n(x) \downarrow u(x), \qquad x \in S(x_0, R). \tag{2.5.2}$$

We can extend $u_n(x)$ to $D(x_0, R)$ by setting

$$u_n(\xi) = \int_{S(x_0, R)} u_n(x)\, K(x, \xi)\, d\sigma_x, \qquad \xi \in D(x_0, R). \tag{2.5.3}$$

The resulting function $u_n(x)$ is by Theorem 1.16 continuous in $C(x_0, R)$ and harmonic in $D(x_0, R)$. Also if x is a point of $S(x_0, R)$ and $\xi \to x$ from $D(x_0, R)$ then it follows from the upper semi-continuity of $u(\xi) - u_n(\xi)$ in $C(x_0, R)$ that

$$\varlimsup_{\xi \to x} (u(\xi) - u_n(\xi)) \leqslant u(x) - u_n(x) \leqslant 0.$$

Thus by the maximum principle we deduce that in $D(x_0, R)$

$$u(\xi) \leqslant u_n(\xi) = \int_{S(x_0, R)} u_n(x)\, K(x, \xi)\, d\sigma_x.$$

We let n tend to ∞ in this inequality. By Fatou's Theorem and in view of (2.5.2) we deduce that (2.5.1) holds.

We note that since, for fixed ξ, $K(x, \xi)$ is bounded above and below by positive constants,

$$\int_{S(x_0, R)} u(x)\, K(x, \xi)\, d\sigma_x$$

is finite or $-\infty$ throughout $D(x_0, R)$ according as

$$\int_{S(x_0, R)} u(x)\, d\sigma_x$$

is finite or $-\infty$. We can essentially discount one of these possibilities by proving

THEOREM 2.6. *Suppose that $u(x)$ is* s.h. *in a domain D and $u(x) \not\equiv -\infty$. Then if $C(x_0, R)$ lies in D*

$$\int_{S(x_0, R)} u(x)\, d\sigma_x > -\infty, \tag{2.5.4}$$

and if E is any compact subset of D, we have

$$\int_E u(x)\, dx > -\infty. \tag{2.5.5}$$

Suppose that E is a compact subset of D such that (2.5.5) is false, so that

$$\int_E u(x)\, dx = -\infty. \tag{2.5.6}$$

Then let $8\delta\sqrt{m}$ be the distance of E from the complement of D. We divide the space R^m into hypercubes Q of the type

$$m_\nu \delta \leqslant x_\nu \leqslant (m_\nu + 1)\,\delta.$$

Since E is compact only a finite number of these hypercubes, $Q_1, \ldots, Q_N$ say, contain points of E. Let

$$Q = \bigcup_{\nu=1}^{N} Q_\nu, \qquad F = Q - E.$$

Then the Q_ν lie in D and so $u(x)$ is bounded above on each Q_ν and so on F.

Thus

$$\int_Q u(x)\,dx = \int_E u(x)\,dx + \int_F u(x)\,dx < \int_E u(x)\,dx + O(1) = -\infty.$$

Also since $Q_\mu \cap Q_\nu$ has zero m-dimensional measure we deduce that

$$\int_Q u(x)\,dx = \sum_{\nu=1}^{N} \int_{Q_\nu} u(x)\,dx,$$

and so for at least one of the hypercubes $Q_\nu = Q'$ we have

$$\int_{Q'} u(x)\,dx = -\infty.$$

Now let ξ be a point distant between $2\delta\sqrt{m}$ and $3\delta\sqrt{m}$ from some point of E in Q'. Then $C(\xi, r)$ lies in D for $r \leqslant 7\delta\sqrt{m} - 3\delta\sqrt{m} = 4\delta\sqrt{m}$. Also in view of Theorem 2.5 we have for $\rho \leqslant 4\delta\sqrt{m}$

$$c_m \rho^{m-1} u(\xi) \leqslant \int_{S(\xi,\rho)} u(x)\,d\sigma_x,$$

and so

$$c_m u(\xi) \int_{\delta\sqrt{m}}^{4\delta\sqrt{m}} \rho^{m-1}\,d\rho \leqslant \int_{\delta\sqrt{m}}^{4\delta\sqrt{m}} d\rho \int_{S(\xi,\rho)} u(x)\,d\sigma_x = \int_{E_1} u(x)\,dx,$$

where $E_1 = D(\xi, 4\delta\sqrt{m}) - C(\xi, \delta\sqrt{m})$. However E_1 contains the hypercube Q' and so

$$\int_{E_1} u(x)\,dx = \int_{E_1 - Q'} u(x)\,dx + \int_{Q'} u(x)\,dx = -\infty.$$

Thus $u(\xi) = -\infty$. Since ξ was any point in a certain hyper-annulus, we deduce that if (2.5.6) holds then $u(x) \equiv -\infty$ on a certain open subset of D.

Consider next the set M of all points ξ in D such that $u(x) \equiv -\infty$ in some hyperball $C(\xi, r)$. Clearly M is open. Also M is closed in D. For let ξ be a limit point of M in D. Then $S(\xi, r)$ meets M for some arbitrarily small r and so, for such r, $u(x) = -\infty$ on an open subset of $S(\xi, r)$. Thus

$$\int_{S(\xi,r)} u(x)\,d\sigma_x = -\infty,$$

and hence, by Theorem 2.5, $u(x) = -\infty$ in $D(\xi, r)$ so that $\xi \in M$. Thus M is open and closed and we have seen that if (2.5.5) is false, then M is not empty. Thus M contains the whole of D, and $u \equiv -\infty$ in D. Again if (2.5.4) is false we see in view of (2.5.1) that $D(x_0, R) \subset M$ and so again M is not empty and $u(x) \equiv -\infty$ in D. This proves Theorem 2.6.

We are now able to prove

THEOREM 2.7. *Suppose that* $u(x)$ *is* s.h. *and* $u(x) \not\equiv -\infty$ *in a domain* V *and let* $C(x_0, R)$ *be a hyperball lying in* V. *Set*

$$v(\xi) = \int_{S(x_0, R)} u(x)\, K(x, \xi)\, d\sigma_x, \quad \xi \in D(x_0, R),$$
$$v(\xi) = u(\xi), \qquad \xi \in V - D(x_0, R).$$

Then $v(\xi)$ *is* s.h. *in* V, $v(\xi)$ *is harmonic in* $D(x_0, R)$ *and* $u(\xi) \leqslant v(\xi)$ *in* $D(x_0, R)$.

We say that v is *the harmonic extension of* u *from* $S(x_0, R)$ *to* $D(x_0, R)$.

The inequality $u \leqslant v$ in $D(x_0, R)$ is (2.5.1). We note that, by Theorem 2.6, $v(\xi)$ is finite in $D(x_0, R)$. Also by using the decreasing sequence $u_n(x)$ satisfying (2.5.2) and defining $u_n(\xi)$ by (2.5.3) we see that $u_n(\xi)$ is harmonic and $u_n(\xi) \to v(\xi)$ in $D(x_0, R)$. Thus, by Harnack's Theorem 1.20, $v(\xi)$ is harmonic in $D(x_0, R)$.

It remains to show that $v(\xi)$ is s.h. in V. For this we have to check the basic properties (i), (ii) and (iii) of Section 2.1. These are obvious in $D(x_0, R)$ (where v is harmonic) and outside $C(x_0, R)$ where $v = u$ locally. Thus we consider $x \in S(x_0, R)$. Here (i) is again obvious since $v(x) = u(x)$. To prove (ii) let ξ tend to x. Since $v(x) = u(x)$ outside $D(x_0, R)$ and $u(x)$ is u.s.c., it is enough to consider $\xi \in D(x_0, R)$. Let $u_n(x)$ be defined by (2.5.2) and (2.5.3). Then $v(\xi) \leqslant u_n(\xi)$ in $D(x_0, R)$ and so by Theorem 1.16

$$\overline{\lim_{\xi \to x}}\, v(\xi) \leqslant \overline{\lim_{\xi \to x}}\, u_n(\xi) \leqslant u_n(x).$$

In view of (2.5.2) we deduce that

$$\overline{\lim_{\xi \to x}}\, v(\xi) \leqslant u(x)$$

as required. This proves the upper semi-continuity. Finally since $v(x) = u(x)$ and $v(\xi) \geqslant u(\xi)$ for all ξ, it is clear that $v(x)$ satisfies (iii) with x, ξ instead of x_0, x. This completes the proof of Theorem 2.7.

2.5.1. We continue by proving an alternative definition for subharmonicity which leads at once to the fact that, if $m = 2$, subharmonicity is invariant under a conformal map in the z plane. This is

THEOREM 2.8. *A function $u(x)$ defined in a domain D in R^m is* s.h. *in D if and only if $u(x)$ satisfies* (i) *and* (ii) *of Section* 2.1 *and in addition*†

(iii′). *Given any domain Δ whose closure $\bar{\Delta}$ is compact and lies in D and a function $v(x)$ harmonic in Δ and continuous in $\bar{\Delta}$, and such that*

$$u(x) \leqslant v(x) \tag{2.5.7}$$

on the frontier of Δ then (2.5.7) *also holds in Δ.*

COROLLARY. *If $u(z)$ is subharmonic in a plane domain D and if $z = t(w)$ is regular for w in D' and maps D' into D then $u[t(w)]$ is* s.h. *in D'.*

(This corollary is the analogue for subharmonic functions of the example in Section 1.5.7). It follows from Theorem 2.4 that if $u(x)$ is s.h. in D then (iii′) holds. It remains to prove the converse. Suppose that $C(x_0, R)$ lies in D and that $u(x)$ satisfies (i), (ii) and (iii′). Let $u(x)$ be a sequence of functions continuous on $S(x_0, R)$ and satisfying (2.5.2) and let (2.5.3) give the harmonic extension $u_n(\xi)$ of $u_n(x)$ to $D(x_0, R)$. Then since

$$u(x) \leqslant u_n(x)$$

on $S(x_0, R)$ it follows from (iii′) that the inequality also holds in $D(x_0, R)$. Thus we deduce on letting n tend to ∞ that

$$u(\xi) \leqslant \int_{S(x_0, R)} u(x)\, K(x, \xi)\, d\sigma_x, \qquad \xi \in D(x_0, R).$$

On setting $\xi = x_0$, we deduce

$$u(x_0) \leqslant \frac{1}{c_m R^{m-1}} \int_{S(x_0, R)} u(x)\, d\sigma_x$$

which is (iii). Thus (iii) holds for all r such that $C(x_0, r)$ lies in D and in particular for all sufficiently small r, so that $u(x)$ is s.h. in D. This proves Theorem 2.8.

To prove the Corollary suppose first that $z = t(w)$ is univalent, i.e. gives a (1, 1) map of D' onto a subdomain of D. Suppose that Δ' is a domain whose closure lies in D', and that $v(w)$ is harmonic in Δ', continuous in $\bar{\Delta}'$, and that

$$u[t(w)] \leqslant v(w) \tag{2.5.8}$$

on the frontier F of Δ'. Then, if Δ is the image of Δ' under $z = t(w)$, $v_1(z) = v[t^{-1}(z)]$ is harmonic in Δ and continuous in $\bar{\Delta}$ and

$$u(z) \leqslant v_1(z) \tag{2.5.9}$$

on the frontier of $\bar{\Delta}$. Since $u(z)$ is s.h. in D, we deduce that (2.5.9) remains

† This property explains the name subharmonic.

true in Δ so that (2.5.8) remains true in Δ'. Thus $u[t(w)]$ satisfies (iii′) in D' and so since (i) and (ii) are clearly satisfied by $u[t(w)]$ we deduce that $u[t(w)]$ is s.h. in D'.

Consider now the general case. We note from the above argument that $u[t(w)]$ is s.h. near a point w_0 of D', provided that $t(w)$ is univalent in a neighbourhood of w_0, i.e. provided that $t'(w_0) \neq 0$. Suppose finally that w_0 is a point of D', where $t'(w_0) = 0$. Then near w_0

$$t(w) = a_0 + a_k(w - w_0)^k + \dots, \qquad a_k \neq 0$$

$$[t(w) - a_0]^{1/k} = a_k^{1/k}(w - w_0) + \dots = \phi(w) \text{ say},$$

where $\phi'(w_0) \neq 0$, i.e. $t(w) = a_0 + [\phi(w)]^k$. We now note that $u(a_0 + Z^k)$ is s.h. near $Z = 0$. For (i) and (ii) are clearly satisfied; it follows from the above argument that $u(a_0 + Z^k)$ is s.h. near z_0 for any sufficiently small z_0 other than zero, so that (iii) holds except at $z_0 = 0$. Finally if R is small

$$\frac{1}{2\pi}\int_0^{2\pi} u[a_0 + (Re^{i\theta})^k]\, d\theta = \frac{1}{2\pi}\int_0^{2\pi} u[a_0 + R^k e^{ki\theta}]\, d\theta$$

$$= \frac{1}{2\pi}\int_0^{2\pi} u[a_0 + R^k e^{i\phi}]\, d\phi \geqslant u(a_0),$$

since $u(Z)$ is s.h. Thus (iii) is satisfied also at $Z = 0$ and so $u(a_0 + Z^k)$ is s.h. near $Z = 0$. Since $\phi(w_0) = 0$, $\phi'(w_0) \neq 0$ it now follows from the previous argument that $u[a_0 + \phi(w)^k]$ is s.h. near w_0, i.e. $u[t(w)]$ is s.h. near w_0 as required.

2.5.2. By investigating the possibility of equality in (2.5.1) we can deduce a simple sufficient condition for harmonicity. We have

THEOREM 2.9. *A function $u(x)$ is harmonic in a domain D if and only if u and $-u$ are* s.h. *in D.*

We noted previously that harmonic functions are s.h. so that the condition is certainly necessary. Next let u be s.h. in D, suppose that $C(x_0, R)$ lies in D and define v as in Theorem 2.7. Then it follows from Theorems 2.4 and 2.7 that $u(\xi) < v(\xi)$ in $D(x_0, R)$ unless $u(\xi) \equiv v(\xi)$ in $D(x_0, R)$, i.e. unless u is harmonic in $D(x_0, R)$. Setting $\xi = x_0$, we deduce that (iii) holds with strict inequality for all sufficiently small r, unless $u(x)$ is harmonic near x_0. If $-u$ is also s.h. near x_0, we deduce that (iii) holds with the inequality reversed for all small r so that equality must hold in (iii) and $u(x)$ is harmonic near x_0. If u and $-u$ are s.h. in D this is true for all x_0 in D, so that u is harmonic in D.

We have seen incidentally that equality is possible in (2.5.1) only if u is harmonic in $D(x_0, R)$. It is not quite evident that equality does in fact always hold in this case and we defer the proof of this result to a later section (see Theorem 2.19).

Example.

If $y = T(x)$ is an orthogonal transformation in R^m mapping a domain D onto a domain D' and if $u(y)$ is harmonic (or s.h.) in D' prove that $u[T(x)]$ is harmonic (or s.h.) in D.

2.6. PERRON'S METHOD† AND THE PROBLEM OF DIRICHLET

While s.h. functions have many of the properties of harmonic functions or in the case $m = 2$ of logarithms of regular functions, they possess a much greater flexibility. Thus while harmonic functions and regular functions are determined by their behaviour in any open set in view of Theorems 1.5 and 1.15, we deduce from Theorem 2.7 that s.h. functions which are not harmonic can always be made harmonic in a fixed hyperball while remaining unchanged outside. This property is one of their main advantages in constructions of various kinds. In this section we exploit it in order to solve the problem of Dirichlet.

Let D be a domain in R^m, suppose that S is the frontier of D, and let $f(\xi)$ be a bounded function defined on S. If D is unbounded we include in S a single "point at infinity" called ∞ and we assume that $f(\infty)$ has been defined. We proceed under suitable conditions to solve the problem of Dirichlet, i.e. to find a function $u(x)$ harmonic in D and approaching $f(\xi)$ as x tends to a point ξ of S from inside D. To do this we define the class $U(f)$ of functions u with the following properties

(a) *u is* s.h. *in D*

(b) $\overline{\lim}\, u(x) \leq f(\xi)$ *as x approaches any point ξ of S from inside D.*

We define

$$v(x) = \sup_{u \in U(f)} u(x). \tag{2.6.1}$$

We shall see that under suitable conditions $v(x)$ provides the desired solution of the problem of Dirichlet. As we saw in Chapter 1 it follows from the maximum principle that the solution is necessarily unique, if it exists.

† Perron [1923].

2.6.1. Harmonicity

We proceed to prove

LEMMA 2.3. *The function $v(x)$ is harmonic in D and if $m \leqslant f(\xi) \leqslant M$ on S, then we have $m \leqslant v(x) \leqslant M$ in D.*

Suppose that $m \leqslant f(\xi) \leqslant M$. Then $u = m$ satisfies (a) and (b) and so belongs to $U(f)$. Thus for every x in D,

$$v(x) \geqslant u(x) = m.$$

Again suppose that $u \in U(f)$. Then it follows from the maximum principle, Theorem 2.3, that $u(x) - M \leqslant 0$. Thus

$$v(x) = \sup_{u \in U(f)} u(x) \leqslant M.$$

We proceed to prove that $v(x)$ is harmonic in D. To see this let Δ be a hyperball, lying with its boundary in D and let x_1, x_2 be two points of Δ. Let $u_{j,n}(x)$ $(j = 1, 2, n = 1$ to $\infty)$ be a sequence of functions in $U(f)$ such that

$$u_{j,n}(x_j) \to v(x_j), \quad \text{as} \quad n \to \infty, j = 1, 2. \tag{2.6.2}$$

We now set

$$v_{j,n}(x) = \sup_{1 \leqslant r \leqslant n} u_{j,r}(x), \tag{2.6.3}$$

$$v_n(x) = \sup\,[v_{1,n}(x), v_{2,n}(x)]. \tag{2.6.4}$$

We now define $V_{j,n}(x)$, $V_n(x)$ to be equal respectively to $v_{j,n}(x)$, $v_n(x)$ in the part of D outside Δ and to be the harmonic extensions of $v_{j,n}(x)$, $v_n(x)$ respectively from the boundary of Δ to Δ. It follows from Example 2, Section 2.1, that the $v_{j,n}(x)$ and $v_n(x)$ are still s.h. in D. They clearly satisfy (b) since the $u_{j,n}(x)$ do and so $v_{j,n}(x)$ and $v_n(x) \in U(f)$. Also, by Theorem 2.7, $V_{j,n}(x)$ and $V_n(x)$ are s.h. in D. These functions satisfy (b) since they are equal to $v_{j,n}(x)$ and $v_n(x)$ respectively outside Δ and so near the boundary of D. Thus $V_{j,n}(x)$ and $V_n(x)$ belong to $U(f)$.

Next we note that $V_n(x)$ and $V_{j,n}(x)$ are harmonic in Δ and are increasing functions of n, bounded above by M for each x in D. Thus it follows from Harnack's Theorem 1.20 that

$$\lim_{n \to \infty} V_{j,n}(x) = V_j(x),$$

$$\lim_{n \to \infty} V_n(x) = V(x),$$

where $V_j(x)$ and $V(x)$ are harmonic in Δ.

Since $V_{j,n}(x)$, $V_n(x) \in U(f)$ it follows from (2.6.1) that

$$V_{j,n}(x) \leqslant v(x), \qquad V_n(x) \leqslant v(x), \qquad j = 1, 2, \quad n = 1 \text{ to } \infty, \quad x \in D,$$

and hence

$$V_j(x) \leqslant v(x), \qquad V(x) \leqslant v(x)$$

for x in D. By our construction and Theorem 2.5 we see that

$$u_{j,n}(x_j) \leqslant v_{j,n}(x_j) \leqslant V_{j,n}(x_j) \leqslant v(x_j); \qquad j = 1, 2, \quad n = 1 \text{ to } \infty$$

(2.6.5) $$V_{j,n}(x) \leqslant V_n(x) \leqslant v(x), \qquad x \in D.$$

Of these two sets of inequalities the first is obvious by construction and the second follows from (2.6.4) and the fact that the larger the function, the larger the harmonic extension, which is a consequence of the positivity of the kernel $K(x, \xi)$ in Theorem 2.7.

In view of (2.6.2) we now deduce that

$$V_j(x_j) = V(x_j) = v(x_j), \qquad j = 1, 2.$$

Also in view of (2.6.5)

$$V_j(x) \leqslant V(x)$$

in Δ. Since $V_j(x) - V(x)$ is harmonic in Δ we deduce from the maximum principle (Theorem 2.1) that

$$V_j(x) = V(x) \quad \text{in} \quad \Delta, \qquad j = 1, 2$$

so that

$$V_1(x) = V_2(x)$$

and in particular that

(2.6.6) $$v(x_2) = V_2(x_2) = V_1(x_2).$$

We now consider x_1 to be a fixed point of Δ and note that the above construction of $V_1(x)$ is quite independent of x_2. By allowing x_2 to vary we deduce from (2.6.6) that

$$v(x) = V_1(x)$$

so that $v(x)$ is harmonic in Δ. Since Δ may be taken to be a neighbourhood of any given point of D, it follows that $v(x)$ is harmonic in D. This proves Lemma 2.3.

2.6.2. Boundary behaviour†

In order to prove that $v(x)$ solves the problem of Dirichlet, we have to show that this function has the right limiting behaviour at the boundary points ξ of D. In order to do this we need the concept of the barrier.

DEFINITION. *Suppose that ζ_0 is a boundary point of a domain D in R^k. We say that D possesses a barrier at ζ_0 if there exists a barrier function $\omega(x)$ at ζ_0 with the following properties.*

(i) *$\omega(x)$ is defined and* s.h. *in $N_0 = N \cap D$, where N is some neighbourhood of ζ_0.*
(ii) *Let $\mu(\delta) = \sup\{\omega(x)\}$, for $x \in N_0$ and $|x - \zeta_0| \geqslant \delta$, when ζ_0 is finite or $|x| \leqslant \delta^{-1}$ when ζ_0 is infinite. Then $\mu(\delta) < 0$ for $\delta > 0$.*
(iii) *$\omega(x) \to 0$ as $x \to \zeta_0$ from inside N_0.*

A boundary point ζ_0 of D is said to be regular or irregular (for the problem of Dirichlet) according as D does or does not possess a barrier at ζ_0.

We can now prove

THEOREM 2.10. *Suppose that D is a domain in R^k, that $f(\zeta)$ is a bounded function defined on the boundary S of D and that ζ_0 is a regular point of S. If $v(x)$ is defined by* (2.6.1) *then $v(x)$ is harmonic in D and*

$$\varliminf_{\zeta\to\zeta_0} f(\zeta) \leqslant \varliminf_{x\to\zeta_0} v(x) \leqslant \varlimsup_{x\to\zeta_0} v(x) \leqslant \varlimsup_{\zeta\to\zeta_0} f(\zeta), \tag{2.6.7}$$

where the outer limits are taken as $\zeta \to \zeta_0$ on S and the inner limits as $x \to \zeta_0$ from inside D. In particular if $f(\zeta)$ is continuous at ζ_0, then $v(x) \to f(\zeta_0)$ as $x \to \zeta_0$ from inside D.

If $f(\zeta)$ is continuous on S and all the points of S are regular then $v(x)$ solves the problem of Dirichlet for $f(\zeta)$ and D. A domain, all of whose frontier points are regular, will be called a regular domain.

Suppose that

$$f(\zeta) < M \quad \text{on} \quad S.$$
$$f(\zeta) < M_0 \leqslant M, \qquad \zeta \text{ on } S, \quad \zeta \in N_1,$$

where N_1 is some neighbourhood of ζ_0. Let $\omega(x)$ be a barrier function at ζ_0 and suppose that N is a neighbourhood of ζ_0 such that $\bar{N} \subset N_1$ and that $\omega(x)$ is defined in $N_0 = \bar{N} \cap D$, where $\bar{N}$ is the closure of N.

† The concepts and results of this section are due to Lebesgue [1924], who also obtained a stronger result than Theorem 2.11 and had given an example of an irregular point, the so called Lebesgue spine.

Let

$$-\eta = \sup \omega(x) \text{ for } x \text{ outside } \overline{N}.$$

In view of (ii) and (iii) it follows that by a suitable choice of $\overline{N}$ we can arrange that η is finite and positive. We now define

$$\omega_1(x) = -\eta, \quad x \in D, \quad x \text{ outside } \overline{N}$$

$$\omega_1(x) = \sup[\omega(x), -\eta] \qquad x \in \overline{N}.$$

Then $\omega_1(x)$ is s.h. in D. This is evident at exterior points of $\overline{N}$, where $\omega_1(x)$ is constant. Also, near any point x_0 in $\overline{N}$, $\omega(x)$ is defined and s.h. and

$$\omega_1(x) = \sup(\omega(x), -\eta)$$

so that $\omega_1(x)$ is s.h. also near x_0 by Example 2, Section 2.1. Hence $\omega_1(x)$ is a barrier function at ζ_0 defined throughout D. We now set

$$v_1(x) = M_0 - \frac{M - M_0}{\eta}\omega_1(x),$$

and proceed to prove that

(2.6.8) $$v(x) \leqslant v_1(x) \text{ in } D.$$

To see this suppose that $u(x) \in U(f)$. Then $u(x)$, $-v_1(x)$ and hence $u(x) - v_1(x)$ are s.h. in D. Also if $\zeta \in S \cap N_1$ then $f(\zeta) < M_0$ and so by (b)

$$u(x) < M_0$$

for all points of D near ζ while $v_1(x) \geqslant M_0$ throughout D. Thus in this case

(2.6.9) $$u(x) \leqslant v_1(x),$$

for all points of x near ζ. On the other hand, if ζ is in S but outside N_1, then ζ is outside $\overline{N}$ and so

$$\omega_1(x) = -\eta, \qquad v_1(x) = M$$

for all x in D near ζ. Thus (2.6.9) holds for x near ζ, when ζ is any point of S. Since $u(x) - v_1(x)$ is s.h. it follows from the maximum principle, Theorem 2.1, that (2.6.9) holds throughout D. Since $u(x)$ is an arbitrary function of $U(f)$ it follows from (2.6.1) and (2.6.9) that (2.6.8) holds. In view of the property (iii) of the barrier function $\omega_1(x)$ we deduce that

$$\overline{\lim_{x\to\zeta_0}}\, v(x) \leqslant \overline{\lim_{x\to\zeta_0}}\, v_1(x) = M_0.$$

Since we can choose M_0 so that

$$M_0 < \overline{\lim_{\zeta\to\zeta_0}}\, f(\zeta) + \varepsilon,$$

where ε is as small as we please, we deduce the right-hand inequality in (2.6.7).

It remains to prove the left-hand inequality of (2.6.7). For this purpose we suppose that

$$f(\zeta) \geqslant m \text{ on } S$$

and

$$f(\zeta) > m_0 \geqslant m \text{ on } S \cap \overline{N}_1.$$

We define $\omega_1(x)$ as above and set

$$u_1(x) = m_0 + \frac{m_0 - m}{\eta}\omega_1(x).$$

Then $u_1(x)$ is s.h. in D, since $\omega_1(x)$ is s.h. in D. Also if $\zeta \in S$ and ζ lies outside $\overline{N}_1$ then we have

$$\omega_1(x) = -\eta, \qquad u_1(x) = m \leqslant f(\zeta)$$

for all points of D near ζ and if ζ lies in $\overline{N}_1$ we have

$$u_1(x) \leqslant m_0 \leqslant f(\zeta)$$

near ζ. Thus $u_1(x)$ satisfies (b) and so $u_1(x) \in U(f)$. Hence

$$v(x) \geqslant u_1(x),$$

and

$$\varliminf_{x \to \zeta_0} v(x) \geqslant \varliminf_{x \to \zeta_0} u_1(x) = m_0.$$

Since we can choose m_0 so that

$$m_0 > \varliminf_{\zeta \to \zeta_0} f(\zeta) - \varepsilon,$$

where ε is as small as we please, we deduce the left-hand inequality of (2.6.7). The middle inequality is obvious.

If $f(\zeta)$ is continuous at ζ_0 then the outer terms and so all the terms in (2.6.7) become equal to $f(\zeta_0)$, so that

$$v(x) \to f(\zeta_0) \quad \text{as} \quad x \to \zeta_0$$

from inside D or from S. If we set $v(x) = f(x)$ on S, the extended function $v(x)$ is continuous at ζ_0. If this is true at all points ζ_0 of S, $v(x)$ is continuous in $\overline{D}$, harmonic in D by Lemma 2.3 and so $v(x)$ solves the problem of Dirichlet. This completes the proof of Theorem 2.10.

2.6.3. Conditions for regularity and construction of the barrier function

In order to use Theorem 2.10 in practice we obtain now a simple geometrical criterion for a boundary point of D to be regular.

Let x_0 be a point of space and let l be the ray of all points given by

$$x = x_0 + t\zeta, \qquad 0 \leqslant t < \infty,$$

where ζ is a point of space. A right circular cone $V(l, \alpha)$ with axis l and vertex x_0 is defined by the set of all points x satisfying the inequality

$$d(x, l) < |x - x_0| \cos \alpha,$$

where $d(x, l)$ is the distance from x to l and we suppose that $0 < \alpha < \pi/2$. We can now prove

THEOREM 2.11. *If D is a domain in R^m and ζ is a boundary point of D, then ζ is regular provided that either*

(a) $m = 2$, *ζ is finite, and there is an arc*

$$z = \zeta + re^{i\theta(r)}, \qquad 0 \leqslant r \leqslant a$$

lying outside D where $\theta(r)$ is continuous in $[0, a]$; or

(b) $m = 2$, $\zeta = \infty$ *and there is an arc*

$$z = \frac{1}{r} e^{i\theta(r)}, \qquad 0 \leqslant r < a$$

lying outside D where $\theta(r)$ is as in a); or

(c) $m > 2$ *and* $\zeta = \infty$; *or*

(d) $m > 2$, *ζ is finite and there is a right circular cone with vertex ζ, all of whose points sufficiently near ζ lie outside D.*

The above criteria, particularly (a) and (b) could be considerably weakened but are adequate for most applications. However we shall see in Chapter 5 that the analogue of (a) is false if $m > 2$; in fact an isolated line segment is irregular for the problem of Dirichlet in R^m in this case.

To prove (a), suppose that $\theta(r) \to \theta_0$ as $r \to 0$. We may suppose that $|\theta_0| \leqslant \pi$ and so that

$$|\theta(r)| < 2\pi, \qquad 0 \leqslant r \leqslant \delta.$$

For z not on γ and $|z| < \delta$, we may define

$$z = \zeta + re^{i\theta}, \qquad \theta(r) < \theta < \theta(r) + 2\pi$$

so that $\theta = \arg (z - \zeta)$ is uniquely defined in the part N_0 of D lying in

$|z-\zeta| < \delta$, and $|\theta| < 4\pi$. We define

$$\omega(z) = -r^{1/9} \cos\left(\frac{\theta}{9}\right)$$

in N_0 and see that $\omega(z)$ is harmonic in N_0 and satisfies the conditions (i), (ii) and (iii) for the barrier function. Similarly in case (b), we define N_0 to be the part of the plane for which $|z| > R_0$, and z is not on γ. By a suitable choice of R_0 and since $\theta(r)$ tends to a limit as $r \to 0$, we may write for z in N_0

$$z = \rho e^{i\theta}, \qquad \rho \geqslant R_0,$$

where $\theta = \arg z$ is uniquely defined and satisfies $|\theta| < 4\pi$. We now define

$$\omega(z) = -\rho^{-1/9} \cos\left(\frac{\theta}{9}\right),$$

and see that $\omega(z)$ is a barrier at ∞.

In the case (c) we set

$$\omega(x) = -|x|^{2-m}.$$

It is evident that all the conditions (i), (ii), and (iii) are satisfied. It is noteworthy that ∞ is always regular for the problem of Dirichlet in R^m if $m \geqslant 3$.

It remains to deal with the case (d). For this we need

LEMMA 2.4. *Let C be a right circular cone of vertex x_0 and let D be the complement of $\bar{C}$ in space. Then there exists a function $u_1(x)$ s.h. in D, continuous in $\bar{D}$ and such that*

$$u_1(x) \leqslant 0 \quad \text{in} \quad \bar{D}, \tag{2.6.10}$$

with equality if and only if $x = x_0$.

We note that under an orthogonal transformation right circular cones go into right circular cones and a given ray can be mapped onto the negative x_1 axis. In view of the example at the end of Section 2.5.2 we may therefore assume that the axis of our cone is the negative x_1 axis, where $(x_1, x_2, \ldots, x_m)$ are the coordinates of x. Thus D is given by

$$x_1 \geqslant \cos(\pi - \delta)\left(\sum_1^m x_r^2\right)^{1/2}. \tag{2.6.11}$$

where $0 < \delta < \frac{\pi}{2}$. We set

$$\rho^2 = \sum_2^m x_r^2,$$

and suppose that $u = g(x_1, \rho) \in C^2$. Then

$$\frac{\partial^2 u}{\partial x_\nu^2} = \frac{x_\nu^2}{\rho^2}\frac{\partial^2 u}{\partial \rho^2} + \left(\frac{1}{\rho} - \frac{x_\nu^2}{\rho^3}\right)\frac{\partial u}{\partial \rho}, \qquad \nu \geqslant 2,$$

so that

$$\nabla^2 u = \sum_{\nu=1}^{m} \frac{\partial^2 u}{\partial x_\nu^2} = \frac{\partial^2 u}{\partial x_1^2} + \frac{\partial^2 u}{\partial \rho^2} + \frac{m-2}{\rho}\frac{\partial u}{\partial \rho}.$$

We now set $x_1 = R\cos\theta$, $\rho = R\sin\theta$ and assume that $u = R^\alpha \phi(\theta)$. Then

$$\nabla^2 u = R^{\alpha-2}[\phi''(\theta) + (m-2)\phi'(\theta)\cot\theta + \alpha(\alpha + m - 2)\phi(\theta)].$$

Let $\phi_m(\theta)$ be the solution of the equations

$$\phi_m''(\theta) + (m-2)\cot\theta\,\phi_m'(\theta) = 1, \qquad \phi_m(0) = \phi_m'(0) = 0,$$

so that

$$\phi_m(\theta) = \int_0^\theta \frac{dt}{(\sin t)^{m-2}}\int_0^t (\sin\tau)^{m-2}\,d\tau.$$

Thus $\phi_m(\theta)$ is analytic at $\theta = 0$ and in fact for $-\pi < \theta < \pi$. We choose a so large that

$$\phi_m(\theta) - a \leqslant -1, \qquad |\theta| \leqslant \pi - \tfrac{1}{2}\delta.$$

We suppose that α is a small positive number and set

$$u_1(x) = R^\alpha[\phi_m(\theta) - a], \qquad 0 \leqslant R < \infty.$$

Then it follows that $u_1(x) \leqslant -R^\alpha$ in the region D given by (2.6.11) and $u_1(0) = 0$, which gives (2.6.10). Also if α is sufficiently small

$$\nabla^2 u_1 = R^{\alpha-2}[1 + \alpha(\alpha + m - 2)(\phi_m(\theta) - a)] > 0$$

for $|\theta| \leqslant \pi - \delta$, so that u_1 is s.h. in the region D. Thus $u_1(x)$ satisfies the conditions of Lemma 2.4 and so $u_1(x)$ is a barrier at ζ.

This completes the proof of Theorem 2.11.

2.7. CONVEXITY THEOREMS

The average of a subharmonic function on a sphere $S(x_0, r)$ has important convexity properties which we now proceed to investigate.

THEOREM 2.12. *Suppose that $u(x)$ is s.h. in the closed annulus $r_1 \leqslant |x - x_0| \leqslant r_2$ in R^m and not identically $-\infty$ there. Set*

$$I(r, u) = \frac{1}{c_m r^{m-1}} \int_{S(x_0, r)} u(x)\, d\sigma(x).$$

Then $I(r, u)$ is a convex function of $\log r$, if $m = 2$ and of r^{2-m} if $m > 2$, for $r_1 \leqslant r \leqslant r_2$, if $0 < r_1 < r_2$.

If $r_1 = 0$, so that $u(x)$ is s.h. in $C(x_0, r_2)$, then $I(r, u)$ is a continuous increasing function of r, for $0 \leqslant r \leqslant r_2$, if we define $I(0, u) = u(x_0)$.

Suppose first that $u(x)$ is harmonic in the annulus $r_1 \leqslant |x - x_0| \leqslant r_2$. It then follows from Green's Theorem 1.9 that

$$\int_{S(x_0, r_2')} - \int_{S(x_0, r_1')} \frac{\partial u}{\partial r} d\sigma = 0. \qquad r_1 \leqslant r_1' < r_2' \leqslant r_2.$$

We may write this equation as

$$r^{m-1} \frac{\partial}{\partial r} \int_{S(x_0, r)} u(x) \frac{d\sigma}{r^{m-1}} = \text{constant}, \qquad r_1 \leqslant r \leqslant r_2,$$

$$r^{m-1} I'(r, u) = \text{constant}, \qquad r_1 \leqslant r \leqslant r_2.$$

Thus in this case there exist constants A, B, such that

$$I(r, u) = A \log r + B, \qquad r_1 \leqslant r \leqslant r_2, \qquad \text{if } m = 2$$

$$I(r, u) = A r^{2-m} + B, \qquad r_1 \leqslant r \leqslant r_2, \quad \text{if } m > 2.$$

The result remains true if u is harmonic for $r_1 < |x - x_0| < r_2$ and continuous for $r_1 \leqslant |x - x_0| \leqslant r_2$, since in this case $I(r, u)$ clearly remains continuous as $r \to r_1$ from above or $r \to r_2$ from below. Thus in this case $I(r, u)$ is a linear function of $\log r$ for $m = 2$ or r^{2-m} for $m > 2$.

We now consider the general case. Let $v_n(x)$ be a sequence of continuous functions such that

$$(2.7.1) \qquad v_n(x) \downarrow u(x), \qquad n \to \infty \quad \text{on} \quad F = S(x_0, r_1) \cup S(x_0, r_2).$$

The sequence exists by Theorem 1.4, since $u(x)$ is u.s.c. on F. Further F is the boundary of the domain $D: r_1 < |x - x_0| < r_2$ and every point ξ of F clearly satisfies the criterion of Theorem 2.11 (a) or (d), so that ξ is a regular boundary point of D. Thus by Theorem 2.10 we can solve the problem of Dirichlet for D with the boundary values $v_n(x)$ so that $v_n(x)$ can be extended to be harmonic in D and continuous in $\bar{D}$.

Set

$$I(r, v_n) = I_n(r), \qquad r_1 \leqslant r \leqslant r_2.$$

Since $v_n(x)$ decreases with the increasing n on $S(x, r_1)$ and $S(x, r_2)$, $I_n(r)$ decreases with increasing n for $r = r_1, r_2$ and by the linearity property $I_n(r)$ decreases with increasing n for $r_1 \leqslant r \leqslant r_2$. Thus

$$I_n(r) \to I(r), \quad \text{as} \quad n \to \infty.$$

We have by (2.7.1) that

$$I(r_1) = I(r_1, u), \qquad I(r_2) = I(r_2, u). \tag{2.7.2}$$

In view of the maximum principle. Theorem 2.4, we deduce that $u(x) \leqslant v_n(x)$ in D for each n and so

$$I(r, u) \leqslant I(r, v_n) = I_n(r), \qquad n = 1, 2, \ldots \quad r_1 < r < r_2$$

and so

$$I(r, u) \leqslant I(r), \qquad r_1 < r < r_2. \tag{2.7.3}$$

If $I(r_1) = -\infty$ or $I(r_2) = -\infty$, we deduce from the linearity of $I_n(r)$ that

$$I(r, u) = I(r) = -\infty, \qquad r_1 < r < r_2.$$

This leads by integration to

$$\int_{r_1 \leqslant |x| \leqslant r_2} u(x)\,dx = \int_{r_1}^{r_2} c_m r^{m-1} I(r, u)\,dr = -\infty,$$

which gives a contradiction unless $u(x) \equiv -\infty$, in view of Theorem 2.6. Thus except in this case $I(r_1)$ and $I(r_2)$ are both finite, $I(r)$ is a linear function of $\log r (m = 2)$ or $r^{2-m}(m > 2)$ and (2.7.2) and (2.7.3) hold. In the above argument we may replace r_1, r, r_2 by numbers r_1', r, r_2', such that $r_1 \leqslant r_1' < r < r_2' \leqslant r_2$. We deduce that, unless $u(x) \equiv -\infty$, $I(r, u)$ is finite for $r_1 \leqslant r \leqslant r_2$ and we have

$$I(r, u) \leqslant I(r), \qquad r_1' < r < r_2',$$

where $I(r)$ is the linear function of $\log r\,(m = 2)$ or $r^{2-m}(m > 2)$ which satisfies

$$I(r_1') = I(r_1', u), \qquad I(r_2') = I(r_2', u).$$

In view of (1.3.3) this proves the convexity of $I(r, u)$.

Suppose finally that u is s.h. in $C(x_0, r_2)$. Then it follows from the definition of subharmonicity ((ii) and (iii) of Section 2.1) that

$$I(r, u) \to u(x_0), \quad \text{as} \quad r \to 0.$$

Suppose that $0 < r_1' < r < r_2' \leqslant r_2$, and set $I(r, u) = J(r)$. Then if $m = 2$ we deduce from the convexity that

$$J(r) \leqslant \frac{\log r_2' - \log r}{\log r_2' - \log r_1'} J(r_1') + \frac{\log r - \log r_1'}{\log r_2' - \log r_1'} J(r_2').$$

We let r_1' tend to zero in this inequality. Then the second term on the right-hand side tends to $J(r_2')$. The first term on the right-hand side tends to zero if $u(x_0) > -\infty$, and is finally negative if $u(x_0) = -\infty$. Thus in either case we deduce that

$$J(r) \leqslant J(r_2'), \qquad 0 < r < r_2' \leqslant r_2.$$

If $m > 2$ this result follows similarly from the inequality

$$J(r) \leqslant \frac{(r_2')^{2-m} - r^{2-m}}{(r_2')^{2-m} - (r_1')^{2-m}} J(r_1') + \frac{r^{2-m} - (r_1')^{2-m}}{(r_2')^{2-m} - (r_1')^{2-m}} J(r_2').$$

Thus in both cases $J(r)$ is an increasing function of r for $0 \leqslant r \leqslant r_2$. This completes the proof of Theorem 2.12.

We can also prove, rather more simply, the analogue of Theorem 2.12 for the maximum

$$B(r, u) = \sup_{x \in S(x_0, r)} u(x).$$

THEOREM 2.13. *With the hypotheses of Theorem 2.12, $B(r, u)$ is a convex function of $\log r$, if $m = 2$ and of r^{2-m} if $m > 2$, for $r_1 \leqslant r \leqslant r_2$, if $0 < r_1 < r_2$.*

If $u(x)$ is s.h. in $C(x_0, r_2)$ then $B(r, u)$ is an increasing function of r for $0 \leqslant r \leqslant r_2$.

The second part of Theorem 2.12 follows at once from the maximum principle, Theorem 2.3. To prove the first part, suppose that

$$r_1 \leqslant r_1' < r < r_2' \leqslant r_2,$$

and let $I(r)$ be the linear function of $\log r$ ($m = 2$) or r^{2-m}($m > 2$) which coincides with $B(r, u)$ for $r = r_1', r_2'$. Then

$$v(x) = I(|x - x_0|)$$

is harmonic for $r_1' \leqslant |x - x_0| \leqslant r_2'$, and

$$u(x) - v(x) \leqslant 0, \qquad |x - x_0| = r_1' \quad \text{and} \quad |x - x_0| = r_2'.$$

Hence by the maximum principle

$$u(x) - v(x) \leqslant 0, \qquad r_1' < |x - x_0| < r_2',$$

so that on $S(x_0, r)$ for $r_1' < r < r_2'$, we have

$$u(x) \leqslant v(x) = I(r),$$

i.e.

$$B(r, u) \leqslant I(r), \qquad r_1' < r < r_2'.$$

This proves the convexity of $B(r, u)$.

2.7.1. Some applications

Theorems 2.12 and 2.13 have an important role in the theory of subharmonic functions. We deduce for instance that if $u(x)$ is s.h. in a neighbourhood of the sphere $S(x_0, r)$ and so in some annulus $r_1 < |x - x_0| < r_2$, where $r_1 < r < r_2$, then

$$I(r, u) = \int_{S(x_0, r)} u(x)\, \mathrm{d}\sigma_x > -\infty,$$

unless $u(x) \equiv -\infty$. This is an extension of (2.5.4) of Theorem 2.6. Also, in view of the properties of convex functions, $I(r, u)$ and $B(r, u)$ are continuous functions of r, possessing left and right derivatives everywhere in $r_1 < r < r_2$ which are equal outside a countable set. Finally

$$r^{m-1} \frac{\mathrm{d}}{\mathrm{d}r} I(r, u) \quad \text{and} \quad r^{m-1} \frac{\mathrm{d}}{\mathrm{d}r} B(r, u)$$

increase with increasing r.

We can deduce at once

THEOREM 2.14. *If $u(x)$ is subharmonic in the plane and not constant then*

$$\alpha = \lim_{r\to\infty} \frac{B(r, u)}{\log r} \tag{2.7.4}$$

exists and $\alpha > 0$. If further u is not harmonic in the plane then

$$\beta = \lim_{r\to\infty} \frac{I(r, u)}{\log r} > 0. \tag{2.7.5}$$

By Theorem 2.13 we deduce that if $B(r) = B(r, u)$ then

$$B'(r) \geqslant 0, \qquad 0 < r < \infty.$$

If $B'(r) \equiv 0$, then $B(r) = u(0)$, $0 < r < \infty$ and so by the maximum principle $u(x) = u(0) =$ constant in the plane. Otherwise $rB'(r) > 0$ for some $r > 0$ and, since $B(r)$ is a convex function of $\log r$,

$$\alpha = \lim_{r\to\infty} rB'(r)$$

exists and $0 < \alpha \leqslant \infty$. If α is finite, then given $\varepsilon > 0$ we have

$$\alpha - \varepsilon < rB'(r) < \alpha, \qquad r > r_0(\varepsilon),$$

$$(\alpha - \varepsilon)\log\frac{r}{r_0} < B(r) - B(r_0) < \alpha\log\frac{r}{r_0}.$$

This yields (2.7.4).

We deduce similarly that β exists, $\beta > 0$ and $\beta = 0$ if

$$I(r, u) = u(0), \qquad 0 < r < \infty.$$

In this case let $v(x)$ be the harmonic extension of u from $S(0, R)$ to $D(0, R)$. By Theorem 2.7 $u(x) - v(x) \leqslant 0$ in $D(0, R)$. Also

$$v(0) = I(R, u) = u(0).$$

Hence by the maximum principle $u(x) - v(x) \equiv 0$ in $D(0, R)$, so that $u(x)$ is harmonic in $D(0, R)$ for every positive R, i.e. $u(x)$ is harmonic in the plane.

We deduce from Theorem 2.14 that if u is not constant u cannot be bounded in the plane, since $B(r, u)$ grows at least as rapidly as $\log r$. This result is false in space of more than two dimensions. In fact $u = -r^{2-m}$ is subharmonic and bounded above in R^m and harmonic except at the origin. Also $\max(-1, u)$ is subharmonic and bounded in R^m. On the other hand a harmonic function in space cannot be bounded. We have

THEOREM 2.15. *If u is harmonic and non-constant in space then*

$$n = \lim_{r\to\infty} \frac{\log B(r, u)}{\log r}$$

is finite if u is a polynomial of degree n in the coordinates and is $+\infty$ otherwise.

The result is a consequence of Example 3, Section 1.5.6.

Some special cases of Theorems 2.12 and 2.13 are worth stating as a separate Theorem.

THEOREM 2.16. *If u is harmonic in a plane annulus $r_1 \leqslant |z| \leqslant r_2$ and $f(z)$ is a, possibly many-valued, function of z there, analytic except for branchpoints and such that $|f(z)|$ is one-valued, then the following quantities are convex functions of $\log r$ for $r_1 < r < r_2$ and, if $r_1 = 0$, increasing functions of r, for $0 \leqslant r \leqslant r_2$:*

$$M(r, f) = \sup_{|z|=r} |f(z)| \quad \textit{and} \quad \log M(r);$$

$$I_\lambda(r, f) = \frac{1}{2\pi}\int_0^{2\pi} |f(re^{i\theta})|^\lambda \, d\theta \quad \textit{and}\dagger \ \log I_\lambda(r, f) \quad \textit{for} \quad 0 < \lambda < \infty;$$

$$I_0(r, f) = \frac{1}{2\pi}\int_0^{2\pi} \log |f(re^{i\theta})| d\theta \quad \textit{and} \quad I_k(r, \log^+ f), \textit{ for } k \geqslant 1;$$

$$I_k(r, u) \quad \textit{for} \quad k \geqslant 1,$$

provided that $f = u + iv$ *is one-valued.*

By example 4 of Section 2.1 $\log|f|$ is s.h. and so are $|f|^\lambda$ for $\lambda > 0$ and $|u|^k$ and $(\log^+|f|)^k$ for $k \geqslant 1$, by Corollaries 2 and 3 of Theorem 2.2. We deduce at once that $M(r, f)$ and

$$\log M(r, f) = \sup_{|z|=r} \log |f(z)|$$

have the required properties in view of Theorem 2.13. Similarly we deduce from Theorem 2.12 that $I_\lambda(r, f)$ for $\lambda > 0$, $I_0(r, f)$ and $I_k(r, \log^+|f|)$ and $I_k(r, u)$ for $k \geqslant 1$ have the required convexity properties.

It remains to show that $\log I_\lambda(r, f)$ is a convex function of $\log r$. For this it is enough to prove that if $L(r) = A \log r + B$ is a linear function of $\log r$ such that

$$\log I_\lambda(r, f) \leqslant L(r)$$

for $r = r_1, r_2$, then the same inequality holds for $r_1 < r < r_2$.

To see this we consider $\phi(z) = z^\mu f(z)$ instead of $f(z)$. Clearly $\phi(z)$ is analytic in our annulus $r_1 < |z| < r_2$ and $|\phi(z)|$ is one valued. Also

$$I_\lambda(r, \phi) = r^{\lambda\mu} I_\lambda(r, f).$$

$$\log I_\lambda(r, \phi) = \log I_\lambda(r, f) + \lambda\mu \log r.$$

We choose μ so that $-\lambda\mu = A$, $\mu = -A/\lambda$, and deduce that

$$I_\lambda(r, \phi) \leqslant e^B,$$

for $r = r_1, r_2$ and since $I_\lambda(r, \phi)$ is a convex function of $\log r$, this inequality remains valid for $r_1 < r < r_2$. This gives

$$\log I_\lambda(r, f) - A \log r \leqslant b, \qquad \log I_\lambda(r, f) \leqslant L(r), \qquad r_1 < r < r_2,$$

and the proof of Theorem 2.16 is complete.

† The result for $I_\lambda(r, f)$ is due to Hardy [1915].

Examples

1. By taking $u = \log r$, show that $I_\lambda(r, u)$ is not in general a convex function of $\log r$, when u is harmonic and $0 < \lambda < 1$.

2. By using the case $\phi(u) = u^{1/\lambda}$ of Jensen's inequality, Theorem 2.1, show that if u is positive, harmonic and not constant in $|x| < r$, then

$$I_\lambda(r, u) < u(0)^\lambda, \qquad 0 < \lambda < 1,$$

so that $I_\lambda(\rho, u)$ is not an increasing function of ρ for $0 < \rho < r$.

3. If $f(z)$ is regular and not zero in $r_1 < |z| < r_2$ and $\lambda < 0$ show that $I_\lambda(r, f)$ is a convex function of $\log r$ for $r_1 < r < r_2$ and that if $r_1 = 0$, $I_\lambda(r, f)$ is an increasing function of r for $0 \leqslant r < r_2$.

Deduce that if $f(z) = z - 1$ and $-1 < \lambda < 0$ then $I_\lambda(r, f)$ increases with r for $0 < r < 1$ and decreases for $1 < r < \infty$, so that $I_\lambda(r, f)$ and $\log I_\lambda(r, f)$ are not convex functions of r in any interval containing $r = 1$ as an interior point.

2.7.2. Harmonic extensions†

In this section we consider some generalizations of Theorem 2.7. We first use the following

DEFINITION 1. *Suppose that D is a bounded regular domain in R^m. Let $f(\zeta)$ be a continuous function defined on the frontier F of D.*

Then if $u(x)$ is continuous in $\bar{D}$, harmonic in D and $u(\zeta) = f(\zeta)$ in F, we say that $u(x)$ is the harmonic extension† of f from F into D.

By Theorem 2.10 the function $u(x)$ always exists. It is unique by Theorem 1.13. We can extend this definition to semi-continuous functions by means of the following

THEOREM 2.17. *Suppose that $f(\zeta)$ is upper semi-continuous in F, $-\infty \leqslant f < +\infty$ and that $f_n(\zeta)$ is a sequence of continuous functions, monotonically decreasing to $f(\zeta)$ as $n \to \infty$ for each ζ in F. Let $u_n(x)$ be the harmonic extension of f_n from F into D. Then $u_n(x)$ decreases to a limit $u(x)$ as $n \to \infty$ which is independent of the choice of the sequence f_n and is either harmonic or identically $-\infty$ in D.*

† Following Poincaré, this process is frequently called balayage, particularly in the context of Theorem 2.18.

DEFINITION 2. *The function $u(x)$ will be called the harmonic extension of f from F into D. If f is lower semi-continuous, the harmonic extension of f from F into D is defined to be $u(x)$, where $-u(x)$ is the harmonic extension of $-f(x)$ from F into D.*

We proceed to prove Theorem 2.17.

It follows from Harnack's Theorem 1.20, that $u(x)$ is harmonic or identically $-\infty$ in D. It remains to show that $u(x)$ is independent of the choice of the sequence f_n. To see this we show that $u(x)$ is the greatest lower bound of all the functions $v(x)$ which are harmonic extensions of continuous functions $g(\zeta)$ from F into D which satisfy $g(\zeta) > f(\zeta)$ in F.

In fact suppose that $g(\zeta)$ is continuous and $g(\zeta) > f(\zeta)$ in F. Then given $\zeta_0 \in F$, there exists $n_0 = n_0(\zeta_0)$, such that

$$f_n(\zeta_0) < g(\zeta_0), \qquad n \geqslant n_0.$$

Since $f_{n_0}(\zeta) - g(\zeta)$ is u.s.c. there exists a neighbourhood N_0 of ζ_0 such that

$$f_{n_0}(\zeta) < g(\zeta), \qquad \zeta \in N_0,$$

and since $f_n(\zeta)$ decreases with increasing n, we deduce that

$$f_n(\zeta) < g(\zeta), \qquad \zeta \in N_0, \qquad n \geqslant n_0.$$

Since F is compact it follows from the Heine–Borel Theorem 1.1, that a finite system of such neighbourhoods, $N_1, N_2, \ldots, N_k$ say, covers F. If $n_1, n_2, \ldots, n_k$ are the associated integers, we deduce that if $n' = \max(n_1, \ldots, n_k)$. Then

$$f_n(\zeta) < g(\zeta), \qquad \zeta \in F, \qquad n \geqslant n'.$$

Thus if $v(x)$ is the harmonic extension of $g(x)$ from F into D we have

$$u_n(x) < v(x), \qquad n \geqslant n',$$

for $x \in F$ and so, by the maximum principle, for $x \in D$. Hence

$$u(x) \leqslant u_n(x) < v(x), \qquad x \in D,$$

and so $u(x)$ is a lower bound for all the functions $v(x)$.

On the other hand if x is fixed and $K > u(x)$ then

$$u_n(x) + \frac{1}{n} < K, \qquad n > n_0$$

Also $u_n(x) + (1/n)$ is the harmonic extension of the function $f_n(x) + (1/n)$ which is continuous and greater than $f(x)$ in F. Thus K is not a lower bound for all the harmonic extensions of continuous functions greater than $f(x)$ and so $u(x)$ is the greatest such lower bound.

We note that if $f(\zeta)$ is continuous in F, we may take $f_n(\zeta) = f(\zeta)$ for each n, so that $u(x)$ is the extension of F in the sense of the previous definition. If f is both lower and upper semi-continuous then f is continuous and so both definitions of the harmonic extension coincide with the definition in the sense of Definition 1, and so Definition 2 is consistent with Definition 1 and not self-contradictory.

We proceed to prove

THEOREM 2.18. *Suppose that u is s.h. in a neighbourhood V of $\bar{D}$ where D is a bounded regular domain in R^m. Let F be the frontier of D and let $v(x)$ be the harmonic extension of $u(x)$ from F into D, for $x \in D$; $v(x) = u(x)$ for all other points of V. Then $v(x)$ is s.h. in V, $v(x) \geqslant u(x)$ in D.*

Suppose that $f_n(\xi)$ is a sequence of continuous functions in F, decreasing to $u(\xi)$ and let $u_n(\xi)$ be the harmonic extensions of $f_n(\xi)$ from F into D. Then

$$u(x) - u_n(x) \text{ is u.s.c. in } \bar{D} \text{ and s.h. in } D$$

and

$$u(x) - u_n(x) \leqslant 0$$

in F and so in D by the maximum principle, Theorem 2.4. By letting n tend to ∞ in this inequality we deduce that

$$u(x) \leqslant v(x)$$

in D, on using the result of Theorem 2.17. It remains to prove that $v(x)$ is s.h. in V. We apply the definition of Section 2.1. It is clear that $v(x) < +\infty$ in D. Also since $v(x) = u(x)$ outside D and $v(x)$ is harmonic or identically $-\infty$ in D, we see that we need only check that $v(x)$ is u.s.c. in F and satisfies the mean value property there.

Let ξ_0 be a point of F, suppose that $K > u(\xi_0)$ and choose n so large that

$$f_n(\xi_0) < K.$$

In view of the fact that ξ_0 is a regular boundary point of D and $f_n(\xi)$ is continuous, it follows from Theorem 2.10 that there exists a neighbourhood N_1 of ξ_0 such that

$$u_n(x) < K, \qquad x \in N_0 \cap D,$$

and so

$$v(x) \leqslant u_n(x) < K, \qquad x \in N_0 \cap D.$$

Also since $u(x)$ is s.h. and so u.s.c. at ξ_0 it follows that there exists a neighbourhood N_1 of ξ_0, such that, if $x \in N_1$ and x is outside D we have

$$v(x) = u(x) < K.$$

Thus $v(x)$ is u.s.c. at ξ_0. Finally, since $u(x)$ is s.h. at ξ_0 we have for all small positive r

$$v(\xi_0) = u(\xi_0) \leqslant \frac{1}{c_m r^{m-1}} \int_{S(\xi_0, r)} u(x)\, d\sigma(x) \leqslant \frac{1}{c_m r^{m-1}} \int_{S(\xi_0, r)} v(x)\, d\sigma(x).$$

Thus $v(x)$ is s.h. in V.

If D is a ball $D(x_0, R)$ then the function $v(x)$ coincides with that of Theorem 2.7. We can use the convexity Theorem 2.12 to prove

THEOREM 2.19. *If D is a ball $D(x_0, R)$ then the function $v(x)$ of Theorem 2.18 is the unique function such that $v(x)$ is harmonic in $D(x_0, R)$, $v(x) = u(x)$ for all points of V not in $D(x_0, R)$ and $v(x)$ is subharmonic in V. Thus $u(x)$ is the harmonic extension of u from $S(x_0, R)$ in the sense of Theorem 2.7.*

We define $v(x)$ as in Theorem 2.18 and suppose that $v_1(x)$ is harmonic in $D(x_0, R)$, $v_1(x) = u(x)$ for all points of V not in $D(x_0, R)$ and $v_1(x)$ is s.h. in V. By Theorem 2.18 $v(x)$ also has these properties, so that we must prove that $v_1(x) = v(x)$.

We show first that $v_1(x) \leqslant v(x)$ in D, and this part of the argument is quite general. In fact let $f_n(\xi)$ be a sequence of continuous functions in F and let $u_n(x)$ be the harmonic extensions of $f_n(\xi)$ from F to D. Suppose further that $f_n(\xi)$ decreases to $u(\xi)$ on F. Set

$$h_n(x) = v_1(x) - u_n(x)$$

Then, if ξ is a point of F, we have, as $x \to \xi$ from inside D,

$$\overline{\lim}\, v_1(x) \leqslant v_1(\xi) = u(\xi) \leqslant f_n(\xi),$$

$$\lim u_n(x) = f_n(\xi),$$

since $v_1(x)$ is s.h. and so u.s.c. at ξ and $u_n(x)$ is the harmonic extension of the continuous function $f_n(\xi)$. Thus

$$\overline{\lim}\, h_n(x) \leqslant 0$$

as $x \to \xi$ from inside D, and hence by the maximum principle

$$h_n(x) \leqslant 0, \qquad v_1(x) \leqslant u_n(x) \quad \text{in} \quad D.$$

By letting n tend to infinity in this we deduce from the fact that

$$u_n(x) \to v(x)$$

that

$$v_1(x) \leqslant v(x) \quad \text{in} \quad D.$$

It now follows from the maximum principle, Theorem 2.3, that $v_1(x) < v(x)$ at every point of D or else that $v_1(x)$ coincides with $v(x)$. We now suppose that $D = D(x_0, R)$ and prove that in this case $v_1(x_0) = v(x_0)$. From this Theorem 2.19 follows.

In fact by Theorem 2.12 and since $v_1(x)$ is subharmonic in V and so in $C(x_0, R)$ it follows that $I(r, v_1)$ is a continuous function of r for $0 \leqslant r \leqslant R$. Also since $v_1(x)$ is harmonic in $D(x_0, R)$, $I(r, v_1)$ is constant for $0 \leqslant r \leqslant R$. Thus

$$v_1(x_0) = I(R, v_1) = I(R, u) = \lim_{n\to\infty} I(R, f_n) = \lim_{n\to\infty} u_n(x_0) = v(x_0).$$

Thus $v_1(x_0) = v(x_0)$ and so $v_1(x) \equiv v(x)$ in $D(x_0, R)$. This proves Theorem 2.19.

2.8. SUBORDINATION

This seems a convenient point to discuss subordination since there is very pretty application of subharmonic functions to this field. The results in this section are due to Littlewood [1924].

Let $f(z)$, $F(z)$ be meromorphic in $|z| < 1$. We say that $f(z)$ is subordinate to $F(z)$ or that $F(z)$ is superordinate to $f(z)$ and write $f(z) \prec F(z)$ if $f(z) = F[\omega(z)]$, where $\omega(z)$ is regular in $|z| < 1$ and

$$|\omega(z)| \leqslant |z|, \qquad |z| < 1. \tag{2.8.1}$$

Thus $\omega(z)$ must satisfy the conditions of Schwarz's Lemma. Most of the useful applications of subordination derive from the following.

THEOREM 2.20. *Suppose that $F(z)$, $f(z)$ are meromorphic in $|z| < 1$ and map $|z| < 1$ into a domain D of the closed complex plane or more generally a Riemann surface. Suppose further that $f(0) = F(0)$ and that the inverse function $z = F^{-1}(w)$ gives a (1, 1) conformal map of D into $|z| < 1$ (if D is simply connected) or more generally can be indefinitely analytically continued throughout D with values lying in $|z| < 1$. Then $f(z) \prec F(z)$.*

The simplest case, when D is a simply connected plane domain is the most useful one, but it is known that given any plane domain D, whose complement contains at least 3 points in the complex plane a function $F(z)$ with the required properties exists. This function is superordinate to all functions $f(z)$ with values lying in D and given $f(0)$ and is in a sense the biggest such function.†

To prove Theorem 2.20 we set

$$\omega(z) = F^{-1}\{f(z)\}$$

† See e.g. Ahlfors and Sario [1960], especially p. 181.

and note that by hypothesis, $\omega(z)$ can be analytically continued throughout $|z| < 1$ with values satisfying $|\omega(z)| < 1$, $\omega(0) = 0$ there. Also, since $|z| < 1$ is simply connected, $\omega(z)$ is regular, i.e. one-valued in $|z| < 1$. Hence for $0 < r < 1$, $\omega(z)/z$ is regular in $|z| \leqslant r$ and satisfies for $|z| = r$

$$|\omega(z)/z| \leqslant 1/r,$$

and hence by the maximum principle this inequality holds also for $|z| \leqslant r$. Taking z fixed and allowing r to tend to one, we deduce (2.81).

The following properties follow very simply from the definition.

Theorem 2.21. *Suppose that* $f(z) = \Sigma a_n z^n \prec F(z) = \Sigma A_n z^n$ *in* $|z| < 1$.

Then $a_0 = A_0$,

$$|a_1| \leqslant |A_1|, \tag{2.8.2}$$

$$|a_2| \leqslant \max(|A_1|, |A_2|), \tag{2.8.3}$$

$$M(r, f) \leqslant M(r, F), 0 < r < 1. \tag{2.8.4}$$

Setting $f(z) = F[\omega(z)]$, where $\omega(z) = \Sigma_1^\infty \omega_n z^n$ satisfies (2.8.1) we have

$$a_0 = A_0, \qquad a_1 = A_1\omega_1, \qquad a_2 = A_2\omega_1^{\ 2} + A_1\omega_2.$$

In view of (2.8.1) we see at once that

$$|\omega_1| \leqslant 1,$$

with equality only when $\omega(z) = ze^{i\lambda}$, $f(z) = F(ze^{i\lambda})$, which proves (2.8.2). Also the function

$$\omega_1(z) = \frac{\omega(z)/z - \omega_1}{1 - \bar{\omega}_1\omega(z)/z} = \frac{\omega_2 z}{1 - |\omega_1|^2} + \ldots$$

satisfies the inequality (2.8.1) so that

$$|\omega_2| \leqslant 1 - |\omega_1|^2.$$

Thus

$$|a_2| \leqslant |A_2|\,|\omega_1|^2 + A_1(1 - |\omega_1|^2) \leqslant \max(|A_1|, |A_2|),$$

which proves (2.8.3). Finally in view of (2.8.1)

$$M(r, f) = \sup_{|z| \leqslant r} F[\omega(z)] \leqslant \sup_{|z| \leqslant r} |F(z)| = M(r, F),$$

which proves (2.8.4). We proceed to prove.

THEOREM 2.22. *Suppose that* $h(z)$ *is* s.h. *in* $|z| < 1$, *and that* $\omega(z)$ *is regular in* $|z| < 1$ *and satisfies* (2.8.1). *Then*

$$\int_0^{2\pi} h[\omega(re^{i\theta})]\, d\theta \leqslant \int_0^{2\pi} h(re^{i\theta})\, d\theta, 0 < r < 1. \tag{2.8.5}$$

As an immediate corollary we deduce

THEOREM 2.23. *If* $f(z) = u(z) + iv(z)$ *is subordinate to* $F(z) = U + iV$, *and if* $\phi(u)$ *is a convex function of* u *in the range of values assumed by* u *in* $|z| < 1$, *then we have*

$$\int_0^{2\pi} \phi[u(re^{i\theta})]\, d\theta \leqslant \int_0^{2\pi} \phi[U(re^{i\theta})]\, d\theta, 0 < r < 1. \tag{2.8.6}$$

If $\psi(R)$ *is a convex increasing function of* $\log R$ *in the range of values assumed by* $R = |F(z)|$ *in* $|z| < 1$, *then we have*

$$\int_0^{2\pi} \psi[|f(re^{i\theta}|)]\, d\theta \leqslant \int_0^{2\pi} \psi(|F(re^{i\theta}|)\, d\theta, \qquad 0 < r < 1. \tag{2.8.7}$$

In particular we may take $\phi(u) = |u|^k$, $k \geqslant 1$; $\psi(R) = R^\lambda$ *for* $\lambda > 0$, $\psi(R) = (\log^+ R)^k$ *for* $k \geqslant 1$, *or* $\psi(R) = \log R$.

To prove Theorem 2.22 we define $H(z)$ to be the harmonic extension of $h(z)$ into $|z| < r$. We may assume without loss in generality that $\omega(z)$ is not identically $ze^{i\lambda}$ for some real λ, since otherwise (2.8.5) holds trivially with equality. Thus $|\omega(z)| < r$, for $|z| \leqslant r$, and so $H[\omega(z)]$ is harmonic in $|z| \leqslant r$, and

$$\int_0^{2\pi} H[\omega(re^{i\theta})]\, d\theta = 2\pi H[\omega(0)] = 2\pi H(0) = \int_0^{2\pi} h(re^{i\theta})\, d\theta$$

in view of Theorem 2.7. Also by that theorem $h(\xi) \leqslant H(\xi)$, $|\xi| < r$ and so

$$\int_0^{2\pi} h[\omega(re^{i\theta})]\, d\theta \leqslant \int_0^{2\pi} H[\omega(re^{i\theta})]\, d\theta = \int_0^{2\pi} h(re^{i\theta})\, d\theta.$$

This proves Theorem 2.22. We deduce from Theorem 2.2 that, with the hypotheses of Theorem 2.23 $\phi[U(z)]$ and $\psi(|F(z)|)$ are subharmonic functions of z in $|z| < 1$, and now Theorem 2.23 follows from Theorem 2.22. If we take $\psi(R) = R^2$, in (2.8.7) we obtain at once

$$\int_0^{2\pi} |f(re^{i\theta})|^2\, d\theta \leqslant \int_0^{2\pi} |F(re^{i\theta})|^2\, d\theta.$$

or, with the notation of Theorem 2.21,

$$\sum_0^\infty |a_n|^2 r^{2n} \leqslant \sum_0^\infty |A_n|^2 r^{2n}, \qquad 0 < r < 1.$$

By letting r tend to one we deduce that

$$\sum_0^\infty |a_n|^2 \leqslant \sum_0^\infty |A_n|^2, \tag{2.8.8}$$

provided that the right-hand side is finite. A more sophisticated approach yields

THEOREM 2.24. *With the hypotheses of Theorem* 2.21 *we have*

$$\sum_0^N |a_n|^2 \leqslant \sum_0^N |A_n|^2, \qquad N = 1, 2, \ldots$$

Let

$$P(z) = \sum_0^N A_n z^n, \qquad p(z) = P[\omega(z)].$$

Then

$$f(z) = \sum_0^\infty A_n[\omega(z)]^n = p(z) + O(z^{N+1}), \quad \text{as} \quad z \to 0.$$

Thus

$$p(z) = \sum_0^N a_n z^n + \sum_{N+1}^\infty b_n z^n$$

say. In view of (2.8.8) we deduce from the fact that $p(z) \prec P(z)$,

$$\sum_0^N |a_n|^2 + \sum_{N+1}^\infty |b_n|^2 \leqslant \sum_0^N |A_n|^2,$$

and this proves Theorem 2.24. It follows in particular that

$$|a_n| \leqslant \sqrt{n} \max(|A_1|, |A_2|, \ldots, |A_n|).$$

and this in fact gives the correct order of magnitude in the most general case†, though more can be proved for special functions $F(z)$.

† See Rogosinski [1943] for examples.

A domain D in R^m is called convex if for any pair of points x_1, x_2 in D the point $tx_1 + (1 - t)x_2$ also lies in D for $0 < t < 1$. We can deduce by induction that if $x_1, x_2, \ldots, x_n$ are points in a convex domain D, then

$$t_1x_1 + t_2x_2 + \ldots + t_nx_n \in D,$$

where the t_j are non-negative numbers, whose sum is one. In particular the centre of gravity

$$\bar{x} = \frac{1}{n}\sum_{\nu=1}^{n} x_\nu$$

belongs to D. Using this notion one can prove

THEOREM 2.25. *If $f(z) \prec F(z)$, where $F(z)$ maps $|z| < 1$ (1, 1) conformally onto a convex domain D, then*

$$|a_n| \leqslant |A_1|, \qquad n = 1, 2 \ldots$$

We set

$$f_n(z) = \frac{1}{n}\sum_{\nu=0}^{n} f(ze^{2\pi i\nu/n}) = a_0 + a_nz^n + a_{2n}z^{2n} + \ldots$$

Clearly $f_n(z)$ assumes only values in D for $|z| < 1$ and hence

$$f_n(z^{1/n}) = a_0 + a_nz + a_{2n}z^2 + \ldots$$

assumes only values in D for $|z| < 1$. Hence by Theorem 2.20 $f_n(z^{1/n}) \prec F(z)$ and so by (2.8.2) applied to this function, we deduce Theorem 2.25.

We mention here one further application of Theorem 2.23, for whose proof we quote a result from elsewhere.

THEOREM 2.26. *Suppose that $f(z) = \Sigma_1^\infty a_nz^n$ is regular in $|z| < 1$ and that all values of $f(z)$ for $|z| < 1$ lie in a simply connected domain not containing the value d. Then*

$$I(r, f) = \frac{1}{2\pi}\int_0^{2\pi} |f(re^{i\theta})|\, d\theta < \frac{4\,|d|\,r}{1 - r}, \qquad 0 < r < 1, \tag{2.8.9}$$

$$|a_n| < 4\,|d|\,en, \qquad n = 2, 3 \ldots \tag{2.8.10}$$

Since $f(z)$ has values in D, $f(z) \prec F(z) = \Sigma_1^\infty A_nz^n$ where $F(z)$ maps $|z| < 1$ (1, 1) conformally onto D. Since $F(z) \neq d$ it follows from Koebe's theorem (see, e.g. Hayman [1958, Theorem 1.2, p. 3]), that $|A_1| \leqslant 4|d|$, and hence from a theorem of Littlewood (ibid. Theorem 1.6, p. 10), that

$$I(r, F) < \frac{4\,|d|\,r}{1 - r}.$$

This gives (2.8.9) in view of (2.8.7) with $\psi(R) = R$. Next the Cauchy integral formula gives

$$|a_n| = \left|\frac{1}{2\pi i}\int_{|z|=r} f(z)\frac{dz}{z^{n+1}}\right| \leqslant \frac{I(r,f)}{r^n},$$

and setting $r = (n-1)/n$, we deduce that

$$|a_n| < 4d(n-1)\left(1+\frac{1}{n-1}\right)^n = 4dn\left(1+\frac{1}{n-1}\right)^{n-1} < 4den,$$

which is (2.8.10).

The function

$$f(z) = \frac{z}{(1-z)^2} = \sum_1^\infty nz^n$$

satisfies the hypotheses of Theorem 2.26 with D the w-plane cut along the negative real axis from $-\infty$ to $-\frac{1}{4}$, so that $d = -\frac{1}{4}$. Also for this function

$$I(r,f) = \frac{r}{1-r^2}.$$

This led Littlewood to conjecture that (2.8.9) and (2.8.10) might be replaced by

$$I(r,f) \leqslant 4\,|d|\,r/(1-r^2)$$

and

$$|a_n| \leqslant 4|d|n, \text{ respectively.}$$

The first of these conjectures has recently been proved by Baernstein [1975]. The second one is still open.

Examples

We assume that $f(z) = \Sigma_0^\infty a_n z^n = u + iv$ is regular in $|z| < 1$ and is subordinate to $F(z) = \Sigma_0^\infty A_n z^n$.

1. If $F(z) \neq 0$ in $|z| < 1$, prove that

$$\inf_{|z|=r} |f(z)| \geqslant \inf_{|z|=r} |F(z)|, \qquad 0 < r < 1.$$

2. If $f(0) = \alpha + i\beta$, and $u(z) > 0$ for $|z| < 1$, prove that

$$|a_n| \leqslant 2\alpha,$$

$$|f(0)|\frac{1-r}{1+r} \leqslant |f(z)| \leqslant |f(0)|\frac{1+r}{1-r}.$$

3. If $f(z)$ assumes in $|z| < 1$ no non-negative real values, prove that $|a_n| \leqslant 4n|a_0|$, (Consider $g(z) = f(z)^{1/2}$ and use the previous example).

4. If $f(z)$ is regular and satisfies $|f(z)| > 1$ for $|z| < 1$, show that $f(z)$ is subordinate to

$$F(z) = \exp\left\{\alpha\frac{1+z}{1-z} + i\beta\right\}$$

for a suitable positive α and real β. Deduce that

$$|a_n| \leqslant |A_n|, n > 1. \text{(consider } \log f(z) \text{ and use Example 2).}$$

5. If $|v| \leqslant l$ in $|z| < 1$, prove that

$$|a_n| \leqslant \frac{4l}{\pi}, n \geqslant 1.$$

6. Show that equality is possible in (2.8.3) for any preassigned function $F(z)$ and a suitable $f(z)$ and find all such functions $f(z)$ for given $F(z)$.

7. Show that equality is possible in (2.8.4) only if $F(z)$ is constant or if $f(z) = F(ze^{i\lambda})$ for real λ.

Chapter 3

Representation Theorems

3.0. INTRODUCTION

One of the most fundamental results in the theory of subharmonic functions is due to F. Riesz [1926, 1930] and states that any such function $u(x)$ can be locally written as the sum of a potential plus a harmonic function, i.e.

$$u(x) = p(x) + h(x).$$

In other words, if $u(x)$ is subharmonic in a domain D in R^m, there exists a positive measure $d\mu$, finite on compact subsets of D, and uniquely determined by $u(x)$, such that if E is a compact subset of D and

$$(3.0.1)\begin{cases} p(x) = \displaystyle\int_E \log|x - \xi|\, d\mu e_\xi, & m = 2, \\ p(x) = -\displaystyle\int_E |x - \xi|^{2-m}\, d\mu e_\xi, & m > 2 \end{cases}$$

then

$$h(x) = u(x) - p(x)$$

is harmonic in the interior of E.

By means of this theorem many of the local properties of subharmonic functions can be deduced from those of potentials such as $p(x)$. The mass distribution $d\mu$ also plays a fundamental role in more delicate questions concerning u. Thus for instance if $m = 2$ and $u(z) = \log|f(z)|$, where f is a regular function of the complex variable z, then $\mu(E)$ reduces to the number of zeros of $f(z)$ on the set E. From this point of view the main difference between this case and that of a general subharmonic function is that in the latter case the "zeros" can have an arbitrary mass distribution instead of occurring in units of one.

In higher dimension we may regard $d\mu$ as the gravitational or electric charge, giving rise to the potential $p(x)$. For this reason the theory of subharmonic functions is frequently called potential theory.

We shall in this chapter prove the representation theorem quoted above after first giving a general discussion of measure, integration and linear functionals, including F. Riesz's famous theorem [1909] that any positive linear functional can be represented by a measure. After proving Riesz's representation theorem we shall deduce a version of the Poisson–Jensen formula, Theorem 3.14, which allows us to express a subharmonic function $u(x)$ in terms of its values on the boundary of a domain D and its Riesz measure in D. This leads in turn to an extension of Theorem 2.19 to more general domains D, a version of Nevanlinna's first fundamental theorem for functions subharmonic in an open ball and a characterization of bounded subharmonic functions in R^m, when $m \geqslant 3$.

3.1. MEASURE AND INTEGRATION

Let A, B be any two sets. We write $A - B$ for the set of all elements x such that $x \in A$ and $x \notin B$. A family of sets R is called a ring if $A \in R$ and $B \in R$ implies that

$$A \cup B \in R \quad \text{and} \quad A - B \in R. \tag{3.1.1}$$

Since $A \cap B = A - (A - B)$ we also have that $A \cap B \in R$ if R is a ring. A ring is called a σ-ring if

$$\bigcup_{n=1}^{\infty} A_n \in R \tag{3.1.2}$$

whenever $A_n \in R$, $n = 1, 2 \ldots$.

It is evident that the intersection of any number of σ-rings (i.e. the class of sets belonging to all the σ-rings) again forms a σ-ring. In any open or compact set X in R^m the intersection of all σ-rings containing the open and closed sets in X is called the σ-ring of Borel sets in X. Any Borel set can be obtained from the open or the closed sets by forming unions and differences a finite or countable number of times.

A function $f(x)$ in X is said to be Borel-measurable if all the subsets $f(x) > a$, $f(x) \geqslant a$, $f(x) < a$ and $f(x) \leqslant a$ of X are Borel sets for different real values of a. If $f(x)$ is continuous all these sets are open or closed, so that continuous functions are always Borel-measurable. It is easy to verify that if f_n, $n = 1, 2, \ldots$ is a sequence of Borel-measurable functions then $f_1 \mp f_2$, $f_1 f_2$ are Borel-measurable and so is f_1/f_2, provided that $f_2 \neq 0$. Further if

$$f_n(x) \to f(x) \quad \text{as} \quad n \to \infty \tag{3.1.3}$$

for each $x \in X$, then $f(x)$ is also Borel-measurable. In particular all semi-continuous functions are Borel-measurable in view of Theorem 1.4.

A set function μ defined on a σ-ring R containing the open and closed sets in X and so all the Borel-sets in X is said to form a measure if the following conditions are satisfied.

$$0 \leqslant \mu(E) \leqslant +\infty \quad \text{for} \quad E \in R. \tag{3.1.4}$$

If E_n is a finite or countable class of mutually nonintersecting sets in R, whose union is E then

$$\mu(E) = \Sigma\, \mu(E_n). \tag{3.1.5}$$

Given a measure μ and a Borel-measurable function $f(x)$ in X the Radon† integral

$$\int_X f(x)\, \mathrm{d}\mu$$

can be defined as follows. We assume first that $f(x)$ is a *simple function*, i.e. that $f(x)$ only assumes a finite number of distinct values y_i, $i = 0$ to n on the subsets E_i of X respectively, where $y_0 = 0$, and $\mu(E_i) < \infty$ for $i > 0$. Then we define

$$\int_X f(x)\, \mathrm{d}\mu = \int_X f(x)\, \mathrm{d}\mu(x) = \sum_{i=1}^{n} y_i \mu(E_i).$$

Next if $f(x) \geqslant 0$ in X

$$I = \int_X f(x)\, \mathrm{d}\mu = \sup \int_X g(x)\, \mathrm{d}\mu,$$

where the supremum is taken over all simple functions $g(x)$, such that $g(x) \leqslant f(x)$ in X. Thus $0 \leqslant I \leqslant +\infty$. Similarly if $f(x) \leqslant 0$, we define

$$\int_X f(x)\, \mathrm{d}\mu = -\int (-f(x))\, \mathrm{d}\mu.$$

Finally for general functions $f(x)$, we write

$$f^+(x) = \max(f(x), 0), f^-(x) = -\min(f, 0),$$

so that

$$f = f^+ - f^-.$$

† Radon [1919].

We write

$$I^+ = \int_X f^+ \, d\mu, \; I^- = \int_X f^- \, d\mu,$$

and define

$$\int_X f \, d\mu = I^+ - I^-,$$

provided that I^+ and I^- are not both infinite. If I^+ and I^- are both finite, we say that f is integrable ($d\mu$).

The integral defined in this way has the usual properties of which the following are the most important from our point of view.

Additivity. If $f(x)$ and $g(x)$ are integrable and a, b are real constants then $af(x) + bg(x)$ is integrable and

$$\int_X (af + bg) \, d\mu = a \int_X f \, d\mu + b \int_X g \, d\mu. \tag{3.1.6}$$

Monotonic convergence. Suppose that $f_n(x)$ is a sequence of Borel-measurable functions and $f_n(x) \to f(x)$ as $n \to \infty$ for each $x \in X$ except perhaps a set X_0 of X, such that $\mu(X_0) = 0$. Then we say that $f_n(x) \to f(x)$ *p.p.*(μ) on X†. If this is so and the sequence $f_n(x)$ is monotonic in n and integrable ($d\mu$) for each fixed n, then

$$\int f_n(x) \, d\mu \to \int f(x) \, d\mu \text{ as } n \to \infty. \tag{3.1.7}$$

A detailed account of the above theory including proofs of all the results quoted will be found for instance in Rudin [1964] chapter 10.

3.2. LINEAR FUNCTIONALS

We have seen how from a measure defined on Borel sets the notion of Lebesgue integration can be built up. We now investigate the converse problem of building up a measure and integration theory from a concept called a positive linear functional which has the basic properties of an integral.

We consider classes of functions $f(x)$ on a space X which will always be a domain or compact subset of R^m. The support of $f(x)$ is the closure in R^m of the set of points where $f(x) \neq 0$, i.e. the set of all points or limit-points of points where $f(x) \neq 0$.

Let $C_0 = C_0(X)$ be the class of all continuous functions on X, whose

† p.p. stands for presque partout (almost everywhere).

support is a compact subset of X. Thus if X is compact, C_0 is simply the class of all continuous functions on X. If X is a domain D, we also define the subset $C_0^\infty(D)$ of all functions belonging to C^∞ and having compact support in D.

Let $\mathscr{F}$ be a class of functions on X such that if $f, g \in \mathscr{F}$ then so do $af + bg$, where a, b are real constants. Such a class is called linear. A positive linear functional or simply functional on a linear class $\mathscr{F}$ is a real function $L(f)$ defined on $\mathscr{F}$ with the following properties.

If $f(x) \in \mathscr{F}$ and $f(x) \geqslant 0$, then

$$0 \leqslant L(f) < \infty. \tag{3.2.1}$$

Further if $f, g \in \mathscr{F}$ and a, b are real then

$$L(af + bg) = aL(f) + bL(g). \tag{3.2.2}$$

We wish to show that (positive) functionals on $C_0^\infty(D)$ can be extended to $C_0(D)$. In order to do this we need the following approximation theorem.

THEOREM 3.1. *If D is a domain in R^m, $f \in C_0(D)$ and $\varepsilon > 0$, there exists $g \in C_0^\infty(D)$ such that*

$$|f(x) - g(x)| < \varepsilon \quad \text{in} \quad D.$$

Let $K(x)$ be a function in R^m with the following properties

$$K(x) = 0, \qquad |x| \geqslant 1,$$
$$K(x) > 0, \qquad |x| < 1,$$

$K(x) \in C^\infty$ and

$$\int K(x)\,\mathrm{d}x = 1, \tag{3.2.3}$$

where the integral is extended over the whole of space or, equivalently, the ball $|x| < 1$. We may take for instance

$$K(x) = C \exp\left[-(1 - |x|^2)^{-1}\right], \qquad |x| < 1$$

where C is a constant of normalization chosen so that (3.2.3) holds. We then define

$$g(x) = \int f(x + \xi)\delta^{-m} K(\xi/\delta)\,\mathrm{d}\xi, \tag{3.2.4}$$

where δ is a sufficiently small positive constant.

In fact we may set $x + \xi = u$ and have

$$g(x) = \delta^{-m} \int f(u) K[(u - x)/\delta]\,\mathrm{d}u.$$

Thus we may differentiate under the sign of integration indefinitely† and deduce that $g(x) \in C^\infty$ in the whole space. Also, since $f \in C_0(D)$, f vanishes outside a set F which is at a positive distance δ_0 from the complement of D. Since $K(\xi/\delta) = 0$ for $|\xi| > \delta$, it follows that $g(x) = 0$ if x is distant more than δ from F. Thus, if $\delta < \delta_0$, $g(x)$ has compact support in D, so that $g(x) \in C_0^\infty(D)$. Finally

$$g(x) - f(x) = \delta^{-m} \int_{|\xi|<\delta} [f(x+\xi) - f(x)]K(\xi/\delta)\,\mathrm{d}\xi.$$

We suppose that δ is so chosen that $|f(x+\xi) - f(x)| < \varepsilon$, for $|\xi| < \delta$. This is possible since $f(x)$ is continuous and of compact support and so is uniformly continuous in the whole space. Then

$$|g(x) - f(x)| < \varepsilon\delta^{-m} \int_{|\xi|<\delta} K(\xi/\delta)\,\mathrm{d}\xi = \varepsilon.$$

This proves Theorem 3.1. We also need

LEMMA 3.1. *If D is a domain in R^m and F is a compact subset of D, there exists a function $g(x) \in C_0^\infty(D)$, such that $g(x) \geqslant 0$ in D, $g(x) > 0$ in F.*

Let $2\delta_0$ be the distance of F from the complement of D, let F_1 be the set of all points distant at most δ_0 from F and let D_1 be the complement of F_1. We define $f(x)$ to be the distance of x from D_1. Clearly $f(x)$ is continuous, $f(x) \geqslant \delta_0 > 0$ in F, $f(x) \geqslant 0$ everywhere and $f(x) = 0$ outside F_1. Thus $f(x) \in C_0(D)$. We now define $g(x)$ by (3.2.4) and note that $g(x) \geqslant 0$, since $f(x) \geqslant 0$. Also if $\varepsilon < \delta_0$, $g(x) > 0$ in F. This proves Lemma 3.1.

We can now prove our first extension Theorem.

THEOREM 3.2. *If D is a domain in R^m and L a functional on $C_0^\infty(D)$, then L can be uniquely extended to $C_0(D)$.*

Suppose that $f(x) \in C_0(D)$. Let $g_1(x)$, $g_2(x)$ be functions in $C_0^\infty(D)$ such that $g_1(x) \leqslant f(x) \leqslant g_2(x)$ in D. We call $g_1(x)$ a lower function and $g_2(x)$ an upper function. Then we define

$$L^-(f) = \sup L(g_1), \qquad L^+(f) = \inf L(g_2),$$

where the supremum is taken over all lower functions and the infimum over all upper functions.

Since $g_1 \leqslant g_2$, we have by (3.2.1) and (3.2.2)

$$L(g_2) = L(g_1) + L(g_2 - g_1) \geqslant L(g_1).$$

† This will be proved under more general hypotheses in Theorem 3.6, section 3.4.1.

Thus $L^-(f) \leqslant L^+(f)$. We proceed to show that $L^-(f) = L^+(f)$. To see this suppose that F is the support of f and let F_1 be a compact subset of D containing F in its interior D_1. Let $h(x)$ be a function in $C_0^\infty(D)$, such that $h(x) \geqslant 0$ everywhere, $h(x) > 0$ in F_1 and let η be the minimum of $h(x)$ in F_1. We approximate $f(x)$ by a function $g(x) \in C_0^\infty(D_1)$, such that for all x, and some positive integer n

$$|f(x) - g(x)| < \frac{\eta}{n}$$

and $f(x) = g(x) = 0$ outside F_1. Thus for all x we have

$$g(x) - \frac{h(x)}{n} \leqslant f(x) \leqslant g(x) + \frac{h(x)}{n},$$

and so writing $g_1 = g - h/n$, $g_2 = g + h/n$, we have

$$L^+(f) - L^-(f) \leqslant L(g_2) - L(g_1) = \frac{2}{n} L(h). \tag{3.2.5}$$

Since n can be made as large as we please, we deduce that $L^-(f) = L^+(f)$. We now write $L(f) = L^-(f) = L^+(f)$ for any continuous function f and note that this definition coincides with the original one when $f \in C_0^\infty(D)$, since we may then take $g_1 = g_2 = f$. Also we evidently have (3.2.1) for the extended definition. Next if a is positive and g_1, g_2 are lower and upper functions for f, then ag_1, ag_2 are lower and upper functions for af. If $a < 0$, ag_2 and ag_1 are lower and upper functions for af. We deduce that for real a and $f \in C_0(D)$

$$L(af) = aL(f).$$

Also if f_1, f_2 are lower and upper functions for f and g_1, g_2 are lower and upper functions for g, where $f, g \in C_0(D)$ then $f_1 + g_1$, $f_2 + g_2$ are lower and upper functions for $f + g$ and also

$$L(f_2 + g_2) - L(f_1 + g_1) = L(f_2 - f_1) + L(g_2 - g_1).$$

The right-hand side can be made as small as we please and we deduce that $L(f + g) = L(f) + L(g)$. Thus our extended functional satisfies (3.2.2).

The extension is unique, since (3.2.1) implies for $f_1 \leqslant f \leqslant f_2$

$$L(f_1) \leqslant L(f) \leqslant L(f_2),$$

so that we must have $L^-(f) \leqslant L(f) \leqslant L^+(f)$ for any extension of the functional L to a wider class of functions f.

The method used for the proof of Theorem 3.2 actually allows us to extend the functional L to a somewhat wider class of functions than merely the continuous ones. The extension obtained in this way corresponds to the

Riemann or Riemann–Stieltjes integral. However, the class obtained is still not sufficiently large. Thus if X reduces to the open interval (0, 1) in R^1 and

$$L(f) = \int_0^1 f(x)\,dx$$

for a function $f(x)$ in $C_0^\infty(D)$ the method leads to the Riemann integral for all Riemann integrable functions. In fact $L^-(f)$ and $L^+(f)$ are respectively the lower and upper Riemann integrals.

3.3. CONSTRUCTION OF LEBESGUE MEASURE AND INTEGRALS—(F. RIESZ'S THEOREM)

We now suppose given a positive linear functional on $C_0(X)$ where X is a domain or a compact subset in R^m and try to extend this uniquely to a wider class of functions. For this purpose we introduce a further axiom.

We assume that if f_n tends monotonically to f and $L(f_n)$ is defined and finite for each $n = 1, 2 \ldots$ then

$$L(f_n) \to L(f) \quad \text{as} \quad n \to \infty. \tag{3.3.1}$$

For Lebesgue integrals this result is true in view of (3.1.7). We also drop the assumption that $L(f) < \infty$ in (3.2.1) for functions $f(x)$ not in $C_0(X)$. We shall show that these additional assumptions allow us to make the required extension. We first show that (3.3.1) is consistent with out previous axioms.

LEMMA 3.2. *Suppose that $f_n \in C_0(X)$, $n = 1, 2, \ldots$ and that the sequence f_n is monotonic, so that $f_n \to f$ and $L(f_n) \to \lambda$ as $n \to \infty$. Then if $g \in C_0(X)$ and $g \leqslant f$ we have $L(g) \leqslant \lambda$, while if $g \geqslant f$ we have $L(g) \geqslant \lambda$.*

Suppose for definiteness that the sequence f_n increases with n. The proof when f_n decreases is similar. Then if $g \geqslant f$, we have $g \geqslant f_n$ for each n and so

$$L(g) \geqslant L(f_n).$$

In view of (3.2.1) and (3.2.2) the sequence $L(f_n)$ is increasing and so tends to a limit λ, such that $\lambda \leqslant L(g)$ as required.

The case when $g \leqslant f$ is a little harder. Let F_1 be the support of f_1, G the support of g and $F = F_1 \cup G$. Then F is a compact subset of D and we have for x outside F

$$g(x) = 0 = f_1(x) \leqslant f_n(x), \qquad n = 1, 2 \ldots. \tag{3.3.2}$$

We now construct a function $h(x) \in C_0(X)$ such that

$$h(x) \geqslant 0 \text{ in } X, \qquad h(x) \geqslant 1 \text{ in } F. \tag{3.3.3}$$

If X is compact we may take $h(x) = 1$. If X is a domain we use Lemma 3.1. Let p be a fixed positive integer and set $\varepsilon = p^{-1}$. Then for each $x \varepsilon F$, we have $f(x) \geqslant g(x)$ and so

$$f_n(x) > g(x) - \varepsilon, \qquad n \geqslant n_0(x).$$

Since $f_n(x)$ and $g(x)$ are continuous there is an open ball $D(x, r)$ with centre x and positive radius such that

$$f_n(\xi) > g(\xi) - \varepsilon, \qquad n = n_0(x), \qquad \xi \in D(x, r).$$

By the Heine–Borel Theorem a finite number $D_1, D_2, \ldots, D_N$ of these balls cover F. If $n_1, n_2, \ldots, n_N$ are the associated indices and $m = \max(n_1, n_2, \ldots, n_N)$, then we deduce that

$$f_n(\xi) > g(\xi) - \varepsilon, \qquad \xi \in F, \qquad n > m.$$

In view of (3.3.2) and (3.3.3) this yields

$$f_n(x) \geqslant g(x) - \frac{1}{p} h(x), \qquad n \geqslant m, \qquad x \in X.$$

Thus

$$L(f_n) \geqslant L(g) - \frac{1}{p} L(h), \qquad n \geqslant m.$$

Hence

$$\lambda \geqslant L(g) - \frac{1}{p} L(h).$$

Since p is any positive integer we have $L(g) \leqslant \lambda$ as required.

We deduce immediately

LEMMA 3.3. *If f_n, f belong to $C_0(X)$ and $f_n \to f$ monotonically then* (3.3.1) *holds.*

For in this case we may take $g = f$ in Lemma 3.2 and obtain $L(f) \leqslant \lambda$ and $L(f) \geqslant \lambda$. Thus we see that in $C_0(X)$ (3.3.1) is a consequence of (3.2.1) and (3.2.2). We can use Lemma 3.2 to define $L(f)$ for semi-continuous functions. To do this we define, as in the last section, for any real function $f(x)$, lower function $f_1(x)$ and an upper function $f_2(x)$ to be functions in $C_0(X)$ such that $f_1(x) \leqslant f(x)$ and $f_2(x) \geqslant f(x)$ respectively. We also write

(3.3.4) $$L^-(f) = \sup L(f_1), \qquad L^+(f) = \inf L(f_2)$$

where the infimum and supremum are taken over all lower and upper functions respectively. The empty set is taken to have supremum $-\infty$ and infimum $+\infty$.

3.3.1

We now prove

LEMMA 3.4. *If $f_n \in C_0(X)$ and $f_n \to f$ monotonically then $L(f_n) \to L^-(f)$ or $L^+(f)$ according as the sequence f_n is increasing or decreasing.*

Suppose first that f_n increases. Then f_n is a lower function for f. Thus

$$L(f_n) \leqslant L^-(f), \qquad n = 1, 2 \dots$$

and so

$$\lambda = \lim_{n \to \infty} L(f_n) \leqslant L^-(f).$$

On the other hand let g be any lower function for f. Then in view of Lemma 3.2 we have $\lambda \geqslant L(g)$. Since this is true for every g we have $\lambda \geqslant L^-(f)$. Thus $\lambda = L^-(f)$. Similarly if f_n decreases we see that $\lambda = L^+(f)$.

Thus if (3.3.1) is to hold we are forced to the following

DEFINITION 3.3.1. *Suppose that f is an* u.s.c. *function on X such that $f(x) \leqslant 0$ outside a compact subset of X. Then we define $L(f) = L^+(f)$. If $f(x)$ is a lower semi-continuous function such that $f(x) \geqslant 0$ outside a compact subset of X we define $L(f) = L^-(f)$.*

We note that the functions considered in the definition are precisely those which are the limits of monotonic sequences from $C_0(X)$. For in view of Theorem 1.3 a monotonic decreasing sequence of continuous function $f_n(x)$ converges to an u.s.c. limit $f(x)$. If in addition $f_n(x)$ vanishes outside a compact subset F_n of X, then $f_n(x) \leqslant 0$ outside F_1 and so $f(x) \leqslant 0$ outside F_1.

Conversely if $f(x)$ is u.s.c. in X, then in view of Theorem 1.4 there exists a decreasing sequence $f_n(x)$ of continuous functions in X which converge to $f(x)$ as $n \to \infty$. If X is compact we may take X to be the compact subset outside which $f_n = 0$ and $f \leqslant 0$. If X is a bounded domain D, suppose that $f(x) \leqslant 0$ outside a compact set E and define $f(x) = 0$ outside D. Then $f(x)$ is u.s.c. in the whole of R^m. Also $f(x)$ is bounded above in R^m and so $\delta(\xi) = 1$ in the proof of Theorem 1.4. As in the proof of the theorem we set

$$M(\xi, h) = \sup_{|x - \xi| \leqslant h} f(x),$$

but we modify (1.2.2) slightly and set

$$f_n(x) = 2n \int_{1/(2n)}^{1/n} M(x, t)\, dt.$$

We prove just as before that $f_n(x)$ decreases with increasing n for fixed x and converges to $f(x)$ as $n \to \infty$, and also that $f_n(x)$ is continuous in x for fixed n. Further, if x is distant less than $1/2n$ from the complement of D, where n is large $M(x, t) = 0$ for $1/(2n) \leqslant t \leqslant 1/n$ so that $f_n(x) = 0$. Thus $f_n(x)$ vanishes outside a compact subset of D.

If D is unbounded we make a topological transformation of R^m onto the open unit ball in R^m and apply the above procedure in the transformed domain. The case of lower semi-continuous functions is dealt with similarly.

We note also that our new definition allows $L(f) = -\infty$ (if $f(x)$ is upper semi-continuous) and $L(f) = +\infty$ (if $f(x)$ is lower semi-continuous).

We can now make our final extension of $L(f)$.

DEFINITION 3.3.2. *Suppose that $f(x)$ is an arbitrary function defined in X. Then we define $L^-(f)$ and $L^+(f)$ as in* (3.3.4) *where the supremum is taken over all upper semi-continuous functions f_1 such that $f_1 \leqslant 0$ outside a compact subset of X and $f_1 \leqslant f$ in X and similarly the infimum is taken over all lower semi-continuous functions f_2 such that $f_2 \geqslant 0$ outside a compact subset of X and $f_2 \geqslant f$ in X. If $L^-(f)$ and $L^+(f)$ are equal we define $L(f)$ to be their common value. If in addition $L(f)$ is finite, we say that f is integrable* (L).

3.3.2

We proceed to prove

THEOREM 3.3. *Let $\mathscr{F}$ be the family of functions which are integrable* (L). *Then $\mathscr{F}$ is linear and the functional $L(f)$ defined as above satisfies* (3.2.1) *and* (3.2.2). *Further if $f_n \in \mathscr{F}$ and the sequence f_n tends monotonically to f and is such that $L(f_n)$ is bounded then $f \in \mathscr{F}$ and* (3.3.1) *holds. The value of $L(f)$ is uniquely determined on $\mathscr{F}$ subject to* (3.2.1), (3.2.2) *and* (3.3.1) *and the original definition of $L(f)$ on $C_0(D)$.*

We can dispose of the uniqueness very simply. For we have seen that (3.3.1) together with the original definitions on $C_0(D)$ determines L on all the lower functions f_1 and upper functions f_2 which are used in the definition of integrability. Thus if $L(f)$ is any extension of L to a linear class containing these semi-continuous functions we must have $L^-(f) \leqslant L(f) \leqslant L^+(f)$ in view of (3.2.1) and (3.2.2). Thus if $L^-(f) = L^+(f)$, then $L(f)$ must be equal to the common value of these two quantities.

Next if $f \geqslant 0$, then $f_1 = 0$ is a lower function and $L(f_1) = 0$. Thus $L(f) \geqslant 0$, so that (3.2.1) holds.

Suppose now that $L(f)$ exists finitely and that a is a positive number. We can then find upper and lower functions f_1 and f_2 such that $f_1 \leqslant f \leqslant f_2$

and $L(f_2) < L(f_1) + \varepsilon$. Then af_2 and af_1 are upper and lower functions for af and

$$L(af_2) - L(af_1) = a\{L(f_2) - L(f_1)\} < a\,\varepsilon.$$

Thus af is integrable. If a is negative af_2 and af_1 are lower and upper functions for af and our conclusion follows as before. Thus $L(af) = aL(f)$.

Finally suppose that f, g are integrable and that f_1, g_1 are lower functions for f, g respectively and f_2, g_2 upper functions for f, g, such that

$$L(f_2) - L(f_1) < \varepsilon, \qquad L(g_2) - L(g_1) < \varepsilon.$$

Then $f_1 + g_1, f_2 + g_2$ are lower and upper functions for $f + g$ and

$$L(f_2 + g_2) - L(f_1 + g_1) = L(f_2) - L(f_1) + L(g_2) - L(g_1) < 2\varepsilon.$$

Thus $f + g$ is integrable and $L(f + g) = L(f) + L(g)$.

Thus $\mathscr{F}$ is a linear class and (3.2.1) and (3.2.2) are both satisfied. We also note that the upper and lower functions f_j of our definition are themselves integrable according to our new definition if $L(f_i)$ is finite. For instance if f_1 is u.s.c. and $f_1 < 0$ outside a compact subset of D, then by definition 3.3.1 there exists a continuous function f_2 of compact support in D, such that, given $K > L(f_1)$

$$f_1 \leqslant f_2 \text{ in } D \text{ and } L(f_2) < K$$

Thus we may take this function as f_2 and for f_1 the function itself in the definition 3.3.2. Lower semi-continuous functions are dealt with similarly.

It remains to deal with (3.3.1). We suppose first that f_n is a decreasing sequence of u.s.c. functions, such that $f_n \leqslant 0$ outside a compact subset of X, and that $f_n \to f$ as $n \to \infty$. Then by Theorem 1.3 f is also u.s.c. Suppose that $g(x) \in C_0(X)$, $g(x) \geqslant f(x)$. Then, just as in the proof of Lemma 3.2, we can show that there exists $h(x) \in C_0(X)$, such that given $\varepsilon > 0$, we can find $n_0 = n_0(\varepsilon)$ such that

$$f_n(x) < g(x) + \varepsilon h(x), \qquad n > n_0.$$

Thus since $g(x), h(x)$ belong to $C_0(X)$, we deduce that

$$L(f_n) < L(g) + \varepsilon L(h), \quad n > n_0.$$

Thus in this case

$$\lambda = \lim_{n \to \infty} L(f_n) \leqslant L(g).$$

Since this is true for every continuous $g(x)$, such that $g(x) \geqslant f(x)$ we deduce, since $f(x)$ is u.s.c., that

$$\lambda \leqslant L(f).$$

Evidently $L(f) \leqslant L(f_n)$ for every n and so $L(f) \leqslant \lambda$. Thus in this case we have $L(f) = \lambda$.

Similarly if f_n is an increasing sequence of lower semi-continuous functions, each of which is non-negative outside a compact set then the limit f has the same property and (3.3.1) holds.

Suppose now that f_n is an increasing sequence of integrable functions. We write $u_1 = f_1$, $u_n = f_n - f_{n-1}$, $n \geqslant 2$, so that

$$f_n = u_1 + u_2 + \ldots + u_n.$$

By what has already been proved the u_n are non-negative and integrable for $n > 1$ and

$$L(f_n) = \sum_{r=1}^{n} L(u_r).$$

We write $\lambda = \lim_{n \to \infty} L(f_n) = \Sigma_{r=1}^{\infty} L(u_r)$. Then we can find a lower semi-continuous function g_r, such that $u_r \leqslant g_r$ and

$$L(g_r) < L(u_r) + \varepsilon 2^{-r}.$$

The function

$$g = \sum_{1}^{\infty} g_r$$

is lower semi-continuous and by what was proved above

$$L(g) = \sum_{1}^{\infty} L(g_r) \leqslant \sum_{1}^{\infty} L(u_r) + \varepsilon = \lambda + \varepsilon.$$

Also g is an upper function for f and so

$$L^{+}(f) \leqslant L(g) \leqslant \lambda + \varepsilon.$$

This is true for every positive ε and so

$$L^{+}(f) \leqslant \lambda = \lim_{n \to \infty} L(f_n).$$

On the other hand we have $f_n \leqslant f$ and so

$$L^{-}(f) \geqslant L^{-}(f_n) = L(f_n)$$

so that

$$L^{-}(f) \geqslant \lambda$$

Thus (3.3.1) holds, and f is integrable if λ is finite. If the sequence f_n decreases with n (3.3.1) is proved similarly. This completes the proof of Theorem 3.3.

We also note that (3.3.1) continues to hold if $L(f)$ is infinite provided that at least one of the functions f_n is integrable. For if, e.g. f_n is monotonic increasing and $L(f_n)$ is finite, $L(f_n)$ can only tend to a finite limit or to $+\infty$. In the latter case we have $L^-(f) \geqslant L^-(f_n)$ for each n and so $L^-(f) = +\infty$. Thus $L^+(f) = L^-(f) = +\infty$ in this case.

3.3.3

It is now easy to construct our measure and to show that it has the required properties. If E is any set then the characteristic function χ_E of E is defined to be equal to 1 in E and zero outside E. A set E is said to be measurable if χ_E is integrable (L) in the sense of definition 3.3.2, or if E is a limit of an expanding sequence of such sets. (Thus certain sets E may be measurable even though $L(\chi_E) = +\infty$). We define the measure μ associated with the functional L by

$$\mu(E) = L(\chi_E).$$

With this definition we have Riesz's Theorem [1909] on (positive linear) functionals.

THEOREM 3.4. *The class of measurable sets is a σ-ring R containing the Borel-subsets of X. Further μ is a measure on R and for any integrable function $f(x)$ we have*

$$L(f) = \int f \, d\mu. \tag{3.3.5}$$

The measure μ is uniquely determined on the Borel-sets provided that (3.3.5) *holds for $f \in C_0(X)$.*

We note that if E is a compact subset of X then χ_E is u.s.c. Also a closed subset of X is a limit of an expanding sequence of compact subsets of X. Thus closed subsets of X are measurable and so are open subsets, since if $A \cup B = X$, $A \cap B = \varnothing$, we have $\chi_A = 1 - \chi_B$.

Next suppose that A, B are measurable sets with compact closures in X, let $\chi_A(x)$, $\chi_B(x)$ be their characteristic functions and let f_1, f_2 be lower and upper functions respectively for χ_A and g_1, g_2 lower and upper functions for χ_B. By writing $\max(f_1, 0)$ instead of f_1 and $\inf(f_2, 1)$ instead of f_2, we may assume that $0 \leqslant f_1 \leqslant \chi_A \leqslant f_2 \leqslant 1$ and similarly $0 \leqslant g_1 \leqslant \chi_B \leqslant g_2 \leqslant 1$. Since f_1, g_1 are u.s.c. and non-negative $f_1 g_1$ is u.s.c., and similarly $f_2 g_2$ is l.s.c. Also $f_1 g_1$ and $f_2 g_2$ are lower and upper functions for $\chi_A \chi_B$ and in particular are measurable. Finally

$$f_2 g_2 - f_1 g_1 = f_2(g_2 - g_1) + g_1(f_2 - f_1) \leqslant g_2 - g_1 + f_2 - f_1.$$

Thus

$$L(f_2g_2 - f_1g_1) \leqslant L[(g_2 - g_1) + (f_2 - f_1)]$$
$$= L(g_2) - L(g_1) + L(f_2) - L(f_1).$$

The right-hand side can be made as small as we please, since χ_A and χ_B are measurable. Thus $\chi_A\chi_B$ is measurable and hence so is $A \cap B$. Next $\chi_A + \chi_B - \chi_A\chi_B$ is measurable and this is the characteristic function of $A \cup B$, so that $A \cup B$ is also measurable.

Also $\chi_A - \chi_A\chi_B$ is measurable and this is the characteristic function of $A - B$, so that $A - B$ is measurable.

Finally let A_n be an expanding sequence of measurable sets, whose closure is compact in D. Then by what we have just proved

$$B_N = \bigcup_{n=1}^{N} A_n$$

is an increasing sequence of measurable sets. Let $\chi_N(x)$ be the characteristic function of B_N. Then $\chi_N(x)$ is an increasing sequence of measurable functions, and so $\chi(x) = \lim_{N\to\infty} \chi_N(x)$ is measurable and

$$L\{\chi(x)\} = \lim_{N\to\infty} L\{\chi_N(x)\}.$$

Thus $B = \bigcup_{n=1}^{\infty} A_n$ is measurable and

$$\mu(B) = \lim_{N\to\infty} \mu(B_N). \tag{3.3.6}$$

Thus R is a σ-ring which contains the Borel sets. Also if A_n is a finite or countable system of disjoint sets in R and $B_N = \bigcup_{n=1}^{N} A_n$, then

$$\chi_{B_N} = \sum_{n=1}^{N} \chi_{A_n},$$

so that, since the χ_{A_n} are all measurable, we deduce that

$$\mu(B_N) = \sum_{n=1}^{N} \mu(A_n).$$

If the system is countable we deduce from (3.3.6) that

$$\mu(B) = \lim \mu(B_N) = \lim \sum_{n=1}^{N} \mu(A_n) = \sum_{1}^{\infty} \mu(A_n)$$

as required.

We next note that (3.3.5) holds if f is the characteristic function of a measurable set. Next in view of the linearity of L and the integral the equation continues to hold for simple functions f and hence for general measurable functions. Finally to prove the uniqueness of μ we note that if (3.3.5) holds for continuous functions, then $L(f)$ is a positive linear functional satisfying (3.2.1) and (3.2.2) on the continuous functions. In view of Lemma 3.4, definition 3.3.1 and (3.1.7) it follows that (3.5.1) continues to hold for the class of functions which are limits of monotonic sequences of continuous functions with compact support, and so for semi-continuous functions and hence finally for the class of all measurable functions with respect to the functional L where measurable is defined as in the Definition 3.3.2. In particular (3.5.1) holds if f is the characteristic function of a Borel set and so μ is determined on the Borel sets. This completes the proof of Theorem 3.4.

3.4. REPEATED INTEGRALS AND FUBINI'S THEOREM

A measure μ defined over a subset X of a Euclidian space will be called a Borel-measure if X is compact and $\mu(X) < +\infty$ or if X is open and $\mu(E) < \infty$ for all compact subsets E of X. The measure whose existence was asserted in Theorem 3.4 is clearly a Borel-measure and so are all the measures with which we shall be dealing in this chapter. We now suppose given a Borel measure μ_1 in a domain D_1 of R^p with coordinates $x = (x_1, x_2, \ldots, x_p)$ and a Borel measure μ_2 in a domain D_2 in R^q with coordinates $y = (x_{p+1}, \ldots, x_{p+q})$. We proceed to define and investigate a Borel measure μ, in the domain $D = D_1 \times D_2$ of R^{p+q} consisting of all points $z = (x_1, \ldots, x_{p+q})$ such that $(x_1, x_2, \ldots, x_p) \in D_1$ and $(x_{p+1}, \ldots, x_{p+q}) \in D_2$.

For this purpose let $f(x) = f(x, y)$ be a continuous function with compact support in D. We write

$$L_1(f) = \int d\mu_1(x) \int f(x, y)\, d\mu_2(y), \tag{3.4.1}$$

$$L_2(f) = \int d\mu_2(y) \int f(x, y)\, d\mu_1(x).$$

We proceed to prove

LEMMA 3.5. *The equations* (3.4.1) *define* $L_1(f)$ *and* $L_2(f)$ *as equal functionals over* $C_0(D)$.

Let F be the support of f in D. Then F is compact and so are the projections F_1 and F_2 of F onto D_1 and D_2 respectively. Thus for each $x \in D_1, f(x, y)$ is a continuous function with compact support in D_2. Thus

$$F(x) = \int f(x, y)\, d\mu_2(y)$$

is defined. Also $F(x)$ is continuous. In fact $f(x, y)$ is continuous and so uniformly continuous on F and so in D. Thus given $\varepsilon > 0$, there exists δ, such that if

(3.4.2) $$|x' - x''| < \delta \quad \text{and} \quad |y' - y''| < \delta,$$

then

(3.4.3) $$|f(x', y') - f(x'', y'')| < \varepsilon.$$

In particular, if $x', x'' \in D_1$ and $|x' - x''| < \delta$, then

$$|F(x') - F(x'')| = |\int \{f(x', y) - f(x'', y)\} \, d\mu_2(y)| < \varepsilon\mu_2(F_2).$$

Thus $F(x)$ is continuous in D_1 with support in F_1 and so $L_1(f)$ is well defined and so is $L_2(f)$ similarly. Also $L_1(f)$ and $L_2(f)$ are finite so that (3.2.1) holds. It is also evident from the definitions that (3.2.2) holds. Thus $L_1(f)$ and $L_2(f)$ are functionals.

We prove next that

$$L_1(f) = L_2(f) = L(f)$$

say. To see this we suppose $\varepsilon > 0$ and choose δ so that (3.4.3) holds subject to (3.4.2). Let δ_0 be the distance of F from the complement of D and let

$$\eta \leqslant \frac{1}{2(p + q)} \min(\delta, \delta_0).$$

By an interval I in R^{p+q} we denote a set defined by inequalities

(3.4.4) $$m_\nu \eta \leqslant x_\nu < (m_\nu + 1)\eta, \ \nu = 1 \text{ to } p + q,$$

where the m_ν are integers. Let I_1 to I_k be the intervals which meet F. In view of the choice of η these intervals all lie in a $(\frac{1}{2}\delta_0)$-neighbourhood F' of F. They are disjoint and their union covers F.

In view of Theorem 3.3 both L_1 and L_2 and their definitions by means of (3.4.1) can be uniquely extended as linear functionals over bounded Borel measurable functions with compact support in D and in particular to the characteristic functions $\chi_s(z)$ of the intervals I_s. If I_s is such an interval given by (3.4.4) we denote by I_s^x, I_s^y the projections of I^s on D_1 and D_2 respectively, i.e. the intervals in R^p and R^q given by (3.4.4) for $\nu = 1$ to p and $\nu = p + 1$ to $p + q$ respectively. With this notation it is evident that

$$L_1(\chi_s) = \mu_1(I_s^x)\mu_2(I_s^y) = L_2(\chi_s) = L_s \text{ say.}$$

Next let b_s and B_s be the lower and upper bounds of $f(x, y)$ in I_s. In view of (3.4.3) and the choice of η we see that

$$|B_s - b_s| < \varepsilon, \qquad s = 1 \text{ to } k.$$

Also for $j = 1, 2$

$$L_j(f) = L_j(f \sum_{s=1}^{k} \chi_s) = \sum_{s=1}^{k} L_j(f\chi_s) \leqslant \sum_{s=1}^{k} B_s L_s.$$

Similarly we have

$$L_j(f) \geqslant \sum_{s=1}^{k} b_s L_s.$$

Thus

$$|L_2(f) - L_1(f)| \leqslant \sum_{s=1}^{k} (B_s - b_s)L_s < \varepsilon \sum_{s=1}^{k} L_s \leqslant \varepsilon L_j(\phi), j = 1 \text{ or } 2,$$

where ϕ is the characteristic function of F'. Since ε is arbitrary it follows that $L_2(f) = L_1(f)$. This completes the proof of Lemma 3.5.

We can now prove Fubini's

THEOREM 3.5. *Suppose that $\mu_1 \mu_2$ are Borel-measures over domains D_1, D_2 in R^p, R^q respectively and let $f(x, y)$ be a Borel-measurable function in $D = D_1 \times D_2$. Then the repeated integrals $L_1(f)$ and $L_2(f)$ defined by (3.4.1) exist and are equal provided that either f has constant sign in D, or more generally if*

$$f \leqslant g \text{ or } f \geqslant -g,$$

where g is nonnegative in D and $L_1(g) = L_2(g) < \infty$.

In view of Lemma 3.5 we can write $L_1(f) = L_2(f) = L(f)$ at least for $f \in C_0(D)$.

Further in view of Theorem 3.3 $L(f)$ can be uniquely extended, to the linear class of functions which are integrable (L), as a linear functional which satisfies (3.2.1), (3.2.2) and (3.3.1). If we admit $+\infty$ as a possible value for $L(f)$ we obtain a unique definition of $L(f)$ for all positive Borel-measurable functions in D.

Suppose that $f_n(x, y)$ is a sequence of non-negative Borel-measurable functions in D tending monotonically to $f(x, y)$. Then

$$F_n(x) = \int f_n(x, y)\, d\mu_1(y)$$

tends for each fixed x monotonically to a nonnegative limit $F(x)$. Thus if $F_n(x)$ is Borel-measurable in D_1 it follows that $F(x)$ is Borel-measurable in D and since the Lebesgue integrals of positive functions satisfy (3.3.1) it follows, if $F_n(x)$ is finite for some n, or $F_n(x)$ increases with n, that

$$F(x) = \int f(x, y)\, d\mu_2(y).$$

Since all characteristic functions of Borel sets can be obtained by repeatedly forming unions or differences of compact sets a finite or countable number of times, we see that the definitions (3.4.1) can be extended to such functions and hence to nonnegative Borel-measurable functions, as in Section 3.1. Thus the unique extension of L to all nonnegative Borel-measurable functions is actually given by either of the definitions in (3.4.1). A similar conclusion holds for non positive Borel measurable functions in D.

Suppose finally that $g \geqslant 0$, and that $L(g) < +\infty$ and $\geqslant -g$. We write

$$f = f + g - g.$$

Then for $j = 1, 2$ we have

$$L_j(f) = L_j(f + g) - L_j(g),$$

where $L_j(f)$ is defined by (3.4.1). In view of what has been proved

$$L_1(f + g) = L_2(f + g) \qquad \text{and} \qquad L_1(g) = L_2(g),$$

so that $L_1(f) = L_2(f)$ as required. This completes the proof of Theorem 3.5.

We remark also that since each of the repeated integrals defined by (3.4.1) is a linear functional it can also be written as the Lebesgue integral with respect to a Borel measure $d\mu$ in D, i.e.

$$L(f) = \int_D f(z)\, d\mu(z).$$

For an interval I in D we see that $\mu(I) = \mu_1(I^x)\mu_2(I^y)$ and it is not hard to see that this relationship determines μ on all Borel sets. The measure μ is called the product measure of μ_1 and μ_2. The process can be repeated for the product of a finite number of measures. As an example the product of k linear Lebesgue-measures on the line gives k-dimensional Lebesgue-measure in R^k.

3.4.1. Convolution transforms

Let $K_\delta(x)$ be a bounded Borel function in space with support in $|x| < \delta$, let $\mu(x)$ be a Borel measure in $|x| < \delta$ and let $f(x)$ be defined in a bounded domain D of R^m. We define the convolution transform

$$F(x) = \int K_\delta(y) f(x + y)\, d\mu(y)$$

and proceed to prove some properties of $F(x)$. If D is any domain we define the δ-interior $D^\circ(\delta)$ of D to be the set of all points distant more than δ from the complement of D. We have

THEOREM 3.6. (i) *If $f(x)$ is* s.h. *in D and $K_\delta(y) \geqslant 0$ then $F(x)$ is* s.h. *in $D^\circ(\delta)$. If $f(x)$ is harmonic in D, then $F(x)$ is harmonic in $D^0(\delta)$.*

(ii) *If $d\mu$ is Lebesgue m-dimensional measure, $K_\delta(y) \in C^p$ and $f(x)$ is integrable over D then $F(x) \in C^p$ in $D^\circ(\delta)$.*

Suppose first that $f(x)$ is continuous in D and so uniformly continuous in $D^\circ(\delta_0)$ for any $\delta_0 > 0$. We assume that δ_0 is fixed. Then given $\varepsilon > 0$, there exists η such that if $|x_1 - x_2| < \eta$, and $x_1, x_2 \in D^\circ(\delta_0)$ we have

$$|f(x_1) - f(x_2)| < \varepsilon.$$

Thus if x_1, x_2 belong to $D^\circ(\delta + \delta_0)$, and $|x_1 - x_2| < \eta$, we deduce that

$$\begin{aligned}|F(x_1) - F(x_2)| &= \left|\int_{|y|<\delta} (f(x_1 + y) - f(x_2 + y))K_\delta(y)\,\mathrm{d}\mu(y)\right| \\ &\leqslant \varepsilon \int_{|y|<\delta} |K_\delta(y)|\,\mathrm{d}\mu(y) = C\varepsilon,\end{aligned}$$

where C is a constant. Thus $F(x)$ is continuous. Suppose next that $f(x)$ is u.s.c. in D and that $K(x) \geqslant 0$. Then by Theorem 1.4 there exists in D a decreasing sequence $f_n(x)$ of continuous functions converging to $f(x)$. We set

$$F_n(x) = \int K_\delta(y)\, f_n(x + y)\,\mathrm{d}\mu(y)$$

and deduce that $F_n(x)$ is continuous in $D^\circ(\delta + \delta_0)$ for any positive δ_0, and that $F_n(x)$ decreases with n, since $K_\delta(y)$ is positive. Thus $F_n(x)$ converges to

$$F(x) = \int K_\delta(y)\, f(x + y)\,\mathrm{d}\mu(y)$$

and so $F(x)$ is the limit of a decreasing sequence of continuous functions and so $F(x)$ is u.s.c.

To conclude the proof of Theorem 3.6(i) it remains to show that if $f(x)$ is s.h. and so satisfies the mean value inequality of the definition (iii) of 2.1, then so does $F(x)$. Suppose then that $x_0 \in D^\circ(\delta + \eta)$ and that $r < \eta$. Then

$$\begin{aligned}&\frac{1}{c_m r^{m-1}} \int_{S(x_0, r)} F(x)\,\mathrm{d}\sigma(x) \\ &= \frac{1}{c_m r^{m-1}} \int_{S(x_0, r)} \mathrm{d}\sigma(x) \int_{|y|<\delta} f(x + y)\, K_\delta(y)\,\mathrm{d}\mu(y) \\ &= \frac{1}{c_m r^{m-1}} \int_{|y|<\delta} K_\delta(y)\,\mathrm{d}\mu(y) \int_{S(x_0, r)} f(x + y)\,\mathrm{d}\sigma(x) \\ &\geqslant \int_{|y|<\delta} K_\delta(y)\,\mathrm{d}\mu(y)\, f(x_0 + y) = F(x_0).\end{aligned}$$

The inversion of the order of integration in the double integral is justified by Fubini's Theorem. In fact the integrand vanishes unless $|y| \leqslant \delta' < \delta$, and $|x - x_0| \leqslant r$. Under these conditions $x + y \in D^\circ(\delta + \eta - \delta' - r) = F_0$ so that $f(x + y)$ is uniformly bounded above by M say and so is $K_\delta(y) f(x + y)$, since $K_\delta(y)$ is nonnegative and bounded. Also $\mu(F_0)$ is finite. Hence if we set $g(x, y) = M$ for $x + y \in F_0$, $g(x, y) = 0$ otherwise, then $g(x, y)$ is integrable with respect to $d\mu$ and $f(x + y) K_\delta(y) \leqslant g(x, y)$. Thus the conditions for Fubini's Theorem are fulfilled and $F(x)$ is s.h. in $D^0(\delta + \eta)$ for every $\eta > 0$, i.e. in $D^0(\delta)$.

Suppose finally that $f(x)$ is harmonic in D. Then $f(x)$ and $-f(x)$ are s.h. in D and hence $F(x)$ and $-F(x)$ are s.h. in $D^0(\delta)$. Hence by Theorem 2.9 $F(x)$ is harmonic in $D^0(\delta)$. This proves Theorem 3.6(i).

We proceed to prove Theorem 3.6(ii). With the hypotheses of that Theorem we may write for $x \in D^0(\delta)$

$$F(x) = \int K_\delta(y) f(x + y) \, dy = \int K_\delta(z - x) f(z) \, dz, \tag{3.4.5}$$

where the integrals are extended over the whole of space but the integrands vanish for $|y| \geqslant \delta$, i.e. $|z - x| \geqslant \delta$. Hence if x_1 and $x_2 \in D^0(\delta + \eta)$, where $\eta > 0$, we have

$$F(x_1) - F(x_2) = \int_{D^\circ(\eta)} f(z)\{K_\delta(z - x_1) - K_\delta(z - x_2)\} \, dz.$$

If $K_\delta(y)$ is continuous and so uniformly continuous in space then, given $\varepsilon > 0$, we may choose η so that if $|y_1 - y_2| < \eta$, we have

$$|K_\delta(y_1) - K_\delta(y_2)| < \varepsilon.$$

Thus if $|x_1 - x_2| < \eta$, we have

$$|F(x_1) - F(x_2)| < \varepsilon \int_D |f(z)| \, dz,$$

so that $F(x)$ is continuous in $D^\circ(\delta)$. Suppose next that $K_\delta(y) \in C^p$. Then we may differentiate partially with respect to the coordinates of x in the second integral in (3.4.5). In fact if l denotes a point on one of the coordinate axes, $|l| = 1$, and $K'_\delta(y)$ is the partial derivative of K in the direction of this axis we can choose η so small that if h is a real number, such that $|h| < \eta$ and y is an arbitrary point, then

$$\left| \frac{K_\delta(y + hl) - K_\delta(y)}{h} - K'_\delta(y) \right| < \varepsilon.$$

This yields for $x_1 \in D^\circ(\delta)$ and h sufficiently small

$$\left| \frac{F(x_1 + hl) - F(x_1)}{h} + \int K'_\delta(z - x) f(z)\, dz \right| < \varepsilon \int_D |f(z)|\, dz.$$

Thus

$$\frac{F(x_1 + hl) - F(x_1)}{h} \to - \int K'_\delta(z - x) f(z)\, dz, \quad \text{as} \quad h \to 0.$$

Thus $F(x)$ possesses partial derivatives of the first order which are continuous in view of what was proved above and since $K'_\delta(y)$ is continuous. If $K_\delta(y) \in C^p$ with $p > 1$ the process can be continued and we can differentiate repeatedly with respect to x in the second integral of (3.4.5) under the sign of integration. This completes the proof of Theorem 3.6.

3.4.2

We can use Theorem 3.6 in order to prove some further results for s.h. functions which are essential for the proof of Riesz's Theorem. We have first

THEOREM 3.7. *If μ is a Borel measure on a compact set E in R^m and $p(x)$ is the potential defined by* (3.0.1) *then $p(x)$ is* s.h. *in R^m and harmonic outside E. In particular $p(x)$ is finite almost everywhere and integrable over any compact set with respect to Lebesgue measure.*

We set

$$f(x) = \log |x|, \qquad m = 2$$

$$f(x) = -|x|^{2-m}, \quad m > 2.$$

Then, as was shown in Section 1.5.1, $f(x)$ is harmonic in space except at the origin. Since $f(x)$ tends to $-\infty$ as $x \to 0$, $f(x)$ clearly remains s.h. at the origin, if we set $f(0) = -\infty$. We define $K_\delta(y) = 1$ in E, $K_\delta(y) = 0$ outside E. Then by Theorem 3.6

$$F(x) = \int K_\delta(\xi) f(x + \xi)\, d\mu e_\xi = \int_E f(x + \xi)\, d\mu e_\xi$$

is s.h. in space and harmonic near any point x for which $x + \xi \neq 0$, when ξ is in E. Thus $F(-x) = p(x)$ is s.h. in space and harmonic outside E. In particular $p(x)$ is finite outside E. The last statement now follows from Theorem 2.6. We have next

THEOREM 3.8. *If $u(x)$ is s.h. in a bounded domain D in R^m, then there exists a sequence $u_n(x)$ s.h. and in C^∞ in $D^\circ(1/n)$ such that, for $x \in D^\circ(1/n_0)$, $u_n(x)$ decreases strictly with n for $n > n_0$ and tends to $u(x)$ as $n \to \infty$.*

If $u(x) = -\infty$, we set $u_n(x) = -n$ in D. Suppose then that $u_n(x)$ is not identically $-\infty$. We define $K(x)$ to be the kernel used for the proof of Theorem 3.1 and set

$$u_n(x) = \int u(x + \xi) n^m K(n\xi)\, d\xi.$$

Evidently $u_n(x)$ is defined in $D^\circ(1/n)$ and, in view of Theorem 3.6, $u_n(x)$ is s.h. there. Further $u_n(x) \in C^\infty$ in $D^\circ(1/n)$ since, for x in $D^\circ(1/n)$, $u(x)$ is integrable over the ball $D(x, 1/n)$ by Theorem 2.6.

It remains to prove that $u_n(x_0)$ decreases to $u(x_0)$ as $n \to \infty$. To see this we write for $0 < r < 1$ and $x_0 \in D^\circ(r)$

$$f_r(x_0) = r^{-m} \int u(x_0 + \xi)\, K(\xi/r)\, d\xi$$

$$= r^{-m} \int_0^r d\rho\, k(\rho/r) \int_{S(x_0, \rho)} u(\eta)\, d\sigma(\eta).$$

Here $k(t)$ denotes the constant value of $K(x)$ on $S(0, t)$ and $\sigma(\eta)$ is areal measure on $S(x_0, \rho)$. In the notation of Theorem 2.12 we may write this as

$$f_r(x_0) = r^{-m} \int_0^r c_m \rho^{m-1}\, k(\rho/r)\, I_\rho(u)\, d\rho.$$

We set

$$\phi(t) = c_m t^{m-1}\, k(t),$$

and obtain

$$f_r(x_0) = \int_0^1 \phi(t)\, I_{rt}(u)\, dt.$$

Here $\phi(t)$ is a positive continuous function of t for $0 < t < 1$ and if $u \equiv 1$, then $f_r(x) \equiv 1$ in view of (3.2.3). Thus

$$\int_0^1 \phi(t)\, dt = 1.$$

It follows from Theorem 2.12 that $I_r(u)$ is an increasing function of r and so $I_{rt}(u)$ is an increasing function of r for each fixed t and so $f_r(x_0)$ is an increasing function of r. Thus $u_n(x_0) = f_{1/n}(x_0)$ is a decreasing function of n. Also by Theorem 2.12

$$I_r(u) \to u(x_0) \quad \text{as} \quad r \to 0.$$

If $u(x_0) = -\infty$, we deduce that, given any positive constant K, we have $I_r(u) < -K$, $r < r_0$ and so

$$f_r(x_0) < -K, \qquad r < r_0.$$

Thus $f_r(x_0) \to -\infty$ as $r \to 0$, and so $u_n(x_0) \to -\infty = u(x_0)$ as $n \to \infty$. If $u(x_0)$ is finite, then given $\varepsilon > 0$ we have

$$u(x_0) \leqslant I_r(u) < u(x_0) + \varepsilon, \qquad 0 < r < r_0.$$

This gives

$$u(x_0) \leqslant f_r(x_0) < u(x_0) + \varepsilon, \qquad 0 < r < r_0.$$

Thus $f_r(x_0) \to u(x_0)$ as $r \to 0$ and so $u_n(x_0) \to u(x_0)$ as $n \to \infty$. The functions $u_n(x)$ are decreasing and tend to $u(x)$ as $n \to \infty$. To make them strictly decreasing with n we replace $u_n(x)$ by $u_n(x) + 1/n$. This completes the proof of Theorem 3.8.

3.5. STATEMENT AND PROOF OF RIESZ'S REPRESENTATION THEOREM

We proceed to prove Riesz's Theorem. We consider functions in R^m where $m \geqslant 2$ and define

(3.5.1)
$$\begin{cases} K(x) = \log |x|, & m = 2 \\ K(x) = -|x|^{2-m}, & m > 2. \end{cases}$$

Then Riesz's Theorem [1926, 1930] becomes

THEOREM 3.9. *Suppose that $u(x)$ is* s.h., *and not identically $-\infty$, in a domain D in R^m. Then there exists a unique Borel-measure μ in D such that for any compact subset E of D*

(3.5.2)
$$u(x) = \int_E K(x - \xi)\, d\mu e_\xi + h(x),$$

where $h(x)$ is harmonic in the interior of E.

There are now many proofs of this deep and fundamental Theorem. The present one follows the ideas of Laurent Schwartz [1950–1951] and is based on the theory of distributions or linear functionals. We shall use three lemmas on which the final proof will be based.

3.5.1

We proceed to construct the measure whose existence is asserted in Theorem 3.9 and do so by way of a functional.

LEMMA 3.6. *Suppose that $u(x)$ is s.h. and not identically $-\infty$ in a domain D in R^m. Then the equation*

$$L_u(v) = \int_D u \nabla^2 v \, dx \tag{3.5.3}$$

defines L_u as a functional on the class of functions $v \in C_0^\infty(D)$.

We assume first that D is an open ball $D = D(x_0, r)$ in R^m. Since $v \in C_0^\infty(D)$, v has compact support in D and so v and all its partial derivatives vanish outside a ball $D' = D(x_0, r')$, where $r' < r$. Also, in view of Theorem 2.6, $u(x)$ is integrable over D' and $\nabla^2 v$ is uniformly bounded there. Thus $L_u(v)$ is well defined and finite. It is clear that $L_u(v)$ is linear for $v \in C_0^\infty(D)$. It remains to show that $L_u(v)$ is positive, i.e. that

$$L_u(v) \geqslant 0, \quad \text{if} \quad v(x) \geqslant 0 \quad \text{in} \quad D.$$

To do this we use Theorem 3.8. Let $u_n(x)$ be a sequence of functions s.h. and in C^∞ in a neighbourhood of the closure $C' = C(x_0, r')$ of D' and decreasing to $u(x)$ in C'. Then since v and all its partial derivatives vanish on the boundary $S' = S(x_0, r')$ of D' we have by Green's Theorem 1.9

$$\int_{D'} (u_n \nabla^2 v - v \nabla^2 u_n) \, dx = 0.$$

Thus if $v \geqslant 0$ in D, we have

$$L_{u_n}(v) = \int_D u_n \nabla^2 v \, dx = \int_{D'} u_n \nabla^2 v \, dx = \int_{D'} v \nabla^2 u_n \, dx \geqslant 0.$$

Also since u_n decreases to u in D' we have that

$$\int_{D'} u_n \, dx \to \int_{D'} u \, dx,$$

i.e.

$$\int_{D'} (u_n - u) \, dx = \int_{D'} |u_n - u| \, dx \to 0, \quad \text{as} \quad n \to \infty.$$

Further $\nabla^2 v$ is continuous and so bounded by M say in C'. Thus

$$\left| \int_{D'} u_n \nabla^2 v \, dx - \int_{D'} u \nabla^2 v \, dx \right| \leqslant \int_{D'} |u_n - u| \, |\nabla^2(v)| \, dx$$

$$\leqslant M \int_{D'} |u_n - u| \, dx \to 0, \quad \text{as} \quad n \to \infty,$$

i.e.

$$L_{u_n}(v) \to L_u(v) \quad \text{as} \quad n \to \infty. \tag{3.5.4}$$

Thus $L_u(v) \geqslant 0$. This completes the proof of Lemmas 3.6 for the case when D is a ball.

To extend the result to the general case, we employ a useful device called *partition of unity*. We suppose that $v \in C_0^\infty(D)$, and that E is a compact subset of D containing the support of v. It follows from the Heine–Borel Theorem that E can be covered by a finite number of open balls $D_\nu = D(x_\nu, r_\nu)$, $\nu = 1$ to N whose closures C_ν are contained in D. In each D_ν we define a function $e_\nu(x) \in C^\infty$ in R^m, positive in D_ν and zero elsewhere. We may set for instance

$$e_\nu(x) = \exp\{-(r_\nu^2 - |x - x_\nu|^2)^{-2}\}, \qquad x \in D_\nu.$$

We then define

$$v_\nu(x) = v(x)\, e_\nu(x) \Big/ \left(\sum_{\nu=1}^{N} e_\nu(x) \right), \qquad x \in D_\nu,$$

$$v_\nu(x) = 0 \qquad \text{elsewhere.}$$

Every boundary point ξ of D_μ in E lies interior to some D_ν with $\nu \neq \mu$ so that $\Sigma\, e_\nu(\xi) \neq 0$. Thus $v_\nu(x) \in C^\infty$ near $x = \xi$. If ξ is a boundary point of D_ν outside E then $v(x)$ and so $v_\nu(x) = 0$ near $x = \xi$. Thus $v_\nu(x) \in C^\infty$ in R^m, $v_\nu(x)$ has support in D_ν.

Also if $v(x) \geqslant 0$, then $v_\nu(x) \geqslant 0$. Finally

$$v(x) = \sum_{\nu=1}^{N} v_\nu(x).$$

Thus if $L_u(v)$ is defined by (3.5.3) then clearly $L_u(v)$ is linear in v and if $v \geqslant 0$ in D then

$$L_u(v) = \sum_{\nu=1}^{N} L_u(v_\nu) \geqslant 0$$

by what was proved above, since $v_\nu(x)$ has support in D_ν. This completes the proof of Lemma 3.6. Also (3.5.4) continues to hold in the general case since this relation is valid for each of the functions v_ν.

We now deduce from Theorem 3.2 that $L_u(v)$ can be uniquely extended as a linear functional to the class of functions $v \in C_0(D)$ and we suppose this extension carried out. Next it follows from Theorem 3.4 that there exists a Borel measure μ, uniquely defined on all Borel subsets of D such that for $v \in C_0(D)$ and in particular for $v \in C_0^\infty(D)$ we have

$$L_u(v) = \int_D v \, d\mu. \tag{3.5.5}$$

We proceed to show that a constant multiple of this measure μ has the properties required in Theorem 3.9. We note incidentally that if u is harmonic in D then it follows directly from Green's Theorem that $L_u(v)$ and hence μ vanishes identically in D. Our next result provides a converse.

3.5.2

In order to deal with the uniqueness part of Theorem 3.9 we prove

LEMMA 3.7. *If u, u' are* s.h. *in D and we have*

$$L_u(v) = L_{u'}(v)$$

for every function $v \in C_0^\infty(D)$, then $u(x) = u'(x) + h(x)$, where $h(x)$ is harmonic in D.

We shall need to consider only functions v whose support is contained in an open ball $D' = D(x_0, r') \subset D$. Suppose first that u is s.h. and belongs to C^∞ in a neighbourhood of $C' = C(x_0, r_2)$ where $r_2 < r'$. Then if $0 < r_1 < r_2$, it follows from Green's Theorem that

$$\int_{S(x_0, r_1)} \frac{\partial u}{\partial r} \, d\sigma = \int_{D(x_0, r_1)} \nabla^2 u \, dx. \tag{3.5.6}$$

If we write

$$I(r, u) = \frac{1}{c_m r^{m-1}} \int_{S(x_0, r)} u(x) \, d\sigma(x),$$

then the equation (3.5.6) may be written

$$\left[r^{m-1} \frac{\partial}{\partial r} I(r, u) \right]_{r = r_1} = \int_{D(x_0, r_1)} \frac{1}{c_m} \nabla^2 u \, dx.$$

We set $t = \log r$, if $m = 2$, $t = -r^{2-m}$, if $m > 2$, and define

$$e_2 = c_2, \; e_m = (m-2)\, c_m, \qquad m > 2.$$

Then our equation becomes

$$\left[\frac{\partial}{\partial t} I(r, u) \right]_{r = r_1} = \frac{1}{e_m} \int_{D(x_0, r_1)} \nabla^2 u \, dx.$$

We integrate this equation from $r = r_1$ to $r = r_2$, where $0 < r_1 < r_2 < r'$ and obtain

$$I(r_2, u) - I(r_1, u) = \int_{r_1}^{r_2} \frac{dr}{r^{(m-1)}} \int_{D(x_0, r)} \frac{1}{c_m} \nabla^2 u \, dx.$$

By Fubini's Theorem we can invert the order of integration in this double integral and deduce that

(3.5.7) $$I(r_2, u) - I(r_1, u) = \int_{D(x_0, r_2)} g(x) \nabla^2 u \, dx,$$

where

$$g(x) = \frac{1}{c_m} \int_{r_1}^{r_2} \frac{dr}{r^{(m-1)}}, \qquad |x - x_0| \leqslant r_1$$

$$g(x) = \frac{1}{c_m} \int_{|x|}^{r_2} \frac{dr}{r^{(m-1)}}, \qquad r_1 < |x - x_0| < r_2.$$

If g were known to belong to C^∞ we could write (3.5.7) as

(3.5.8) $$I(r_2, u) - I(r_1, u) = L_u(g).$$

We show next that this equation is in fact valid even for general s.h. functions u, which need not belong to C^∞.

Suppose first still that $u \in C^\infty$ and that, for all $v \in C_0^\infty(D')$, $L_u(v)$ is defined by (3.5.3). Then for such v we have by Green's Theorem

(3.5.9) $$L_u(v) = \int v \nabla^2 u \, dx.$$

However this equation defines $L_u(v)$ as a positive linear functional on all functions $v \in C_0(D')$. Since by Theorem 3.2 the extension from $C_0^\infty(D')$ to $C_0(D')$ is unique, we see that (3.5.9) is valid for all functions $v \in C_0(D')$ and in particular for g. Thus (3.5.8) holds.

Next if u is an arbitrary function s.h. and not identically $-\infty$ in D' let u_n be a sequence of functions s.h. and in C^∞ in D' and such that u_n decreases to u as $n \to \infty$. Then it follows from (3.3.1) that for $j = 1, 2$

(3.5.10) $$I(r_j, u_n) \to I(r_j, u), \quad \text{as} \quad n \to \infty.$$

We proceed to show that

(3.5.11) $$L_{u_n}(g) \to L_u(g), \quad \text{as} \quad n \to \infty.$$

To see this we construct g_1, g_2 in $C_0^\infty(D')$, such that

(3.5.12) $$g_1 \leqslant g \leqslant g_2, \qquad L_u(g_2) < L_u(g_1) + \varepsilon.$$

The possibility of constructing such functions g_1 and g_2 was shown in (3.2.5). It also follows from (3.5.4) and the Definition (3.5.3) which is valid for g_1 and g_2 that

$$L_{u_n}(g_j) \to L_u(g_j), \quad \text{as} \quad n \to \infty.$$

We deduce from this and (3.5.12) that for all sufficiently large n we have

$$L_u(g) - \varepsilon < L_{u_n}(g_j) < L_u(g) + \varepsilon, \qquad j = 1, 2.$$

Since $L_{u_n}(v)$ is a positive linear functional for each n we deduce from this and (3.5.12) that

$$L_u(g) - \varepsilon < L_{u_n}(g) < L_u(g) + \varepsilon$$

This yields (3.5.11). Since (3.5.8) was shown to be valid for the functions u_n in $C_0^\infty(D)$ instead of u, it follows from (3.5.10) and (3.5.11) that (3.5.8) is valid for general s.h. functions u.

Suppose now finally that u, u' satisfy the hypotheses of Lemma 3.7. Then $L_u(v)$ and $L_{u'}(v)$ define the same positive linear functional on $C_0^\infty(D')$ and hence, by the unique extension Theorem 3.2 on $C_0(D')$. In particular we may take $v = g$ in the hypotheses of Lemma 3.7 and deduce from (3.5.8) that

$$I(r_2, u') - I(r_1, u') = I(r_2, u) - I(r_1, u).$$

In particular

$$I(r) = I(r, u') - I(r, u)$$

is constant for $0 < r \leqslant r'$. We may apply the same conclusion with any ball $D(x_1, \rho) \subset D$ instead of $D' = D(x_0, r')$. Thus for a fixed $x_1 \in D$

$$I(x_1, \rho) = \frac{1}{c_m \rho^{m-1}} \int_{S(x_1, \rho)} [u'(x) - u(x)] \, d\sigma(x)$$

is constant provided that $\rho < \delta(x_1)$, where $\delta(x_1)$, is the distance of x_1 from the complement of D.

We now define

(3.5.13) $$h(x_1) = I(x_1, \rho), \qquad 0 < \rho < \delta(x_1).$$

Then it follows from Theorem 2.12 that

(3.5.14) $$u'(x_1) = u(x_1) + h(x_1)$$

unless $u'(x)$, $u(x_1)$ are both $-\infty$. In this ambiguous case the equation (3.5.14) still holds, since in this case both sides are $-\infty$. Thus the equation is always valid.

It remains to prove that $h(x)$ is harmonic in D. To see this we note that $h(x)$ is finite in D and in view of (3.5.14) we have for $x_1 \in D$, $\rho < \delta(x_1)$

(3.5.15) $$\frac{1}{c_m \rho^{m-1}} \int_{S(x_1, \rho)} h(x) \, d\sigma(x)$$

$$= \frac{1}{c_m \rho^{m-1}} \int_{S(x_1, \rho)} [u'(x) - u(x)] \, d\sigma(x) = h(x_1).$$

Also in view of (3.5.14) and since $u(x)$, $u'(x)$ are s.h. and not identically infinite it follows that $h(x)$ is integrable with respect to Lebesgue measure over any compact subset of D. Let $k_\delta(\rho)$ be continuous for all positive ρ, zero for $\rho \geqslant \delta$ and positive for $0 < \rho < \delta$, and such that

$$\int_0^\delta c_m \rho^{m-1} k_\delta(\rho)\, d\rho = 1.$$

Then we set

$$H(x) = \int_{D(x,\delta)} h(\xi)\, k_\delta(|x - \xi|)\, d\xi = \int_0^\delta k_\delta(\rho)\, d\rho \int_{S(x,\rho)} h(\xi)\, d\sigma(\xi) = h(x).$$

In view of Theorem 3.6 $H(x)$ is continuous in $D_0(\delta)$ and hence so is $h(x)$. Since δ is arbitrary $h(x)$ is continuous in D. Since (3.5.15) holds $h(x)$ and $-h(x)$ are both s.h. in D and so, by Theorem 2.9, $h(x)$ is harmonic in D. This proves Lemma 3.7,

3.5.3

Our last subsidiary result is

LEMMA 3.8. *Let $K(x)$ be the kernel defined by (3.5.1), let μ be a Borel measure over a compact subset of D and suppose that*

$$p(x) = \frac{1}{e_m} \int K(x - \xi)\, d\mu(\xi)$$

where $e_m = c_m d_m$ and $d_2 = 1$, $d_m = m - 2$, $m > 2$. Then

$$L_p(v) = \int_D v\, d\mu.$$

It is enough to prove the result for $v \in C_0^\infty(D)$ in view of Theorem 3.2. We suppose first that v has support in $D' = D(x_0, r')$, where the closure of D' lies in D. By Theorem 3.7 $p(x)$ is s.h. in R^m and in particular in D and we have from (3.5.3)

$$L_p(v) = \frac{1}{e_m} \int_{D'} p(x)\, \nabla^2 v(x)\, dx \tag{3.5.16}$$

$$= \frac{1}{e_m} \int_{D'} \nabla^2 v(x)\, dx \int_D K(x - \xi)\, d\mu(\xi).$$

We proceed to invert the order of integration in this double integral. This may be justified as follows. We write

$$v_1(x) = \max(\nabla^2 v(x), 0), \quad -v_2(x) = \inf(\nabla^2 v(x), 0),$$

so that $v_1(x)$, $v_2(x)$ are nonnegative continuous functions. Also $v_1(x)\,K(x-\xi)$ is bounded above by M_0 say and

$$\int_{D'} M_0\,\mathrm{d}x \int_D \mathrm{d}\mu(\xi) < +\infty,$$

since $\mu(D) < \infty$. Thus by Theorem 3.5 the integrals corresponding to v_1 and v_2 separately may be inverted. Thus the corresponding integral for v may be inverted, provided that the integrals for v_1 and v_2 remain finite. To see this suppose that $v_1 \leqslant M_1$. Then if $K^-(x) = -\inf(K(x), 0)$

$$\int_{D'} v_1(x)\,\mathrm{d}x \int_D K(x-\xi)\,\mathrm{d}\mu(\xi) \geqslant -\int_{D'} \mathrm{d}x \int_D M_1 K^-(x-\xi)\,\mathrm{d}\mu(\xi)$$

$$= -\int_D \mathrm{d}\mu(\xi) \int_{D'} M_1 K^-(x-\xi)\,\mathrm{d}x > -\infty,$$

since $K(x)$ is locally integrable near the origin. Thus the inversion of the order of integration in (3.5.16) is justified. We deduce that

$$L_p(v) = \frac{1}{e_m}\int_D \mathrm{d}\mu(\xi) \int_{D'} \nabla^2 v(x)\,K(x-\xi)\,\mathrm{d}x. \tag{3.5.17}$$

We now write

$$I(\xi) = \int_{D'} \nabla^2 v(x)\,K(x-\xi)\,\mathrm{d}x.$$

To evaluate this integral, we take first $\xi \in D' = D(x_0, r')$ and evaluate

$$I_\varepsilon(\xi) = \left(\int_{D(x_0,r')} - \int_{D(\xi,\varepsilon)}\right) \nabla^2 v\,K(x-\xi)\,\mathrm{d}x,$$

where ε is a small positive number. In view of Green's Theorem 1.9 this can be written as

$$I_\varepsilon(\xi) = \int_{S(x_0,r')} \left(v\frac{\partial K}{\partial n} - K\frac{\partial v}{\partial n}\right)\mathrm{d}\sigma(x) + \int_{S(\xi,\varepsilon)} \left(v\frac{\partial K}{\partial n} - K\frac{\partial v}{\partial n}\right)\mathrm{d}\sigma(x)$$

$$+ \left(\int_{D(x_0,r')} - \int_{D(\xi,\varepsilon)}\right) v\nabla^2 K(x-\xi)\,\mathrm{d}x.$$

Since $K(x-\xi)$ is harmonic except at ξ, the last integral vanishes. Since v and its partial derivatives vanish on $S(x_0, r')$ the integral over this surface also vanishes. Thus

$$I_\varepsilon(\xi) = \int_{S(\xi,\varepsilon)} \left(v\frac{\partial K}{\partial \varepsilon} - K\frac{\partial v}{\partial \varepsilon} \right) d\sigma(x).$$

Here the second integral tends to zero as $\varepsilon \to 0$. Also

$$\frac{\partial K}{\partial \varepsilon} \sim d_m \varepsilon^{1-m}, \quad \text{as} \quad \varepsilon \to 0,$$

where $d_2 = 1$, $d_m = m - 2$, $m > 2$. Thus

$$I_\varepsilon(\xi) = \{v(\xi) + o(1)\}\, d_m \varepsilon^{1-m} \int_{S(\xi,\varepsilon)} d\sigma(x) = c_m d_m v(\xi) + o(1) \quad \text{as} \quad \varepsilon \to 0.$$

Hence

$$I(\xi) = e_m v(\xi). \tag{3.5.18}$$

Suppose next that ξ is exterior to D'. Then $K(x - \xi)$ is harmonic on the closure of D' and a direct application of Green's Theorem to D' yields

$$I(\xi) = \int_{S(x_0, r')} \left(v\frac{\partial K}{\partial n} - K\frac{\partial v}{\partial n} \right) d\sigma(x) + \int_{D'} v\nabla^2 K(x - \xi)\, dx = 0.$$

If ξ lies on the boundary $S(x_0, r')$ of D' the same conclusion holds. For since the support of v is a compact subset of D', we may replace r' by a slightly smaller number. Thus (3.5.18) holds generally for functions v, whose support lies in a ball contained in D. Since a general function $v \in C_0^\infty(D)$ can by the method of Section 3.5.1 be expressed as a finite sum of functions $v_n \in C_0^\infty(D_n)$, where the D_n are balls contained in D, we deduce (3.5.18) for all $v \in C_0^\infty(D)$. Now Lemma 3.8 follows from this and (3.5.17).

3.5.4. Proof of Riesz's Theorem

Suppose now that $u(x)$ is s.h. and not identically $-\infty$ in a domain $D \subset R^m$. Then we define the functional $L_u(v)$ for $v \in C_0^\infty(D)$ by (3.5.3). By Lemma 3.6 this is a positive linear functional which is uniquely extensible to the class $C_0(D)$ and is given in terms of a Borel measure μ, uniquely determined by u and D, by the formula (3.5.5).

If E is any compact subset of D we construct the potential

$$p(x) = \frac{1}{e_m} \int_E K(x - \xi)\, d\mu(\xi)$$

and deduce from Lemma 3.8 that if Δ denotes the interior of E and $v \in C_0(\Delta)$, then

$$L_p(v) = \int v \, d\mu = L_u(v)$$

in view of (3.5.5). It now follows from Lemma 3.7 that

$$u(x) = p(x) + h(x),$$

where $h(x)$ is harmonic in Δ.

Conversely suppose that Δ is any domain, whose closure is a compact subset of D and that $\mu_1(x)$ is a measure finite in Δ such that

$$u(x) = p_1(x) + h_1(x),$$

where

$$p_1(x) = \frac{1}{e_m} \int K(x - \xi) \, d\mu_1(\xi),$$

and $h_1(x)$ is harmonic in Δ. Then it follows from (3.5.3) that for any $v \in C_0^\infty(\Delta)$ we have

$$L_u(v) = L_{p_1}(v) + L_{h_1}(v) = L_{p_1}(v) = L_p(v).$$

Hence in view of Lemma 3.8 we have

$$\int v \, d\mu = \int v \, d\mu_1.$$

Since this is true for any $v \in C_0^\infty(\Delta)$ the result also holds by the unique extension Theorem 3.2 for any $v \in C_0(\Delta)$ and hence, in view of Theorem 3.4, $\mu(e) = \mu_1(e)$ for any Borel subset e of Δ. This proves the uniqueness part of Theorem 3.9 and completes the proof of that theorem.

We note that we have proved Theorem 3.9 with $e_m^{-1} \, d\mu$ instead of $d\mu$ where $d\mu$ is the measure defined by (3.5.5). If $u \in C_0^2(D)$ then it follows from Green's Theorem that in (3.5.5)

$$d\mu = \nabla^2 u \, dx,$$

so that in this case (3.5.2) becomes

$$u(x) = \frac{1}{e_m} \int_E K(x - \xi) \nabla^2 u \, dx + h(x),$$

In the sequel the measure μ occurring in (3.5.2) and not the differently normalized measure in (3.5.5) will be called the *Riesz measure* of u.

3.6. HARMONIC MEASURE

In this section we take up again the notion of harmonic extension first defined in Section 2.7.2. With the material now at our disposal we can deepen our understanding of this subject. We proceed to prove

THEOREM 3.10.† *Suppose that D is a bounded regular domain in R^m with frontier F. Then for every x in D and Borel set e on F there exists a number $\omega(x, e)$ with the following properties*

(i) *For fixed $x \in D$ $\omega(x, e)$ is a Borel measure on F and $\omega(x, F) = 1$.*
(ii) *For fixed $e \subset F$, $\omega(x, e)$ is a harmonic function of x in D.*
(iii) *If $f(\xi)$ is a semi-continuous function defined on F then*

$$u(x) = \int_F f(\xi)\, d\omega(x, e_\xi) \tag{3.6.1}$$

is the harmonic extension of $f(\xi)$ to D. The measure $\omega(x, e) = \omega(x, e, D)$ will be called the harmonic measure of e at x with respect to D.

To prove Theorem 3.10 we write $L_x(f)$ for the harmonic extension $u(x)$ of f to the point x in D.

Then, for fixed x, $L_x(f)$ is defined for continuous functions f on E and by the maximum principle $L_x(f) \geqslant 0$ if $f \geqslant 0$. Also $L_x(f)$ is clearly linear in f. Thus $L_x(f)$ is a positive linear functional on the class of continuous functions f on F. Hence it follows from Theorem 3.4 that there exists a measure $\omega(x, e)$ uniquely determined by F, x and D such that (iii) holds for continuous f. In view of Theorem 2.17 and the property (3.3.1) of linear functionals it follows that $u(x) = L_x(f)$ is still the harmonic extension of f to D when f is semi-continuous. Also for any bounded Borel measurable function f on F we define the harmonic extension of f from F onto D to be given by (3.6.1). Since the extension of $f = 1$ is the function $u(x) = 1$, it follows that $\omega(x, F) = 1$, which proves (i). From this it also follows that $u(x)$ can be defined by (3.6.1) for bounded Borel-measurable f.

It remains to show that (ii) holds. We proceed to prove more generally that, when f is bounded and Borel measurable, $u(x)$ is harmonic in x. For any Borel set $e \subset F$ we may then take for f the characteristic function of e and deduce (ii).

If f is upper semi-continuous it follows from Theorem 2.17 that $u(x)$ is harmonic or identically $-\infty$ and similarly if f is lower semi-continuous $u(x)$ is harmonic or identically $+\infty$. The infinite case is excluded if f is bounded

† Brelot [1939a]. Brelot's solution of the problem of Dirichlet for general f, which are integrable $d\omega$ yields the same conclusion but alsong a slightly different road.

since in this case $u(x)$ also lies between the same bounds for each x. Suppose finally that f is bounded and Borel measurable. Let x_0 be a fixed point of D. Then f is integrable with respect to $\omega(x_0, e)$ and so we can, in view of the Definition 3.3.2, find u.s.c. functions $f_n(\xi)$ on F such that

$$L_{x_0}(f_n) > L_{x_0}(f) - \frac{1}{n}, \tag{3.6.2}$$

and $f_n \leqslant f$ on E. We may suppose further that the sequence f_n is monotonic increasing since otherwise we may replace f_n by

$$g_n = \max_{\nu = 1 \text{ to } n} f_\nu.$$

Let $u_n(x)$ be the harmonic extension of f_n from F to D. Since f_n is u.s.c., it follows from Theorem 2.17 that $u_n(x)$ is harmonic in x and, since $u_n(x) = L_x(f_n)$ is a positive linear functional, $u_n(x)$ increases with n for each fixed x in D. Thus by Harnack's Theorem 1.20 $u_n(x)$ converges to a harmonic function $u(x)$ in D and in view of (3.6.2) we have

$$L_{x_0}(f) = u(x_0).$$

Also since $f_n \leqslant f$ on F it follows that for $x \in D$

$$u_n(x) = L_x(f_n) \leqslant L_x(f)$$

and so

$$u(x) \leqslant L_x(f), \qquad x \in D.$$

We may similarly choose a decreasing sequence of lower semi-continuous functions $g_n(\xi)$, such that $g_n(\xi) \geqslant f(\xi)$ on F and that

$$L_{x_0}(g_n) \to L_{x_0}(f) \quad \text{as} \quad n \to \infty.$$

Then if $v_n(x)$ is the harmonic extension of g_n into D it follows that $v_n(x)$ decreases to a harmonic limit $v(x)$ in D, such that $v(x_0) = L_{x_0}(f)$ and

$$v(x) \geqslant L_x(f), \qquad x \in D.$$

Thus

$$v(x) - u(x) \geqslant 0$$

in D with equality at $x = x_0$ and hence we deduce from the maximum principle that $v(x) = u(x)$ in D and so that

$$u(x) = L_x(f) = v(x),$$

so that $L_x(f)$ is harmonic in D as a function of x.

Finally it is possible to define $L_x(f)$ for nonnegative unbounded functions f by setting $f_n = \inf(f, n)$ and

$$L_x(f) = \lim_{n\to\infty} L_x(f_n).$$

Since f_n increases with n so does $L_x(f_n)$ and hence by Harnack's Theorem $L_x(f)$ is harmonic in x or identically $+\infty$. If f is a general function and $f^+ = \max(f, 0)$, $f^- = -\min(f, 0)$ we have $f = f^+ - f^-$ and define

$$L_x(f) = L_x(f^+) - L_x(f^-)$$

provided both these quantities are finite for some (and hence all) x in D. In this case say that f is integrable with respect to harmonic measure. If $L_x(f^+) = \infty$ and $L_x(f^-)$ is finite we write $L_x(f) = +\infty$, and if $L_x(f^+)$ is either finite and $L_x(f^-) = \infty$ we say $L_x(f) = -\infty$. Thus in all cases $L_x(f)$ is either harmonic in x or identically $-\infty$ or identically $+\infty$ or indeterminate for each x. This justifies us in calling $u(x)$ given by (3.6.1) the harmonic extension of f for general Borel measurable f, such that the integral exists. We have incidentally proved (ii) and completed the proof of Theorem 3.10.

It follows from Theorem 3.10 that, for any Borel set on F, $\omega(x, e)$ is a harmonic nonnegative function of x. Hence by the maximum principle $\omega(x, e) > 0$ in D, or $\omega(x, e) = 0$ in D.

In the former case we say that e has positive harmonic measure and in the latter that e has harmonic measure zero. Clearly two functions f_1 and f_2 which differ only on a subset e of F, having harmonic measure zero, have identical harmonic extensions from F to D.

3.6.1

The following result is in fact only a slight generalisation of Theorem 2.18, but it seems to fit naturally into the present framework.

THEOREM 3.11. (Principle of harmonic measure).† *Suppose that $u(x)$ is* s.h. *in a bounded regular domain D in R^m with frontier F and that*

$$\overline{\lim}\, u(x) \leqslant f(\xi), \text{ as } x \to \xi \text{ from inside } D$$

for each ξ in F, where f is bounded above and Borel-measurable on F. Then

$$u(x) \leqslant L_x(f)$$

where $L_x(f)$ is the harmonic extension from f to D.

† For a slightly less general formulation, with f a characteristic function, see F. and R. Nevanlinna [1922].

The condition that f and hence u is bounded above is essential. For instance if D is the unit disk in R^2 and we set for $z = re^{i\theta}$

$$u(z) = \frac{1 - r^2}{1 - 2r\cos\theta + r^2},$$

then we may take $f(1) = +\infty$, $f(e^{i\theta}) = 0$, $0 < \theta < 2\pi$. The harmonic extension of this function f is zero, but $u(z)$ is harmonic and $u(z) > 0$ in $|z| < 1$.

The most usual application of Theorem 3.11 is to the case when $f(\xi) = 1$ on a certain closed subset of F and $f(\xi) = 0$ on the complementary open set. The conclusion is that in this case

$$u(x) \leqslant \omega(x, e),$$

where $\omega(x, e)$ is the harmonic measure of e in D. For this reason the result is frequently called the principle of harmonic measure.

We proceed to prove Theorem 3.11. To do this we set

$$g(\zeta) = \overline{\lim}\, u(x)$$

where the upper limit is taken as x tends to ζ from inside D. Then $g(\zeta)$ is u.s.c. on F. In fact if ζ_0 is a point of F and $\varepsilon > 0$, then $u(x) < g(\zeta_0) + \varepsilon$ at all points x of D belonging to a neighbourhood N of ζ_0 in R^m and hence

$$g(\zeta) \leqslant g(\zeta_0) + \varepsilon$$

at all points of F in N. In particular $g(\zeta)$ is Borel measurable and bounded above and so $L_x(g)$ is defined in D and is finite or identically $-\infty$.

Let $g_n(\zeta)$ be a sequence of continuous functions on F decreasing to $g(\zeta)$ on F and let $u_n(x)$ be the harmonic extensions of $g_n(\zeta)$ to D. The sequence g_n exists since $g(\zeta)$ is u.s.c. Then since D is a regular domain

$$u_n(x) \to g_n(\zeta)$$

as $x \to \zeta$ from inside D. Hence

$$\overline{\lim}\, \{u(x) - u_n(x)\} \leqslant g(\zeta) - g_n(\zeta) \leqslant 0$$

as $x \to \zeta$ from inside D, and this is true for each ζ in F. Since $u(x) - u_n(x)$ is s.h. in D we deduce from the the maximum principle (Theorem 2.3) that $u(x) \leqslant u_n(x)$ in D. Thus for each n

$$u(x) \leqslant L_x(g_n).$$

Since g_n decreases to g as $n \to \infty$, we deduce from (3.3.1) that $L_x(g_n) \to L_x(g)$ and so that

$$u(x) \leqslant L_x(g).$$

Since L_x is a positive linear functional and $g \leqslant f$ we deduce that $L_x(g) \leqslant L_x(f)$

and this yields Theorem 3.11. We deduce an interesting uniqueness theorem for harmonic functions.

THEOREM 3.12. *If $u(x)$ is bounded and harmonic in a regular bounded domain D in R^m and $g(x)$ is bounded on F and*

$$u(x) \to g(\zeta)$$

as $x \to \zeta$ from inside D for all ζ on the boundary F of D apart from a set of harmonic measure zero, then $u(x) = L_x(g)$.

We set

$$g_1(\zeta) = \overline{\lim}\, u(x), \qquad g_2(\zeta) = \underline{\lim}\, u(x),$$

both limits being taken as $x \to \zeta$ from inside D. Then g_1 is u.s.c. and g_2 is l.s.c. so that both functions are Borel-measurable and bounded and hence integrable. Also $g_1 = g = g_2$ outside a set of harmonic measure zero and so

$$L_x(g_1) = L_x(g_2).$$

Hence g is also integrable and $L_x(g) = L_x(g_1) = L_x(g_2)$. Also by Theorem 3.11 we have

$$u(x) \leqslant L_x(g_2)$$

and by applying the theorem to $-u(x)$ we have

$$u(x) \geqslant L_x(g_1).$$

Thus Theorem 3.12 follows.

Theorem 3.12 provides us with a convenient way of calculating harmonic measure. It is enough to construct a bounded harmonic function in D which has boundary values 1 at all proints of a certain set e of frontier points of D and 0 at complementary points apart from a set of harmonic measure zero. If D is an open ball $D(x_0, r)$ in R_m it follows from (1.5.1) that for continuous functions $f(\zeta)$ we have (3.6.1) with

$$\omega(x, e) = \frac{1}{c_m} \int_e \frac{r^2 - |x - x_0|^2}{r|x - \zeta|^m} \, d\sigma_\zeta,$$

where $d\sigma_\zeta$ denotes surface area on $S(x_0, r)$. In particular harmonic measure at x_0 is proportional to the surface area of e, or the solid angle which e subtends at the origin. If $m = 2$ harmonic functions are invariant under conformal mapping and so if D is a simply-connected domain and e is a set on the boundary of D we can calculate $\omega(x_0, e)$ by mapping D onto $|z| < 1$ in such a way that x_0 corresponds to the origin. If e' is the image set of e on $|z| = 1$, then $\omega(x_0, e) = l/(2\pi)$, where l is the length of e'.

In higher dimensions the method of conformal mapping is not available and it is not in general possible to determine harmonic measure explicitly.

Example

If D is a half space $x_1 > 0$ in space R^m, where x is the point $(x_1, x_2, \ldots, x_m)$, show that if $u(x)$ is harmonic and bounded in D and continuous on the frontier $F: x_1 = 0$ of D, then

$$u(x) = \frac{1}{c_m} \int_F \frac{2x_1 u(\xi) \, d\sigma(\xi)}{|x_1 - \xi|^m},$$

where $d\sigma(\xi)$ denotes Lebesgue $(m-1)$-dimensional measure on F.

(Hint show that the theory of this section extends to unbounded domains, provided that $\zeta = \infty$ is taken as a single boundary point).

3.7. THE GREEN'S FUNCTION AND THE POISSON–JENSEN FORMULA

The (Classical) Green's Function was defined in Section 1.5.1 and its uniqueness was proved in Theorem 1.14. We proceed to prove the existence of the Green's function and then to use it for a general representation Theorem for subharmonic functions.

THEOREM 3.13. *Suppose that D is a regular domain in R^m. Then the Green's function $g(x, \xi, D)$ exists, provided that, if $m = 2$, D has an exterior point.*

We have to distinguish slightly the cases $m = 2$ and $m > 2$. Suppose first that $m > 2$ and that ξ is a point of D. We set

$$f(x) = -|x - \xi|^{2-m}$$

for x on the frontier F of D. Let $u(x)$ be the harmonic extension of f from F into D. Since $f(x)$ is continuous and bounded on F, including ∞, $u(x)$ is uniquely defined by these properties in view of Theorem 2.10. Clearly

$$g(x, \xi, D) = u(x) + |x - \xi|^{2-m}$$

satisfies the requirements for the Green's function. The uniqueness was proved in Theorem 1.14.

Next suppose that $m = 2$ and that D has an exterior point x_0 say. We set

$$f(x) = \log \left| \frac{x - \xi}{x - x_0} \right|$$

and note that $f(x)$ is continuous and bounded at the frontier F of D, including

∞. If $u(x)$ is the harmonic extension of $f(x)$ into D then

$$g(x, \xi, D) = u(x) + \log\left|\frac{x - x_0}{x - \xi}\right|$$

is the required Green's function.

3.7.1

We can now prove a rather general form of the Poisson–Jensen formula.† This is

THEOREM 3.14. *Suppose that D is a bounded regular domain in R^m whose frontier F has zero m-dimensional Lebesgue measure, and that $u(x)$ is* s.h. *and not identically $-\infty$ on $D \cup F$.*

Then we have for $x \in D$

$$u(x) = \int_F u(\xi)\, d\omega(x, e_\xi) - \int_D g(x, \xi, D)\, d\mu e_\xi, \tag{3.7.1}$$

where $\omega(x, e)$ is the harmonic measure of e at x, $g(x, \xi, D)$ is the Green's function of D and $d\mu$ is the Riesz measure of u in D.

Example

Suppose that $u(z)$ is subharmonic in the plane disk $|z| \leqslant r$ with Riesz measure $d\mu$ in $|z| < r$. Then if we apply Theorem 3.14 to the subharmonic function $u(z)$ we obtain

$$u(\rho e^{i\theta}) = \frac{1}{2\pi}\int_0^{2\pi} u(re^{i\phi}) \frac{(r^2 - \rho^2)\, d\phi}{r^2 - 2r\rho\cos(\theta - \phi) + \rho^2} + \int \log\left|\frac{r(z - \xi)}{r^2 - \bar{z}\xi}\right| d\mu e_\xi. \tag{3.7.2}$$

For by Theorems 1.16 and 3.10 the first integral gives the harmonic extension of $u(re^{i\phi})$ into $|z| < r$. If $u(z) = \log|f(z)|$, where $f(z)$ is regular in $|z| \leqslant r$, then $\mu(e)$ is equal to the number of zeros of $f(z)$ on e and the second integral becomes

$$\sum_{\nu=1}^{n} \log\left|\frac{r(z - a_\nu)}{r^2 - \bar{a}_\nu z}\right|,$$

where a_ν, $\nu = 1$ to n, are the zeros of $f(z)$ in $|z| < r$. The form of the Green's

† R. Nevanlinna [1929], for the special case (0.1). For s.h. functions in a plane disk see Hayman [1952]. A less restrictive result is proved in Theorem 5.27.

function in this case was given in Theorem 1.10. Thus we obtain the classical Poisson–Jensen formula for regular functions as a special case. The meromorphic version also follows easily when we express a meromorphic function as the ratio of two regular functions.

Other formulae in which the disk $|z| \leqslant r$ is replaced by a different simply-connected or multiply connected domain D in the plane also follow easily, provided that the Green's function of D and harmonic measure in D can be calculated. The hypothesis that $u(z)$ is s.h. on the closure $\bar{D}$ of D is more convenient than that frequently given since we need no continuity assumptions for $u(z)$. The m-dimensional version of Theorem 3.14 will also be useful. We write down the special case of (3.7.1) when D is a ball $D(0, r)$ in R^m where $m \geqslant 3$, and obtain

$$u(x) = \frac{1}{c_m} \int_{S(0,r)} \frac{(r^2 - |x|^2)\,d\sigma(\xi)}{r|x - \xi|^m} - \int_D g(x, \xi)\,d\mu e_\xi. \tag{3.7.3}$$

Here by Theorem 1.10 we have for x, ξ in D

$$g(x, \xi) = |x - \xi|^{2-m} - \{|\xi|\,|x - \xi'|/r\}^{2-m},$$

where ξ' is the inverse point of ξ with respect to $S(0, r)$.

3.7.2

We proceed to prove Theorem 3.14. Theorem 3.9 allows us to reduce the general case to the special case when $u(x) = K(x - \eta)$, and we prove Theorem 3.14 first in this special case.

LEMMA 3.9. *If D is the domain of Theorem* 3.14, *$K(x)$ is defined by* (3.5.1), *and x, η are points of D and R^m respectively, then*

$$\int_F K\,(\xi - \eta)\,d\omega(x, e_\xi, D) = \begin{cases} K(x - \eta) + g(x, \eta, D), & \eta \in D \\ K(x - \eta), & \textit{otherwise}. \end{cases} \tag{3.7.4}$$

Suppose first that η is an exterior point of D, so that η does not lie on $D \cup F$. Then $K(\xi - \eta)$ is a harmonic function of ξ on $D \cup F$ and so (3.7.4) follows from Theorem 3.10 in this case.

Next if η is a point of D we write

$$h(\xi) = K(\xi - \eta) + g(\xi, \eta, D).$$

Then $h(\xi)$ remains harmonic at η and so throughout D and, since $g(\xi, \eta, D)$ vanishes continuously as ξ approaches F from inside D, $h(\xi)$ is continuous for ξ on F if we define $h(\xi) = K(\xi - \eta)$ there. Thus we may replace $K(\xi - \eta)$

by $h(\xi)$ in the left-hand side of (3.7.4) and deduce from Theorem 3.10 that the right-hand side becomes $h(x)$ in this case. Thus Lemma 3.9 has been proved except when $\eta \in F$.

This final case is less easy and it is only for this case that we require F to have Lebesgue measure zero in R^m. We now consider x fixed and η variable and write

$$J(\eta) = \int_F K(\xi - \eta)\, d\omega(x, e_\xi, D).$$

Since $\omega(x, e_\xi, D)$ is a Borel measure on F it follows from Theorem 3.7 that $J(\eta)$ is s.h. in R^m. Let η_0 be a point of F. Then it follows from the properties (ii) and (iii) of subharmonicity in Section 2.1, that

$$I(r, J) = \frac{1}{c_m r^{m-1}} \int_{S(\eta_0, r)} J(\eta)\, d\sigma(\eta) \to J(\eta_0) \quad \text{as} \quad r \to 0.$$

Here $\sigma(\eta)$ denotes $(m - 1)$-dimensional measure on $S(\eta_0, r)$. Hence also

$$\text{(3.7.5)} \qquad \frac{m}{c_m r^m} \int_{D(\eta_0, r)} J(\eta)\, d\eta = \frac{m}{r^m} \int_0^r I(\rho, J) \rho^{m-1}\, d\rho \to J(\eta_0), \quad \text{as} \quad r \to 0.$$

Here $d\eta$ denotes m-dimensional Lebesgue measure in R^m. Since F has zero m-dimensional measure we may alter the integrand on F arbitrarily without changing the integral. In particular we may replace $J(\eta)$ by $K(x - \eta)$ on F. Since

$$K(x - \eta) \to K(x - \eta_0) \quad \text{as} \quad \eta \to \eta_0$$

and

$$g(x, \eta, D) \to 0 \quad \text{as} \quad \eta \to \eta_0 \text{ from inside } D,$$

we deduce from the fact that Lemma 3.9 holds except on F, that we have as $r \to 0$

$$\int_{D(\eta_0, r)} J(\eta)\, d\eta = \int_{D(\eta_0, r)} \{K(x - \eta_0) + o(1)\}\, d\eta = \frac{c_m r^m}{m} \{K(x - \eta_0) + o(1)\}.$$

On comparing this result with (3.7.5) we deduce that

$$J(\eta_0) = K(x - \eta_0).$$

This completes the proof of Lemma 3.9.

3.7.3

Suppose now that u satisfies the hypotheses of Theorem 3.14. Then $u(\xi)$

is s.h. on a compact set E containing $D \cup F$ in its interior Δ. Thus by Theorem 3.9 we may write

$$u(\xi) = \int_E K(\xi - \eta)\, d\mu e_\eta + h(\xi), \tag{3.7.6}$$

where $\mu(e_\eta)$ denotes the Riesz measure of u in E so that $\mu(E) < \infty$, $K(\xi)$ is defined by (3.5.1), and $h(\xi)$ is harmonic in Δ. For any function $f(\xi)$ which is bounded above and Borel measurable on F we write

$$L(f) = \int_F f(\xi)\, d\omega(x, e_\xi, D).$$

In particular

$$L\{u(\xi)\} = \int_F d\omega(x, e_\xi, D) \int_E K(\xi - \eta)\, d\mu e_\eta + L\{h(\xi)\}.$$

Since $h(\xi)$ is harmonic on $D \cup F$ it follows from Theorem 3.10, that

$$L\{h(\xi)\} = h(x).$$

Also $K(\xi - \eta)$ is bounded above for $\xi \in F$ and $\eta \in E$ and $\omega(x, F, D)$ and $\mu(E)$ are both finite. Thus it follows from Fubini's Theorem 3.5 that we may invert the order of integration in the double integral. This gives

$$\begin{aligned} L\{u(\xi)\} &= \int_E d\mu e_\eta\, L\{K(\xi - \eta)\} + h(x) \\ &= \int_E K(x - \eta)\, d\mu e_\eta + \int_D g(x, \eta, D)\, d\mu e_\eta + h(x) \end{aligned}$$

by Lemma 3.9. Using (3.7.6) we deduce Theorem 3.14. We shall use the rest of the chapter to give some applications of this basic result.

3.8. HARMONIC EXTENSIONS AND LEAST HARMONIC MAJORANTS

We proceed to prove a generalization of Theorem 2.18. This is

THEOREM 3.15. *With the hypotheses of Theorem* 3.14 *the harmonic extension of* $u(x)$ *from* F *into* D *is the least harmonic majorant of* $u(x)$ *in* D.

It follows from Theorem 2.18 that if $v_0(x)$ is the harmonic extension of $u(x)$ from F into D, so that by Theorem 3.10

$$v_0(x) = \int_F u(\xi)\, d\omega(x, e_\xi),$$

then $v_0(x)$ is a harmonic majorant of $u(x)$ in D, i.e. $v_0(x)$ is harmonic in 0, and $v_0(x) \geqslant v(x)$ there.

Next we note that the Green's function $g(x, \xi, D)$ is continuous in x, ξ jointly for $\xi \in D$, $x \in D \cup F$ and $x \neq \xi$. We assume $m \geqslant 3$. The case $m = 2$ is similar. Then $K(x - \xi) = -|x - \xi|^{2-m}$ is continuous for x and ξ satisfying these conditions and in particular uniformly for $x \in F$ and ξ on a compact subset e of D. Hence the harmonic extension $h(x, \xi)$ of $K(x - \xi)$ from F into D as a function of x is also continuous in x and ξ jointly, uniformly for ξ in e and x in D, since a slight change in ξ causes a slight change in $K(x - \xi)$ on F and so, by the maximum principle, a slight change in $h(x, \xi)$ for x in D. Thus

$$g(x, \xi, D) = h(x, \xi) + |x - \xi|^{2-m}$$

is continuous for x outside e and $\xi \in e$ if we set $g = 0$ for $x \in F$.

In particular

$$g(x, \xi, D) \to 0,$$

uniformly as x approaches F for $\xi \in e$. Let μ be the Riesz measure in D and set

$$p(x) = \int_e g(x, \xi, D)\, \mathrm{d}\mu e_\xi.$$

Then $p(x) \to 0$ as x approaches F. In particular, given $\varepsilon > 0$, we can find a compact subset e_1 of D, containing e and such that

$$p(x) < \varepsilon, \qquad x \in D - e_1.$$

Thus if $v(x)$ is any harmonic majorant of $u(x)$ in D, we have

$$u(x) + p(x) < u(x) + \varepsilon \leqslant v(x) + \varepsilon, \qquad x \in D - e_1.$$

Since

$$u(x) + p(x) = v_0(x) + \int_{D-e} g(x, \xi, D)\, \mathrm{d}\mu e_\xi$$

is still s.h. in D, it follows from the maximum principle that

$$u(x) + p(x) \leqslant v(x) + \varepsilon, \qquad x \in e_1,$$

so that

$$v(x) \geqslant u(x) + p(x) - \varepsilon, \qquad x \in D.$$

Since ε is arbitrary we deduce that

$$v(x) \geqslant u(x) + p(x) = v_0(x) - \int_{D-e} g(x, \xi, D)\, \mathrm{d}\mu e_\xi.$$

We may allow e to expand through a sequence e_n to D and then

$$\int_{D-e_n} g(x, \xi, D)\, \mathrm{d}\mu e_\xi \to 0.$$

Thus

$$v(x) \geqslant v_0(x), \qquad x \in D.$$

Hence $v_0(x)$ is the least harmonic majorant of $u(x)$ in D. The following special case of Theorem 3.14 has independent interest.

THEOREM 3.16. *Suppose that $u(x)$ satisfies the hypotheses of Theorem* 3.14 *and that $u(x)$ is harmonic in D. Then $u(x)$ is in D the harmonic extension of $u(\xi)$ from F into D.*

In fact, since $u(x)$ is harmonic in D, $\mathrm{d}\mu = 0$ in D. This gives

$$u(x) = \int_F u(\xi)\, \mathrm{d}\omega(x, e_\xi).$$

This result also enables to extend significantly Theorem 2.19. We have

THEOREM 3.17. *If D, F and $u(x)$ satisy the hypotheses of Theorem* 3.14 *and $v(x)$ is* s.h. *on $D \cup F$ and $v(x)$ is harmonic in D, $v(x) = u(x)$ on F, then $v(x)$ coincides in D with the least harmonic majorant of $u(x)$ in D and with the harmonic extension of $u(x)$ from F into D.*

In fact by Theorem 3.15 the least harmonic majorant of u in D is the same as the harmonic extension of u from F into D. Since $u = v$ on F this latter coincides with the harmonic extension of v from F into D and by Theorem 3.16 this coincides with $v(x)$ in D.

3.9. NEVANLINNA THEORY†

We shall consider functions subharmonic in a closed ball $C(0, r)$ in R^m and not identically $-\infty$. It is convenient to suppose that $u(0) \neq -\infty$. If this condition is not satisfied we can replace $u(x)$ in $D(0, \varepsilon)$ by the harmonic extension of the values of $u(x)$ on $S(0, \varepsilon)$ and obtain a function $v(x)$ s.h. in $C(0, r)$, harmonic in $D(0, \varepsilon)$ and coinciding with $u(x)$ elsewhere. These conditions determine $v(x)$ uniquely by Theorem 3.17 (or even Theorem 2.19). Then

† See e.g. R. Nevanlinna [1929] for the special case (0.1).

the Poisson–Jensen formula (3.7.2) or (3.7.3) yields

$$u(0) = \frac{1}{c_m r^{m-1}} \int_{S(0,r)} u(x)\, d\sigma(x) - \int_{D(0,r)} g(0,\xi)\, d\mu e_\xi. \tag{3.9.1}$$

Here we have by Theorem 1.10

$$g(0,\xi) = \log\left|\frac{r}{\xi}\right|, \qquad m = 2$$

$$g(0,\xi) = |\xi|^{2-m} - r^{2-m}, \qquad m > 2.$$

We proceed to modify (3.9.1) slightly. We write

$$u^+(x) = \max(u(x), 0), u^-(x) = -\min(u(x), 0)$$

and set†

$$T(r,u) = \frac{1}{c_m r^{m-1}} \int_{S(0,r)} u^+(x)\, d\sigma(x). \tag{3.9.2}$$

$$m(r,u) = \frac{1}{c_m r^{m-1}} \int_{S(0,r)} u^-(x)\, d\sigma(x). \tag{3.9.3}$$

Next let $n(t)$ be the Riesz mass of the closed ball $C(0,t)$. Then we see that

$$\int_{D(0,r)} g(0,\xi)\, d\mu e_\xi = d_m \int_0^r \frac{n(t)\, dt}{t^{m-1}}, \tag{3.9.4}$$

where $d_2 = 1, d_m = m - 2, m > 2$.

Suppose first that $m = 2$. Then

$$\int_{D(0,r)} g(0,\xi)\, d\mu e_\xi = \int_0^r \log\frac{r}{t}\, dn(t) = -\lim_{\varepsilon\to 0}\left[\log\frac{r}{\varepsilon}\, n(\varepsilon) + \int_\varepsilon^r \frac{n(t)\, dt}{t}\right].$$

Also, since $u(0)$ is finite, the integrals

$$\int_{D(0,r)} g(0,\xi)\, d\mu e_\xi = \int_0^r \log\frac{r}{t}\, dn(t)$$

are both finite and since $n(t)$ increases with t, we deduce that

$$n(\varepsilon)\log\frac{r}{\varepsilon} \leqslant \int_0^\varepsilon \log\frac{r}{t}\, dn(t) \to 0, \quad \text{as} \quad \varepsilon \to 0.$$

† The fact that m is also used for the dimension of the space throughout the book will not, we hope, lead to any confusion.

Thus (3.9.4) follows in this case. If $m > 2$ we see similarly that

$$\int_{D(0,r)} g(0, \xi)\, d\mu e_\xi = \int_0^r (t^{2-m} - r^{2-m})\, dn(t) = (m-2) \int_0^r \frac{n(t)\, dt}{t^{m-1}}.$$

Thus (3.9.4) holds.

We now define

$$N(r, u) = d_m \int_0^r \frac{n(t)\, dt}{t^{m-1}}, \tag{3.9.5}$$

and see that (3.9.1) may be written as

$$T(r, u) = m(r, u) + N(r, u) + u(0). \tag{3.9.6}$$

This is Nevanlinna's first fundamental Theorem for s.h. functions. The function $T(r, u)$ is called the Nevanlinna characteristic of u. In a very significant way it determines the growth of $u(x)$.

We have immediately an interesting convexity Theorem.

THEOREM 3.18. *If $u(x)$ is s.h. in $C(0, r)$ and $u(0) \neq -\infty$, then $T(\rho, u)$ and $N(\rho, u)$ are increasing functions of ρ for $0 \leqslant \rho \leqslant r$. They are also convex functions of $\log \rho$ if $m = 2$ and of ρ^{2-m} for $m > 2$.*

Since $u^+(x) = \max\, (u(x), 0)$ is s.h. the results concerning $T(r)$ are immediate consequences of Theorem 2.12. Those concerning $N(r, u)$ follow from (3.9.5). In fact we have

$$r^{m-1} \frac{d}{dr} N(r, u) = d_m n(r)$$

except at the countable set of discontinuities of $n(r)$, where the result remains true if we take the right derivative on the left-hand side. Clearly the right-hand side is an increasing function of r.

We are also interested in the maximum of $u(x)$ on hyperspheres and define

$$B(r, u) = \sup_{|x| = r} u(x).$$

The quantities $B(r,u)$ and $T(r, u)$ have comparable growth. This follows from the following inequality.

THEOREM 3.19. *If $u(x)$ is s.h. in $C(0, r)$ then we have for $0 < \rho < r$*

$$T(\rho, u) \leqslant B(\rho, u^+) \leqslant \frac{r^{m-2}(r+\rho)}{(r-\rho)^{m-1}} T(r, u).$$

F

The left-hand inequality is obvious. To prove the right-hand inequality we apply the Poisson–Jensen formula to $u^+(x)$ in $D = D(0, r)$. The contribution from the Riesz mass is non-positive. Thus we obtain

$$u^+(\xi) \leqslant \frac{1}{c_m}\int_{S(0,r)} \frac{r^2 - |\xi|^2}{r|r - \xi|^m} u^+(x)\,d\sigma_x$$

$$\leqslant \frac{r^2 - \rho^2}{r(r-\rho)^m}\frac{1}{c_m}\int_{S(0,r)} u^+(x)\,d\sigma_x = \frac{r^{m-2}(r+\rho)}{(r-\rho)^{m-1}} T(r, u).$$

Choosing ξ so that $u^+(\xi) = B(\rho, u^+)$ we deduce the right-hand inequality of Theorem 3.19.

3.10. BOUNDED SUBHARMONIC FUNCTIONS IN R^m

It follows from Theorem 2.14 that the only functions which are s.h. and bounded above in the plane are constants. However, if $m > 2$ we saw in Section 2.7.1 that s.h. functions which are bounded above exist in R^m. We proceed to investigate these functions. Our main result is

THEOREM 3.20. *Given any Borel measure μ in R^m, where $m \geqslant 3$, let $n(t)$ be the measure of the closed ball $C(0, t)$. Then a necessary and sufficient condition that μ is the Riesz measure of a function $u(x)$* s.h. *in R^m is that*

$$\int_1^\infty \frac{n(t)\,dt}{t^{m-1}} < \infty. \tag{3.10.1}$$

If this condition is satisfied, the unique s.h. *function with Riesz measure μ in R^m and least upper bound C is given by*

$$u(x) = C - \int \frac{1}{|x - \xi|^{m-2}}\,d\mu e_\xi. \tag{3.10.2}$$

We prove first that (3.10.1) is necessary. In fact if $u(x) \leqslant C$ it is evident that

$$T(r, u) \leqslant \max(C, 0), \qquad 0 < r < \infty.$$

It follows from (3.9.6) that

$$T(r, u) - T(1, u) = N(r, u) - N(1, u) + m(r, u) - m(1, u).$$

This result remains true with suitable definitions even if $u(0) = -\infty$. Thus if $r > 1$

$$d_m \int_1^r \frac{n(t)\,dt}{t^{m-1}} = N(r, u) - N(1, u) \leqslant m(1, u) + \max(C, 0) = O(1),$$

as $r \to \infty$, if $u(x)$ is not identically $-\infty$, which we assume. Thus (3.10.1) holds.

Suppose next that (3.10.1) holds and that $u(x)$ is given by (3.10.2) with $C = 0$. Suppose that $|x| \leqslant r$, and write

$$u(x) = -\int_{|\xi| \leqslant r} |x - \xi|^{2-m}\, d\mu e_\xi - \int_{|\xi| > r} |x - \xi|^{2-m}\, d\mu e_\xi$$
$$= u_1(x) + u_2(x).$$

Then $u_1(x)$ is s.h. in space by Theorem 3.7. Also for any integer $N > r$

$$u_N(x) = -\int_{r < |\xi| < N} |x - \xi|^{2-m}\, d\mu e_\xi$$

is harmonic in $|x| < r$ by the same Theorem and clearly $u_N(x)$ decreases with increasing N. Thus by Harnack's Theorem 1.20

$$u_2(x) = \lim_{N \to \infty} u_N(x)$$

is harmonic in $|x| < r$, provided that the limit is finite at some point in $|x| < r$, for instance the origin. Now we have

$$u_N(0) = -\int_r^N |\xi|^{2-m}\, d\mu e_\xi = -\int_r^N t^{2-m}\, dn(t)$$
$$= -(m-2)\int_r^N \frac{n(t)\, dt}{t^{m-1}} - [t^{2-m} n(t)]_r^N.$$

By (3.10.1) the integral remains bounded as $N \to \infty$. Also since the integral converges and $n(t)$ increases with t we may, given $\varepsilon > 0$, choose r so large that for $r > r_0$

$$\int_r^\infty \frac{n(t)\, dt}{t^{m-1}} < \varepsilon,$$

hence also

$$n(r)\int_r^\infty t^{1-m}\, dt = \frac{n(r) r^{2-m}}{(m-2)} < \varepsilon.$$

Thus

$$\frac{n(r)}{r^{m-2}} \to 0, \quad \text{as} \quad r \to \infty. \tag{3.10.3}$$

Thus we deduce that

$$u_2(0) = \lim_{N \to \infty} u_N(0) = -(m-2)\int_r^\infty \frac{n(t)\, dt}{t^{m-2}} + \frac{n(r)}{r^{m-2}} > -\infty. \tag{3.10.4}$$

Thus $u_2(x)$ is harmonic in $|x| < r$ and so $u(x) = u_1(x) + u_2(x)$ is s.h. and also, since $u(x) - u_1(x)$ is harmonic in $|x| < r$, $u(x)$ has Riesz measure μ in $|x| < r$ and so in space.

Next it follows from (3.10.3) and (3.10.4) that given $\varepsilon > 0$, we may choose r_0 so large that for $r > r_0$

$$u_2(0) > -\varepsilon$$

Also for fixed r, and $|x| > 2r$

$$u_1(x) = \int_{|\xi| \leqslant r} |x - \xi|^{2-m} \, d\mu e_\xi \leqslant |\tfrac{1}{2}x|^{2-m} n(r) \to 0$$

as $x \to \infty$. Since $u_2(x)$ is s.h. we have for any $t > 0$

$$-\varepsilon < u_2(0) \leqslant \frac{1}{c_m t^{m-1}} \int_{S(0,t)} u_2(x) \, d\sigma(x).$$

Hence we can find x on $S(0, t)$ such that

$$u_2(x) > -\varepsilon,$$

and hence, if t is large enough, also

$$u_1(x) > -\varepsilon.$$

Thus

$$u(x) = u_1(x) + u_2(x) > -2\varepsilon.$$

Thus the least upper bound of $u(x)$ in R^m cannot be negative, and since $u(x) \leqslant 0$ everywhere, the least upper bound is zero. Thus $u(x)$ given by (3.10.2) has the properties required by Theorem 3.20.

It remains to show that $u(x)$ is unique, subject to these conditions. Suppose then that contrary to this there exists another function $u_0(x)$, nonpositive in R^m and having Riesz measure μ there. Set

$$h(x) = u_0(x) - u(x).$$

Then, by Theorem 3.9, $h(x)$ is harmonic in R^m. Also since $u_0(x) \leqslant 0$

$$h^+(x) = \max(h(x), 0) \leqslant u^-(x).$$

Thus

$$T(r, h) \leqslant \frac{1}{c_m r^m} \int_{S(0,r)} u^-(x) \, d\sigma(x) = m(r, u).$$

We suppose that $u(0) > -\infty$, otherwise we consider $u(x_0 + x)$ instead of

$u(x)$ for a suitable x_0. Then in view of (3.9.6) and since $u(x) \leqslant 0$, we deduce that

$$T(r, u) = 0,$$

and so

$$m(r, u) = O(1) \quad \text{as} \quad r \to \infty.$$

Thus

$$T(r, h) = O(1) \quad \text{as} \quad r \to \infty.$$

Taking $\rho = \frac{1}{2}r$ in Theorem 3.19, we deduce that $h(x)$ is bounded above in R^m and hence, in view of Theorem 2.15, that $h(x)$ is constant. This completes the proof of Theorem 3.20.

3.10.1

It follows from Theorem 3.20 that the upper bound of the function $u(x)$ given by (3.10.2) in R^m is C. We conclude the chapter by proving a somewhat sharper result.

THEOREM 3.21. *Let $u(x)$ be* s.h. *in R^m with a finite least upper bound C. Then given $q > m - 2$, we have that*

$$u(x) \to C \quad \text{as} \quad x \to \infty \tag{3.10.5}$$

outside a finite or countable set of closed balls $C(x_k, r_k)$ such that $\Sigma(r_k/|x_k|)^q < \infty$.

In particular (3.10.5) *holds as $x \to \infty$ along almost all fixed rays through the origin*†.

We need a lemma which is the analogue of a corresponding result of Cartan [1928] for s.h. functions in the plane.

LEMMA 3.10 (Cartan's Lemma). *Suppose that μ is a measure in R^m such that $\mu(R^m) = \mu_0 < \infty$ and suppose that $0 < p < q < \infty$. Then if $h > 0$ and*

$$I(x) = \int_{R^m} |x - \xi|^{-p}\, d\mu e_\xi, \tag{3.10.6}$$

we have

$$I(x) < h, \tag{3.10.7}$$

outside the union of a finite or countable set of closed balls $C(x_n, r_n)$ such that

$$\sum_n r_n^q < A_1(\mu_0/h)^{q/p}, \tag{3.10.8}$$

where the constant A_1 depends on p, q only.

† Deny [1948] has proved that the result holds along all rays except a set of capacity zero.

For each fixed positive integer ν, we construct a maximal number of mutually disjoint closed balls $C_{k,\nu} = C(x_{k,\nu}, \frac{1}{2}r_\nu)$, $k = 1$ to k_ν such that

$$r_\nu = (\mu_0/h)^{1/p}\, 2^{-\nu/s}, \tag{3.10.9}$$

where $s = \frac{1}{2}(p + q)$, and

$$\mu(C_{k,\nu}) \geqslant \mu_0 2^{-\nu}. \tag{3.10.10}$$

Clearly there can be at most 2^ν disjoint balls $C_{k,\nu}$ for each ν, since the total measure is μ_0. Also

$$\sum_{\nu=1}^{\infty} \sum_{k=1}^{k_\nu} r_\nu^q \leqslant \sum_{\nu=1}^{\infty} 2^\nu (\mu_0/h)^{q/p}\, 2^{-\nu q/s} \tag{3.10.11}$$

$$= (\mu_0/h)^{q/p} \sum_{\nu=1}^{\infty} 2^{-\nu(q/s - 1)}$$

$$= A_2 (\mu_0/h)^{q/p},$$

where the constant A_2 depends on q, s and so on p, q only.

Suppose now that x is a point outside all the balls $C(x_{k,\nu}, r_\nu)$. Then if $n(r)$ denotes the measure of the ball $C(x, r)$, we have

$$n(\tfrac{1}{2}r_\nu) < \mu_0 2^{-\nu}, \quad \nu = 1 \quad \text{to} \quad \infty. \tag{3.10.12}$$

For if this result is false for some ν the ball $C = C(x, \frac{1}{2}r_\nu)$ does not meet any of the balls $C(x_{k,\nu}, \frac{1}{2}r_\nu)$ for $k = 1$ to k_ν and we obtain a contradiction to our hypothesis that the set of balls satisfying (3.10.10) is maximal, i.e. that no ball not meeting the others can be added to the set. In particular $n(0) = 0$. Thus

$$I(x) = \int_{|\xi - x| \geqslant \frac{1}{2}r_1} |x - \xi|^{-p}\, d\mu e_\xi + \sum_{\nu=1}^{\infty} \int_{\frac{1}{2}r_{\nu+1} \leqslant |\xi - x| < \frac{1}{2}r_\nu} |x - \xi|^{-p}\, d\mu e_\xi$$

$$\leqslant 2^p \mu_0 \left\{ r_1^{-p} + \sum_{\nu=1}^{\infty} 2^{-\nu} r_{\nu+1}{}^{-p} \right\}$$

$$= 2^p \mu_0 \sum_{\nu=1}^{\infty} 2^{1-\nu} r_\nu^{-p} = 2^{p+1} h \sum_1 2^{-\nu + \nu p/s} = A_3 h,$$

where the constant A_3 depends on p and s, i.e. on p and q only. If we write $A_3 h$ instead of h and use (3.10.11) we obtain Lemma 3.10 with

$$A_1 = A_2 A_3^{q/p}.$$

3.10.2

We can now prove Theorem 3.21. We suppose that ν is a positive integer and that

$$2^{\nu} < |x| \leqslant 2^{\nu+1}.$$

Let $u(x)$ be the function given by (3.10.2) and write

$$C - u(x) = I_1(x) + I_2(x) + I_3(x),$$

where

$$I_1(x) = \int_{|\xi| \leqslant 2^{\nu-1}} |x - \xi|^{2-m} \, d\mu,$$

$$I_2(x) = \int_{2^{\nu-1} < |\xi| < 2^{\nu+2}} |x - \xi|^{2-m} \, d\mu,$$

and

$$I_3(x) = \int_{|\xi| \geqslant 2^{\nu+2}} |x - \xi|^{2-m} \, d\mu.$$

Then

(3.10.13) $$I_1(x) \leqslant |\tfrac{1}{2}x|^{2-m} \int_{|\xi| \leqslant 2^{\nu-1}} d\mu \leqslant \frac{n(2^{\nu-1})}{(2^{\nu-1})^{m-2}} \to 0$$

as $\nu \to \infty$, in view of (3.10.3). Again

$$I_3(x) \leqslant \int_{|\xi| \geqslant 2^{\nu+2}} |\tfrac{1}{2}\xi|^{2-m} \, d\mu = 2^{m-2} \int_{2^{\nu+2}}^{\infty} t^{2-m} \, dn(t)$$

$$= 2^{m-2} \left[\frac{n(t)}{t^{m-2}} \right]_{2^{\nu+2}}^{\infty} + 2^{m-2}(m-2) \int_{2^{\nu+2}}^{\infty} \frac{n(t) \, dt}{t^{m-1}}$$

In view of (3.10.1) and (3.10.3) we see that

(3.10.14) $$I_3(x) \to 0 \quad \text{as} \quad x \to \infty.$$

It remains to consider $I_2(x)$. Let μ_ν be the total measure of the region

(3.10.15) $$2^{\nu-1} < |\xi| < 2^{\nu+2}.$$

Let ε_ν be a positive number. Then we deduce from Lemma 3.10 that

(3.10.16) $$I_3(x) < \varepsilon_\nu$$

outside a set $E_\nu(\varepsilon)$ balls of radii r_n, such that, for $q > m - 2$, we have

(3.10.17) $$\Sigma\, r_n^q < A_1(\mu_\nu/\varepsilon_\nu)^{q/(m-2)}.$$

We set

$$\eta_\nu = \mu_\nu \, 2^{(2-m)\nu},$$

and note that

$$\sum_{\nu=1}^{\infty} \eta_\nu \leqslant \sum_{\nu=1}^{\infty} 2^{(2-m)\nu} \left[\int_{2^{\nu-1}}^{2^\nu} + \int_{2^\nu}^{2^{\nu+1}} + \int_{2^{\nu+1}}^{2^{\nu+2}}\right] dn(t)$$

(3.10.18)
$$\leqslant C \int_1^\infty \frac{dn(t)}{t^{m-2}} < \infty,$$

where C is a constant depending on m. In terms of η_ν we may write (3.10.17) as

$$\sum \left(\frac{r_n}{2^\nu}\right)^q < A_1\, \eta_\nu^{q/(m-2)}\, \varepsilon_\nu^{-q/(m-2)}.$$

We set

$$\varepsilon_\nu^{q/(m-2)} = A_1 \eta_\nu^{q/(m-2)-1},$$

so that $\varepsilon_\nu \to 0$ as $\nu \to \infty$, and note that we have (3.10.16) in the region (3.10.15) outside a set of balls whose radii r_n satisfy

$$\sum \left(\frac{r_n}{2^\nu}\right)^q < \eta_\nu.$$

In view of (3.10.18) it follows that $\eta_\nu \to 0$, so that, we may assume that $\eta_\nu < 4^{-q}$, $\nu > \nu_0$. We ignore all the balls for $\nu \leqslant \nu_0$ and those balls for $\nu > \nu_0$ which do not meet the region (3.10.15). If $C(x_{n,\nu}, r_{n,\nu})$ are the remaining balls corresponding to (3.10.15) then it follows that $r_{n,\nu} < 2^{\nu-2}$, and since the balls meet (3.10.15) we must have

$$|x_{n,\nu}| > 2^{\nu-2},$$

so that

$$\sum \left(\frac{r_{n,\nu}}{|x_{n,\nu}|}\right)^q \leqslant 4^q \sum \left(\frac{r_n}{2^\nu}\right)^q \leqslant 4^q \eta_\nu.$$

Thus as $x \to \infty$ in any manner outside the exceptional balls corresponding to the region (3.10.15) we have for $2^\nu < |x| \leqslant 2^{\nu+1}$

$$C - u(x) < \varepsilon_\nu + o(1) \quad \text{as} \quad \nu \to \infty.$$

Also we have for the exceptional balls

$$\sum_{\nu=\nu_0}^{\infty} \sum_n \left(\frac{r_{n,\nu}}{|x_{n,\nu}|}\right)^q \leqslant 4^q \sum_{\nu=\nu_0}^{\infty} \eta_\nu < \infty,$$

in view of (3.10.18). This proves the main statement of Theorem 3.21.

We may in particular take $q = m - 1$. Then the solid angle subtended by $C(x_k, r_k)$ at the origin is proportional to $(r_k/|x_k|)^{m-1}$ and so Theorem 3.21 tells us that the solid angles ω_ν subtended by the exceptional balls C_ν have convergent sum. We may leave out a finite number of the exceptional balls, without altering the conclusion of the theorem and so ensure that

$$\sum \omega_\nu < \varepsilon.$$

For all rays, not meeting the remaining exceptional balls for large $|x|$, (3.10.5) holds without restriction and, since ε is arbitrary, we deduce that (3.10.5) holds without restriction on all rays through the origin except those meeting the unit sphere $S(0, 1)$ on a set of zero $(m - 1)$—dimensional surface area. This completes the proof of Theorem 3.21.

Example

The conclusion of Theorem 3.21 no longer holds if $q = m - 2$. For instance in the case $m = 3$ we may set for $x = (x_1, x_2, x_3)$

$$u(x) = -\int_0^\infty \frac{dt}{\{(x_1 + t)^2 + x_2^2 + x_3^2\}^{1/2}(t + 1)}.$$

This function satisfies the condition of Theorem 3.20, with all the measure concentrated on the negative x_1 axis and

$$n(t) = \int_0^t \frac{d\tau}{\tau + 1} = \log(t + 1).$$

But $u(x) = -\infty$ on the whole negative x_1 axis and this axis cannot be included in a set of balls $C_k = C(\xi_k, r_k)$ for which

$$\sum \frac{r_k}{|\xi_k|} < \infty.$$

In fact if E is the set in which the C_k meet the part $x_1 < -1$ of the x_1 axis we have

$$\int_E \frac{dx_1}{x_1} = O\left\{\sum \frac{r_k}{|\xi_k|}\right\} < \infty.$$

CHAPTER 4

Functions Subharmonic in Space

4.0. INTRODUCTION

In the last chapter we developed the representation theorem of F. Riesz which expresses a subharmonic function locally as the sum of a harmonic function plus a potential. In this chapter we shall consider the analogous results for functions subharmonic in the whole of R^m. When $m = 2$ and

$$u(z) = \log|f(z)|,$$

where $f(z)$ is an entire function we obtain in particular the classical representation theorems of Weierstrass and Hadamard of $f(z)$ as a product in terms of its zeros.

Just as in the above mentioned special cases, the order of the function will be seen to play a big role, and in particular we shall be able to obtain some sharp estimates for functions of order less than one, by using the above representation theorems. These refer specifically to the relations between the quantities $N(r)$, $T(r)$ and $B(r)$.

In the last part of the chapter we consider how to generalize the notion of asymptotic values and tracts to subharmonic functions in R^m. The estimates obtained although not as sharp as the Theorem of Ahlfors at least give the right order of magnitude for the number of such tracts for a function with given order μ in R^m. We shall also be able to generalize a classical Theorem of Iversen.

4.1. THE WEIERSTRASS REPRESENTATION THEOREM

Let $K(x)$ be the function defined by (3.5.1). Then $K(x)$ is harmonic in R^m except at $x = 0$, when $m \geqslant 2$, and so $K(x - \xi)$ is harmonic in R^m except at $x = \xi$. In particular, if $\xi \neq 0$, $K(x - \xi)$ is harmonic near the origin and so has a multiple power series expansion in $x_1, x_2, \ldots, x_m$, convergent in a neighbourhood of the origin. We write

$$K(x - \xi) = \sum_{0}^{\infty} a_\nu(x, \xi), \tag{4.1.1}$$

where, for fixed ν and $\xi \neq 0$, $a_\nu(x, \xi)$ is a homogeneous polynomial in x_1 to x_m of degree ν. We now set

$$K_q(x, \xi) = K_q(x, \xi, m) = K(x - \xi) - \sum_{\nu=0}^{q} a_\nu(x, \xi). \tag{4.1.2}$$

We shall need to obtain some estimates for $K_q(x, \xi)$, but first proceed to estimate the polynomials $a_\nu(x, \xi)$.

LEMMA 4.1. *The polynomials $a_\nu(x, \xi)$ are harmonic in x for fixed ξ, and continuous in x, ξ jointly for $|\xi| \neq 0$. If $|x| = \rho$, $|\xi| = r > 0$, we have the sharp inequality*

$$|a_\nu(x, \xi)| \leqslant \frac{b_{\nu,m}\rho^\nu}{r^{m+\nu-2}}, \tag{4.1.3}$$

where $b_{\nu,m} = 1/\nu$ *if* $m = 2$, $\nu \geqslant 1$;

$$b_{\nu,m} = (\nu + m - 3)(\nu + m - 4)\ldots(\nu + 1)/(m - 1)!, \quad m \geqslant 3, \nu \geqslant 0.$$

Suppose first that $m = 2$. Then we write $z = x_1 + ix_2$, $\zeta = \xi_1 + i\xi_2$, so that z, ζ are complex variables and

$$K(x - \zeta) = \log|z - \zeta| = \mathrm{R}\left\{\log|\zeta| + \log\left(1 - \frac{z}{\zeta}\right)\right\}$$
$$= \mathrm{R}\left\{\log|\zeta| - \sum_{\nu=1}^{\infty} \frac{1}{\nu}\left(\frac{z}{\zeta}\right)^\nu\right\}.$$

Thus in this case we have for $\nu \geqslant 1$

$$|a_\nu(x, \zeta)| = \left| -\frac{1}{\nu}\mathrm{R}\left(\frac{z}{\zeta}\right)^\nu\right| \leqslant \frac{1}{\nu}\left(\frac{\rho}{r}\right)^\nu,$$

as required. We also see that equality holds if z/ζ is positive.

In the case $m \geqslant 3$ we write for any real vectors x, ξ, and a complex h

$$\phi(h) = -K(\xi - hx) = \left\{\sum_{n=1}^{m} (\xi_n - hx_n)^2\right\}^{-(m/2)+1} = -\sum_{0}^{\infty} h^\nu a_\nu(x, \xi).$$

Then $\phi(h)$ is an analytic function of the complex variable h at least for small h, if we choose the square root so as to make $\phi(0)$ positive. We have

$$\sum_{n=1}^{m} (\xi_n - hx_n)^2 = r^2 - 2th + \rho^2 h^2,$$

where $t = \Sigma_{n=1}^{m} \xi_n x_n$, so that $|t| \leqslant r\rho$. Thus

$$\phi(h) = \{r^2(1 - \alpha h)(1 - \beta h)\}^{1-(m/2)},$$

where α, β are $(t \mp i\sqrt{(r^2\rho^2 - t^2)})/r^2$. In particular

$$|\alpha| = |\beta| = \frac{\rho}{r}.$$

We now note that if $p = (m - 2)/2 > 0$,

$$(1 - \alpha h)^{-p} = \sum_{0}^{\infty} c_{\nu, p}(\alpha h)^{\nu},$$

where

$$c_{\nu, p} = \frac{p(p + 1) \dots (p + \nu - 1)}{\nu!} > 0.$$

Thus, on multiplying, we see that the coefficients of $\phi(h)$ are maximal for $|\alpha| = |\beta| = \rho/r$ if $\alpha = \beta = \rho/r$. In this case they reduce to the coefficients of

$$r^{2-m}\left(1 - \frac{\rho}{r}h\right)^{2-m} = \sum_{\nu=0}^{\infty} b_{\nu, m}\, \rho^{\nu} r^{2-m-\nu} h^{\nu}.$$

This completes the proof of (4.1.3). Equality holds if $x = \lambda\xi$, where $\lambda > 0$.

Also $K(\xi - hx)$ is real analytic in h and $x_1, \dots, x_m$ jointly for $|h| < r/\rho$. Thus in any partial derivative the orders of differentiation can be reversed. In particular

$$\nabla^2 \left(\frac{\partial}{\partial h}\right)^{\nu} K(\xi - hx) = \left(\frac{\partial}{\partial h}\right)^{\nu} \nabla^2 K(\xi - hx) = 0,$$

since $K(\xi - hx)$ is harmonic in x_1 to x_m for fixed h. Thus all the derivatives $\phi^{(\nu)}(h)$ of $\phi(h)$ are harmonic in x_1 to x_m, and setting $h = 0$, we see that the polynomials $a_\nu(x, \xi)$ are harmonic. This completes the proof of Lemma 4.1.

4.1.1.

We now estimate $K_q(x, \xi)$.

LEMMA 4.2. *If* $|\xi| = r > 0$, *then* $K_q(x, \xi) - K(x, \xi)$ *is harmonic in* R^m. *We set*

$|x| = \rho$ *and have the following estimates*

(4.1.4) $$|K_q(x, \xi)| \leqslant 4^{m+q} \frac{\rho^{q+1}}{r^{m+q-1}} \text{ if } \rho \leqslant \tfrac{1}{2}r.$$

If $q = 0$, $m = 2$ *we have*

(4.1.5) $$K_0(x, \xi) \leqslant \log(1 + \rho/r),$$

while in all other cases

(4.1.6) $$K_q(x, \xi, m) \leqslant 4^{m+q} \frac{\rho^q}{r^{m+q-2}} \inf\left\{1, \frac{\rho}{r}\right\}.$$

We have if $m = 2$

$$K_0(x, \xi) = \log|z - \xi| - \log|\xi| = \log|1 - z/\xi| \leqslant \log(1 + |z|/|\xi|).$$

This proves (4.1.5). Next suppose that $\rho \leqslant \lambda r$, $0 \leqslant \lambda < 1$. Then, by (4.1.3)

$$|K_q(x, \xi)| = \left|\sum_{\nu=q+1}^{\infty} a_\nu(x, \xi)\right| \leqslant \sum_{\nu=q+1}^{\infty} b_{\nu,m}\rho^\nu r^{2-m-\nu}.$$

We note that $b_{\nu+1,m}/b_{\nu,m}$ decreases with increasing ν for $m \geqslant 3$. Thus for $\nu = q + l$, where $l > 0$,

$$b_{\nu,m} \leqslant b_{q+1,m}b_{l,m}/b_{1,m}.$$

Thus for $m \geqslant 3$

(4.1.7) $$|K_q(x, \xi)| \leqslant \frac{b_{q+1,m}}{b_{1,m}} \frac{\rho^q}{r^{m+q-2}} \sum_{l=1}^{\infty} b_{l,m}\left(\frac{\rho}{r}\right)^l$$
$$\leqslant \frac{(m + q - 2)!\,\{(1 - \lambda)^{2-m} - 1\}\,\rho^{q+1}}{\lambda(q + 1)!\,(m - 2)!\,r^{m+q-1}}.$$

In case $m = 2$, we obtain

(4.1.8) $$|K_q(x, \xi)| \leqslant \frac{1}{q + 1}\left(\frac{\rho}{r}\right)^q \sum_{l=1}^{\infty}\left(\frac{\rho}{r}\right)^l \leqslant \frac{1}{(q + 1)(1 - \lambda)}\left(\frac{\rho}{r}\right)^{q+1}.$$

We note that

$$\frac{(m + q - 2)!}{q!(m - 2)!}$$

is a coefficient in the binomial expansion of $(1 + 1)^{m+q-2}$, and so

$$\frac{(m + q - 2)!}{(q + 1)!\,(m - 2)!} \leqslant \frac{(m + q - 2)!}{q!(m - 2)!} \leqslant 2^{m+q-2}.$$

We set $\lambda = \frac{1}{2}$, and deduce from (4.1.7) that for $m \geqslant 3$

$$|K_q(x, \xi)| \leqslant 2^{m+q-2} 2^{m-1} \frac{\rho^{q+1}}{r^{m+q-1}} = 2^{2m+q-3} \frac{\rho^{q+1}}{r^{m+q-1}}.$$

For $m = 2$, $q \geqslant 1$, this remains valid by (4.1.8). This proves (4.1.4) and also (4.1.6) for $\rho \leqslant \frac{1}{2}r$.

Suppose next that $\rho > \frac{1}{2}r$. Then if $m = 2$, $q \geqslant 1$

$$K_q(x, \xi) \leqslant \log\left(1 + \frac{\rho}{r}\right) + \sum_{\nu=1}^{q} \frac{1}{\nu}\left(\frac{\rho}{r}\right)^{\nu} \leqslant 2\frac{\rho}{r} + \sum_{\nu=2}^{q}\left(\frac{\rho}{r}\right)^{\nu} \tag{4.1.9}$$

$$\leqslant \left(\frac{\rho}{r}\right)^q \left[1 + \left(\frac{r}{\rho}\right) + \dots + \left(\frac{r}{\rho}\right)^{q-2} + 2\left(\frac{r}{\rho}\right)^{q-1}\right]$$

$$\leqslant 2^{q+1}\left(\frac{\rho}{r}\right)^q.$$

If $m > 2$, we deduce from (4.1.3) that

$$K_q(x, \xi) \leqslant K(x - \xi) + \sum_{\nu=0}^{q} b_{\nu,m} \frac{\rho^\nu}{r^{m+\nu-2}} \leqslant \frac{b_{q,m}\rho^q}{r^{m-2+q}}(1 + 2 + 2^2 + \dots + 2^q)$$

$$\leqslant 2^{q+1} b_{q,m} \frac{\rho^q}{r^{m+q-2}}.$$

Also

$$\sum_{q=0}^{\infty} b_{q,m}\left(\frac{1}{2}\right)^q = \left(1 - \frac{1}{2}\right)^{2-m},$$

so that

$$b_{q,m} \leqslant 2^{q+m-2}.$$

Thus

$$K_q(x, \xi) \leqslant 2^{2q+m-1} \frac{\rho^q}{r^{m+q-2}},$$

In view of (4.1.9) this remains valid for $m = 2$. Thus we obtain (4.1.6) also in this case. This completes the proof of Lemma 4.2.

4.1.2.

We can now state and prove Weierstrass' Theorem.†

† For the case (0.1) see Weierstrass [1876].

THEOREM 4.1. *Suppose that μ is a Borel measure in R^m, let $n(t)$ be the measure of $D(0, t)$ and let $q(t)$ be a positive integer-valued increasing function of t, continuous on the right, and so chosen that*

$$\int_1^\infty \left(\frac{t_0}{t}\right)^{q(t)+m-1} \mathrm{d}n(t) < \infty \tag{4.1.10}$$

for all positive t_0. Then there exist functions $u(x)$, s.h. in R^m and with Riesz measure μ, and all such functions take the form

$$u(x) = \int_{|\xi|<1} K(x-\xi)\,\mathrm{d}\mu e_\xi + \int_{|\xi|\geq 1} K_{q(|\xi|)}(x, \xi)\,\mathrm{d}\mu e_\xi + v(x), \tag{4.1.11}$$

where $v(x)$ is harmonic in R^m. The second integral in (4.1.11) converges absolutely near ∞ and uniformly for $|x| \leqslant \rho$ and any fixed positive ρ.

It is evident that given $n(t)$, it is possible to choose $q(t)$, so as to satisfy (4.1.10). We may for instance take for $q(t)$ the integral part of $\log(n(t)+1)$. Then we have for $t \geqslant e^2 t_0$

$$\left(\frac{t_0}{t}\right)^{q(t)+m-1} \leqslant \exp\{-2\log(n(t)+1)\} = \frac{1}{(n(t)+1)^2}.$$

Thus for this choice of $q(t)$ (4.1.10) holds.

Suppose now that, with the hypotheses of Theorem 4.1,

$$q(t) = q = \text{constant}, \qquad t_1 \leqslant t < t_2,$$

where $1 \leqslant t_1 < t_2 < \infty$ and

$$u_q(x) = \int_{t_1 \leqslant |\xi| < t_2} K_q(x, \xi)\,\mathrm{d}\mu e_\xi.$$

Then $u_q(x)$ is s.h. in R^m and has Riesz measure μ in the set $t_1 \leqslant |\xi| < t_2$ and zero elsewhere. For this is true of the function

$$p(x) = \int_{t_1 \leqslant |\xi| < t_2} K(x-\xi)\,\mathrm{d}\mu e_\xi$$

by Theorem 3.7. Also

$$u_q(x) - p(x) = \sum_{\nu=0}^{q} \int_{t_1 \leqslant |\xi| < t_2} -a_\nu(x, \xi)\,\mathrm{d}\mu e_\xi.$$

In view of Lemma 4.1 the right-hand side is a uniform limit of harmonic polynomials of degree at most q and so is itself a harmonic polynomial of degree at most q. Also by Lemma 4.2 we have for $|x| \leqslant \rho$, $|\xi| = r$, if $1 \leqslant \rho \leqslant \frac{1}{2}t_1$,

$$|u_q(x)| \leqslant 4 \int_{t_1 \leqslant |\xi| < t_2} \left(\frac{4\rho}{r}\right)^{m+q-1} d\mu e_\xi$$

$$= 4 \int_{t_1 \leqslant t < t_2} \left(\frac{4\rho}{t}\right)^{m+q-1} dn(t).$$

We now consider two cases. Suppose first that $q(t)$ is unbounded. Then let $q(1) = q_1$, and for $q > q_1$, let r_q be the least value of r, such that $q(r) \geqslant q$. If $q(t)$ is bounded above and q_2 the maximum value of $q(t)$, we define r_q as before for $q \leqslant q_2$ and set $r_q = r_{q_2} + q - q_2$ for $q > q_2$. Then in either case $q(t)$ is constant for $r_q \leqslant t < r_{q+1}$ and

$$r_{q_1} = 1, \qquad r_q \to \infty, \qquad \text{as} \quad q \to \infty.$$

We now define

$$u_q(x) = \int_{r_q \leqslant |\xi| < r_{q+1}} K_{q(|\xi|)}(x_1 \xi)\, d\mu e_\xi.$$

Then the above analysis shows that for $|x| \leqslant \rho$, $1 \leqslant \rho < \frac{1}{2} r_q$, we have

$$|u_q(x)| < 4 \int_{r_q \leqslant t < r_{q+1}} \left(\frac{4\rho}{t}\right)^{q(t)+m-1} dn(t).$$

For fixed ρ this inequality thus holds for all sufficiently large q. Thus the series

$$\sum_{q=q_0}^{\infty} u_q(x)$$

converges uniformly and absolutely for $|x| \leqslant \rho$ in view of (4.1.10). The sum

$$\sum_{q_0}^{\infty} u_q(x)$$

is harmonic in $|x| \leqslant \rho$ if $r_{q_0} > 2\rho$, being a uniformly convergent sequence of harmonic functions. This follows easily from Theorem 1.11. Thus

$$u(x) = \int_{|\xi| < 1} K(x - \xi)\, d\mu e_\xi + \sum_{q=q_1}^{q_2 - 1} u_q(x) + \sum_{q_2}^{\infty} u_q(x)$$

is s.h. in $|x| \leqslant \rho$ and has Riesz measure μ there, as required. Finally if $u_1(x)$ is any other function with Riesz measure μ in R^m, then $v(x) = u(x) - u_1(x)$ is harmonic in any compact subset of R^m, and so in the whole of R^m, by Theorem 3.9. This completes the proof of Theorem 4.1.

4.2. HADAMARD'S REPRESENTATION THEOREM

Theorem 4.1 is not very useful in many practical cases in view of the arbitrary functions $q(t)$ and $v(x)$. The most interesting cases are those in which we can take q to be constant. We accordingly make the following

DEFINITION 4.1. *If $S(r)$ is any positive increasing function of r for $r \geqslant r_0$, then the "order" λ and "lower order" μ of $S(r)$ are defined by*

$$\lambda = \overline{\lim_{r\to\infty}} \frac{\log S(r)}{\log r}, \quad \mu = \underline{\lim_{r\to\infty}} \frac{\log S(r)}{\log r}.$$

if $0 < \lambda < \infty$, we also say that $S(r)$ has "minimal", "mean" or "maximal" type, according as

$$C = \overline{\lim_{r\to\infty}} \frac{S(r)}{r^\lambda}$$

is zero, finite and positive, or infinite. We say that $S(r)$ has "convergence" or "divergence" class according as

$$\int_{r_0}^{\infty} \frac{S(r)\,dr}{r^{\lambda+1}} \text{ converges or diverges.}$$

DEFINITION 4.2. *If $u(x)$ is* s.h. *in R^m and not bounded above we define the "order" and "type class" of $u(x)$ to be that of either of the functions $T(r, u)$ or $B(r, u)$ defined in Section* 3.9.

The functions $B(r, u)$ and $T(r, u)$ are increasing functions of r by Theorems 2.13 and 3.18 respectively. Thus we have only to show that the definitions in terms of $T(r, u)$ and $B(r, u)$ are equivalent. This is an almost immediate deduction from Theorem 3.19. In fact we have by that Theorem, provided that $u(x)$ is positive somewhere on $|x| = \rho$

(4.2.1) $$T(\rho, u) \leqslant B(\rho, u) \leqslant 3\cdot 2^{m-2}\, T(2\rho, u),$$

on setting $r = 2\rho$. From these inequalities we deduce the equality of the orders and lower orders of $T(r)$ and $B(r)$, and if λ is finite and positive we also deduce that $T(r)$ and $B(r)$ have minimal, mean and maximal type or convergence class and divergence class together. (For details see Hayman [1964, Theorem 1.7, p. 18].)

We also prove

LEMMA 4.3. *If $S(r)$ satisfies the hypotheses of Definition* (4.1) *and for some $k > 0$*

(4.2.2) $$\int_{r_0}^{\infty} \frac{S(r)\,dr}{r^{k+1}} < \infty,$$

then $\lambda \leqslant k$ and if $\lambda = k$, $S(r)$ has minimal type.

We say $S(r)$ has at most order k and minimal type. Conversely it is obvious that if $\lambda < k$ then (4.2.2) holds, since in this case we may choose $\varepsilon > 0$, so that $\lambda + \varepsilon < k$ and

$$S(r) < r^{\lambda+\varepsilon}, \qquad r > r_1.$$

To prove Lemma 4.3, suppose that (4.2.2) holds and choose r_1 so large that

$$\int_{r_1}^{\infty} \frac{S(r)\,dr}{r^{k+1}} < \varepsilon.$$

Since $S(r)$ increases with r we deduce that for all large r_1

$$S(r_1)\int_{r_1}^{\infty} \frac{dr}{r^{k+1}} < \varepsilon, \quad \text{i.e.} \quad S(r_1) < \frac{\varepsilon r_1^k}{k}.$$

Since ε may be chosen as small as we please, Lemma 4.3 is proved.

4.2.1.

We also need an estimate for the integrals in (4.1.11), when $q(t)$ is constant. This is

LEMMA 4.4. *Suppose that μ is a Borel measure in R^m, that $n(t)$ is the measure of $D(0, t)$, that $N(r)$, defined by (3.9.5) has at most convergence class of order $q + 1$, and that $n(1) = 0$. Then if*

$$u(x) = \int_{|\xi| \geq 1} K_q(x, \xi)\, d\mu e_\xi, \tag{4.2.3}$$

the integral converges absolutely at infinity and uniformly for $|x| \leq \rho$, and we have, for $\rho \geq 1$,

$$u(x) \leq 4^{m+q}(q+2)\left\{q\rho^q \int_1^{\rho} \frac{N(r)\,dr}{r^{q+1}} + (q+1)\rho^{q+1}\int_{\rho}^{\infty} \frac{N(r)\,dr}{r^{q+2}}\right\}. \tag{4.2.4}$$

In view of Theorem 4.1, to prove the absolute convergence of the integral in (4.2.3) it is sufficient to show that (4.1.10) holds, i.e. that

$$\int_1^{\infty} \frac{1}{t^{q+m-1}}\, dn(t) < \infty. \tag{4.2.5}$$

Now if $R > 1$

$$\int_1^R \frac{dn(t)}{t^{q+m-1}} = \frac{n(R)}{R^{q+m-1}} + (q+m-1)\int_1^R \frac{n(t)\,dt}{t^{q+m}}, \tag{4.2.6}$$

since $n(1) = 0$. Similarly, since $N(1) = 0$,

(4.2.7)
$$\int_1^R \frac{n(t)\,\mathrm{d}t}{t^{q+m}} = \frac{1}{d_m}\int_1^R \frac{\mathrm{d}N(t)}{t^{q+1}} = \frac{1}{d_m}\frac{N(R)}{R^{q+1}} + (q+1)\int_1^R \frac{N(t)\,\mathrm{d}t}{t^{q+2}}.$$

By hypothesis $N(r)$ has at most convergence class and so at most minimal type of order $q+1$ by Lemma 4.3. Thus the right-hand side of (4.2.7) is bounded as $R \to \infty$ and hence so is the left-hand side. Thus $n(r)$ has at most convergence class of order $q+m-1$, and so at most minimal type of that order. Thus (4.2.5) holds, and $u(x)$, defined by (4.2.3) is s.h. in R^m with Riesz measure μ, in view of Theorem 4.1.

It remains to prove (4.2.4). We suppose $\rho \geqslant 1$, and deduce from (4.1.6) that for $|x| = \rho$, we have unless $q = 0$ and $m = 2$

$$u(x) = \int_{|\xi|\geqslant 1} K_q(x, \xi)\,\mathrm{d}\mu e_\xi$$
$$\leqslant 4^{m+q}\left\{\int_1^\rho \frac{\rho^q}{r^{m+q-2}}\mathrm{d}n(r) + \int_\rho^\infty \frac{\rho^{q+1}\,\mathrm{d}n(r)}{r^{m+q-1}}\right\}.$$

We integrate by parts twice as in (4.2.6) and (4.2.7). Then unless $q = 0$ and $m = 2$, we have

$$\int_1^\rho \frac{\rho^q\,\mathrm{d}n(r)}{r^{m+q-2}} + \int_\rho^\infty \frac{\rho^{q+1}\,\mathrm{d}n(r)}{r^{m+q-1}}$$
$$= (q+m-2)\rho^q \int_1^\rho \frac{n(r)\,\mathrm{d}r}{r^{m+q-1}} + (q+m-1)\rho^{q+1}\int_\rho^\infty \frac{n(r)\,\mathrm{d}r}{r^{m+q}}$$
$$\leqslant \frac{(q+m-1)}{d_m}\left\{\rho^q \int_1^\rho \frac{\mathrm{d}N(r)}{r^q} + \rho^{q+1}\int_\rho^\infty \frac{\mathrm{d}N(r)}{r^{q+1}}\right\}$$
$$= \frac{(q+m-1)}{d_m}\left\{q\rho^q \int_1^\rho \frac{N(r)\,\mathrm{d}r}{r^{q+1}} + (q+1)\rho^{q+1}\int_\rho^\infty \frac{N(r)\,\mathrm{d}r}{r^{q+2}}\right\}.$$

Since $d_m \geqslant m-2$, this yields (4.2.4). If $q = 0$, $m = 2$, we deduce similarly from (4.1.5) that

$$u(x) \leqslant \int_1^\infty \log\left(1 + \frac{\rho}{r}\right)\mathrm{d}n(r) = \rho\int_1^\infty \frac{n(r)\,\mathrm{d}r}{r(r+\rho)}$$
$$\leqslant \int_1^\rho \frac{n(r)\,\mathrm{d}r}{r} + \rho\int_\rho^\infty \frac{n(r)\,\mathrm{d}r}{r^2} = \rho\int_\rho^\infty \frac{N(r)\,\mathrm{d}r}{r^2}.$$

Thus (4.2.4) still holds, and the inequality is true in all cases for $|x| = \rho$. The inequality remains true for $|x| < \rho$ by the maximum principle.

4.2.2.

We can now state and prove Hadamard's Theorem.

THEOREM† 4.2. *Suppose that* μ *is a Borel measure in* R^m, *that* $n(t)$ *denotes the measure of the set* $1 \leqslant |x| < t$ *and that* $N(r)$ *is defined by* (3.9.5). *Then if* $N(r)$ *has at most order* $q+1$, *convergence class, we may take* $q(\xi) = q =$ *constant in Theorem* 4.1. *In particular if* $u(x)$ *has at most order* $q+1$, *convergence class, then in addition* $v(x)$ *is a harmonic polynomial of degree at most* q.

With the hypotheses of Theorem 4.2 we may apply Lemma 4.4 to

$$u_0(x) = u(x) - \int_{|\xi|<1} K(x-\xi)\, d\mu e_\xi,$$

and deduce that we can take $q(\xi) = q =$ constant in Theorem 4.1. Suppose next that $u(x)$ has at most order $q+1$, convergence class. If $m = 2, |x| = \rho > 2$

$$\int_{|\xi|<1} K(x-\xi)\, d\mu e_\xi = \int_{|\xi|<1} \log|x-\xi|\, d\mu e_\xi \geqslant 0.$$

If $m > 2$

$$\int_{|\xi|<1} K(x-\xi)\, d\mu e_\xi \to 0 \quad \text{as} \quad x \to \infty.$$

Thus in either case the order of $u_0(x)$ cannot exceed that of $u(x)$. Thus $u_0(x)$ has at most order $q+1$, convergence class and hence so does $N(r)$, in view of (3.9.6). Thus we may apply our previous conclusion.

It remains to estimate $v(x)$ in this case. We have

$$v(x) = u_0(x) - \int_{|\xi|\geqslant 1} K_q(x, \xi)\, d\mu e_\xi.$$

Suppose now that $|x| = \rho > 1$, and set

$$u_1(x) = \int_{|\xi|\geqslant 1} K_q(x, \xi)\, d\mu e_\xi.$$

Then it follows from (4.2.4) that, for $|x| = \rho > 1$,

$$u_1(x) \leqslant \rho^q \int_1^\rho \frac{o(r^{q+1})\, dr}{r^{q+1}} + \rho^{q+1} o(1) = o(\rho^{q+1}).$$

† Hadamard [1893] proved this in the case (0.1). For plane s.h. functions the result is due to Heins [1948] for $q = 0$ and Kennedy [1956] in general.

Thus, with the notation of Section 3.9, we have

$$m(r, v) = \frac{1}{c_m r^{m-1}} \left\{ \int_{S(0,r)} v^-(x)\, d\sigma(x) \right\}$$

$$\leqslant \frac{1}{c_m r^{m-1}} \left\{ \int_{S(0,r)} u_0^-(x)\, d\sigma(x) + \int_{S(0,r)} u_1^+(x)\, d\sigma(x) \right\}$$

$$= m(r, u_0) + T(r, u_1) = o(r^{q+1}).$$

Since $v(x)$ is harmonic $N(r, v) = 0$, so that by (3.9.6)

$$T(r, v) = o(r^{q+1})$$

and hence in view of (4.2.1)

$$B(r, v) = o(r^{q+1}), \qquad B(r, -v) = o(r^{q+1}).$$

It now follows (for instance from Example 3 of Section 1.5.6) that $v(x)$ is a polynomial of degree at most q. This concludes the proof of Theorem 4.2.

4.3. RELATIONS BETWEEN $T(r)$ AND $B(r)$

We are now in a position to investigate the orders of magnitude of the various functions which we have used to estimate the growth of s.h. functions in space. We have clearly

$$T(r, u) \leqslant B(r, u)$$

unless the right-hand side is negative. This will happen for large r only if $m > 2$ and u takes the form given in Theorem 3.20. In the opposite direction we can prove.†

THEOREM 4.3. *If $u(x)$ is s.h. of finite lower order μ in R^m, then*

$$\varliminf_{r\to\infty} \frac{B(r, u)}{T(r, u)} \leqslant K(\mu, m),$$

where $K(\mu, m) \leqslant 3\{2e(\mu - 1)/(m - 1)\}^{m-1}$, $\mu \geqslant m$, $K(\mu, m) \leqslant (2me/\mu)^{\mu}$, $0 < \mu < m$ *and* $K(0, m) = 1$.

We have by Theorem 3.19 for $r = \lambda\rho$, where $\lambda > 1$

$$B(\rho) \leqslant \frac{\lambda^{m-2}(1 + \lambda)}{(\lambda - 1)^{m-1}} T(r).$$

† For the best possible form of this result see Govorov [1969] and Petrenko [1969], when $u = \log|f|$, where f is entire, and Dahlberg [1972] for the general case.

Suppose now that $\alpha > \mu$ and that we have for all sufficiently large positive integers $k \geqslant k_0$

$$T(\lambda^k) \geqslant \lambda^\alpha T(\lambda^{k-1}).$$

Then we should have for all $k \geqslant k_0$

$$T(\lambda^k) \geqslant \lambda^{\alpha(k-k_0)} T(\lambda^{k_0}).$$

If $\lambda^k \leqslant \rho \leqslant \lambda^{k+1}$, we deduce that

$$T(\rho) \geqslant T(\lambda^k) \geqslant \lambda^{\alpha(k+1)} \frac{T(\lambda^{k_0})}{\lambda^{\alpha(k_0+1)}} \geqslant \rho^\alpha \frac{T(\lambda^{k_0})}{\lambda^{\alpha(k_0+1)}} = A\rho^\alpha,$$

say. This holds for all sufficiently large ρ, giving a contradictión to our assumption that $u(x)$ has lower order $\mu < \alpha$. Thus we must have for some arbitrarily large $\rho = \lambda^k$

$$T(\lambda\rho) \leqslant \lambda^\alpha T(\rho).$$

Thus we obtain for such values of λ

$$B(\rho) \leqslant \frac{\lambda^{\alpha+m-2}(1+\lambda)}{(\lambda-1)^{m-1}} T(\rho).$$

Hence

$$\varliminf_{r\to\infty} \frac{B(r)}{T(r)} \leqslant \frac{\lambda^{\alpha+m-2}(1+\lambda)}{(\lambda-1)^{m-1}}.$$

Since α may be any number greater than μ, we may replace α by μ. Suppose first that $\mu \geqslant m$. Then we set $\lambda = (\alpha + m - 2)/(\alpha - 1)$. In this case

$$\frac{\lambda^{\alpha+m-2}(1+\lambda)}{(\lambda-1)^{m-1}} = \frac{(\alpha+m-2)^{\alpha+m-2}(2\alpha+m-3)}{(m-1)^{m-1}(\alpha-1)^\alpha}$$
$$\leqslant 3\left(\frac{\alpha-1+m-1}{m-1}\right)^{m-1}\left(\frac{\alpha-1+m-1}{\alpha-1}\right)^{\alpha-1}$$
$$\leqslant 3\left\{\frac{2e(\alpha-1)}{m-1}\right\}^{m-1}.$$

Next if $\mu < m$, we set $\lambda = 1 + m/\alpha$. Then

$$\frac{\lambda^{\alpha+m-2}(1+\lambda)}{(\lambda-1)^{m-1}} = \frac{(\alpha+m)^{\alpha+m-2}(2\alpha+m)}{m^{m-1}\alpha^\alpha} \leqslant \frac{(\alpha+m)^{\alpha+m}}{m^m\alpha^\alpha}$$
$$= \left(1+\frac{\alpha}{m}\right)^m \cdot \left(\frac{\alpha+m}{\alpha}\right)^\alpha \leqslant \left(\frac{2me}{\alpha}\right)^\alpha.$$

For $\mu > 0$ we may set $\alpha = \mu$ in the above inequalities, and Theorem 4.3 is proved. For $\mu = 0$, we may take α as small as we please and since

$$\left(\frac{2me}{\alpha}\right)^{\alpha} \to 1 \quad \text{as} \quad \alpha \to 0,$$

the result follows also in this case.

4.3.1. Two examples

The case $\mu = 0$ of Theorem 4.3 yields the sharp relation

$$\varliminf_{r\to\infty} \frac{B(r, u)}{T(r, u)} = 1.$$

In other cases the inequality for $K(\mu, m)$ is not sharp. It appears however that $K(\mu, m)$ has the correct order of magnitude as $\mu \to \infty$ for fixed m. For instance if $m = 2$, we may set $x_1 + ix_2 = \rho e^{i\theta}$, suppose $\alpha \geqslant \frac{1}{2}$ and set

$$u(x_1, x_2) = \rho^{\alpha} \cos \alpha\theta, \qquad |\theta| < \frac{\pi}{2\alpha},$$

$$= 0, \qquad |\theta| > \frac{\pi}{2\alpha}$$

The resulting function is clearly s.h., of order α and

$$B(\rho, u) = \rho^{\alpha}$$

$$T(\rho, u) = \frac{1}{\pi} \int_0^{\pi/(2\alpha)} \rho^{\alpha} \cos \alpha\theta \, d\theta = \frac{\rho^{\alpha}}{\pi\alpha}.$$

Thus in this case

$$\frac{B(\rho, u)}{T(\rho, u)} = \pi\alpha.$$

We can also construct corresponding examples for $m > 2$. To do this we recall the analysis of Section 2.6.3. We set

$$\rho^2 = \sum_2^m x_n^2, \qquad R^2 = \sum_1^m x_n^2,$$

$$x_1 = R \cos \theta, \qquad \rho = R \sin \theta,$$

$$u = R^{\alpha}\phi(\theta).$$

Then $\nabla^2 u = R^{\alpha-2}[\phi''(\theta) + (m-2)\phi'(\theta) \cot \theta + \alpha(\alpha + m - 2)\phi(\theta)]$.

We set $\lambda > 2$ and

$$\phi(\theta) = 1 + \cos\lambda\theta, \qquad |\theta| \leqslant \pi/\lambda,$$
$$\phi(\theta) = 0, \qquad \pi/\lambda \leqslant |\theta| \leqslant \pi.$$

The resulting function u is then clearly s.h. in R^m, provided that

$$L(\phi) = \phi''(\theta) + (m-2)\phi'(\theta)\cot\theta + \alpha(\alpha+m-2)\phi(\theta) > 0, \quad 0 < \theta < \pi/\lambda. \tag{4.3.1}$$

We proceed to show that (4.3.1) holds if α is sufficiently large. In fact suppose that $\pi - \delta \leqslant \lambda\theta < \pi$, where $\delta = \pi/[4(m-2)]$. Then

$$\phi''(\theta) + (m-2)\phi'(\theta)\cot\theta = -\lambda^2\left[\cos\lambda\theta + (m-2)\frac{\sin\lambda\theta}{\lambda\sin\theta}\cos\theta\right]$$
$$> \lambda^2[\cos\delta - (m-2)\sin\delta] > 0,$$

since

$$\tan\delta < \frac{4}{\pi}\delta = \frac{1}{(m-2)}.$$

Thus $L(\phi) > 0$ in this range. On the other hand if

$$0 < \lambda\theta \leqslant \pi - \delta$$

we have

$$|\phi''(\theta) + (m-2)\phi'(\theta)\cot\theta| \leqslant (m-1)\lambda^2,$$

$$\alpha(\alpha+m-2)\phi(\theta) \geqslant \alpha^2(1-\cos\delta) = 2\alpha^2\sin^2\frac{\delta}{2} \geqslant \frac{2\alpha^2\delta^2}{\pi^2} \geqslant \frac{\alpha^2}{8(m-2)^2}.$$

Thus $L(\phi) > 0$ in the whole range provided that

$$\alpha^2 > 8(m-2)^2(m-1)\lambda^2.$$

In particular we may take $\alpha = 3\lambda m^{3/2}$. The resulting function is s.h. and has order α in R^m. However only the part $|\theta| \leqslant \pi/\lambda$ of each sphere $S(0, R)$ contributes to $T(R, u)$, since $u = 0$ elsewhere. The surface area of this region is less than $A(m)(\pi R/\lambda)^{m-1}$. Thus

$$T(R, u) \leqslant B(R,u)\frac{1}{c_m}A(m)\left(\frac{\pi}{\lambda}\right)^{m-1} \leqslant A_1(m)\,B(R, u)\alpha^{1-m}.$$

Here $A_1(m)$ is a constant depending on m only. Thus $K(\mu, m)$ has the right order of magnitude in Theorem 4.3 as $\mu \to \infty$ for fixed m.

4.4. RELATIONS BETWEEN $N(r)$ AND $T(r)$

It is well known that integral functions of finite order λ, which is not a positive integer, assume all values (and in particular zero) infinitely often, The corresponding result for s.h. functions is that the only such functions which are harmonic and of finite order in R^m are the harmonic polynomials, whose order is equal to their degree. This is an immediate consequence of Theorem 4.2. We can however also extend the subtler theorems which concern the minimum number of zeros which integral functions of non-integral or zero order must have. The following result is very simple.

THEOREM 4.4. *Suppose that* $u(x)$ *is* s.h. *and of order* λ *in* R^m, *where* λ *is finite, positive and non-integral. Then the function* $N(r, u)$ *also has order* λ *and the same type-class (maximal, minimal or mean type) as* $u(x)$.

We suppose that $q < \lambda < q + 1$, where q is a positive integer or zero, and further that

$$N(r, u) < cr^{\lambda_1}, \qquad r_0 < r < \infty, \tag{4.4.1}$$

where $c > 0$ and $q < \lambda_1 \leqslant \lambda$. Then by Theorem 4.1

$$u(x) = \int_{|\xi| \geqslant 1} K_q(x, \xi)\, d\mu e_\xi + v(x),$$

where for $|x| = r$

$$v(x) = O(r^q) + \int_{|\xi| < 1} K(x - \xi)\, d\mu e_\xi.$$

The last term is negative if $m > 2$, and reduces to

$$\int_{|\xi| < 1} \log|x - \xi|\, d\mu e_\xi = O(\log|x|), \quad \text{as} \quad x \to \infty,$$

if $m = 2$. Thus in any case,

$$u(x) \leqslant \int_{|\xi| \geqslant 1} K_q(x, \xi)\, d\mu e_\xi + o(r^{\lambda_1}).$$

Again by Lemma 4.4 we have for $\rho > 1$, $|x| = \rho$

$$u(x) \leqslant A(q, m)\left\{\rho^q \int_1^\rho \frac{N(r)\, dr}{r^{q+1}} + \rho^{q+1} \int_\rho^\infty \frac{N(r)\, dr}{r^{q+2}}\right\}$$

$$\leqslant A(q, m)\left\{c\rho^q \int_1^\rho r^{\lambda_1 - q - 1}\, dr + c\rho^{q+1} \int_\rho^\infty r^{\lambda_1 - q - 2}\, dr + o(\rho^{\lambda_1})\right\}$$

in view of (4.4.1). Thus

$$B(\rho, u) \leqslant A(q, m)c\rho^{\lambda_1}\left[\frac{1}{\lambda_1 - q} + \frac{1}{q + 1 - \lambda_1} + o(1)\right].$$

This gives a contradiction unless $\lambda_1 = \lambda$, so that $N(r)$ cannot have smaller order than $B(r, u)$. Also we obtain a contradiction if $u(x)$ has maximal type, so that in this case $N(r)$ must have maximal type. If $N(r)$ has minimal type, we may take $\lambda_1 = \lambda$ and c as small as we please in (4.4.1) and deduce that $B(r, u)$ and so $u(x)$ also has minimal type. Since the order and type class of $N(r)$ cannot exceed those of $B(r)$ and $T(r)$ by (3.9.6), we have proved Theorem 4.4.

We can prove a result which is in some ways stronger and in order to do this we define the *deficiency* $\delta(u)$ of u by the equation

$$\delta(u) = \varliminf_{r\to\infty} \frac{m(r, u)}{T(r, u)} = 1 - \varlimsup_{r\to\infty} \frac{N(r, u)}{T(r, u)}. \tag{4.4.2}$$

If $m = 2$ and $u(z) = \log|f(z)|$, where $f(z)$ is an integral function, then $\delta(u)$ reduces to the deficiency of the value zero with respect to $f(z)$ in Nevanlinna Theory. We proceed to prove

THEOREM 4.5. *Suppose that $u(x)$ is* s.h. *in R^m, of finite order λ where $q < \lambda < q + 1$. Then*†

$$\delta(u) \leqslant A(m, \lambda) = 1 - \frac{(q + 1 - \lambda)(\lambda - q)}{\lambda(q + 2)4^{m+q}}, \qquad \lambda \neq 0.$$

We need a lemma on Pólya peaks. (For a fuller account of this concept, which will deal also with lower orders, see Edrei [1969]).

LEMMA 4.5. *Suppose that $\phi_1(r)$, $\phi_2(r)$, $\phi(r)$ are positive continuous functions of r for $r_0 < r < \infty$, such that $\phi_2(r)/\phi_1(r)$ increases with r and*

$$\varlimsup_{r\to\infty} \frac{\phi(r)}{\phi_1(r)} = \infty, \qquad \varliminf_{r\to\infty} \frac{\phi(r)}{\phi_2(r)} = 0. \tag{4.4.3}$$

Then there exists a sequence $r = r_n$, tending to infinity with n, such that

$$\frac{\phi(\rho)}{\phi_1(\rho)} \leqslant \frac{\phi(r)}{\phi_1(r)}, \; r_0 \leqslant \rho \leqslant r, \qquad \frac{\phi(\rho)}{\phi_2(\rho)} \leqslant \frac{\phi(r)}{\phi_2(r)}, \; r \leqslant \rho < \infty. \tag{4.4.4}$$

† This inequality is not sharp. For $0 \leqslant \lambda < 1$, we prove the corresponding sharp bounds in the next section. If $\lambda > 1$ the best known bounds are due to Miles and Shea [1973] if $m = 2$, and Rao and Shea [1976] if $m > 2$.

We wish to construct r satisfying (4.4.4) and such that $r \geqslant C$, where C is a preassigned positive number. To do this let

$$M_1 = \sup_{r_0 \leqslant r \leqslant C} \frac{\phi(r)}{\phi_1(r)}$$

and let R_1 be the smallest number such that $R_1 \geqslant C$ and

$$\frac{\phi(R_1)}{\phi_1(R_1)} = M_1.$$

Since (4.4.3) holds, R_1 must exist. Let R_2 be such that $R_2 \geqslant R_1$ and

$$\frac{\phi(R_2)}{\phi_2(R_2)} = \sup_{R \geqslant R_1} \frac{\phi(R)}{\phi_2(R)},$$

and let R_3 be such that $R_1 \leqslant R_3 \leqslant R_2$ and

$$\frac{\phi(R_3)}{\phi_1(R_3)} = \sup_{R_1 \leqslant R \leqslant R_2} \frac{\phi(R)}{\phi_1(R)}. \tag{4.4.5}$$

Then

$$\frac{\phi(R_3)}{\phi_1(R_3)} \geqslant \frac{\phi(R_1)}{\phi_1(R_1)} = M_1 \geqslant \frac{\phi(r)}{\phi_1(r)}, \qquad r_0 \leqslant r \leqslant R_1.$$

Also we have by construction for $R_1 \leqslant R \leqslant R_3$

$$\frac{\phi(R_3)}{\phi_1(R_3)} \geqslant \frac{\phi(R)}{\phi_1(R)}. \tag{4.4.6}$$

Thus (4.4.6) holds for $r_0 \leqslant R \leqslant R_3$. Again we have for $R_3 \leqslant R \leqslant R_2$

$$\frac{\phi(R)}{\phi_2(R)} = \frac{\phi(R)}{\phi_1(R)} \cdot \frac{\phi_1(R)}{\phi_2(R)} \leqslant \frac{\phi(R_3)}{\phi_1(R_3)} \cdot \frac{\phi_1(R_3)}{\phi_2(R_3)} = \frac{\phi(R_3)}{\phi_2(R_3)},$$

in view of (4.4.5) and the fact that ϕ_2/ϕ_1 increases. Also for $R \geqslant R_2$

$$\frac{\phi(R)}{\phi_2(R)} \leqslant \frac{\phi(R_2)}{\phi_2(R_2)} \leqslant \frac{\phi(R_3)}{\phi_2(R_3)}.$$

Thus we have for $R \geqslant R_3$

$$\frac{\phi(R)}{\phi_2(R)} \leqslant \frac{\phi(R_3)}{\phi_2(R_3)}.$$

Thus $r = R_3$ has the required properties and Lemma 4.5 is proved.

We can now complete the proof of Theorem 4.5. We suppose that q is an integer, $q \geqslant 0$, $q < \lambda < q + 1$, and we apply Lemma 4.5 with $\phi_1(r) = r^{\lambda - \varepsilon}$, $\phi_2(r) = r^{\lambda + \varepsilon}$, $\phi(r) = B(r, u)$. We assume that

$$N(r, u) \leqslant KB(r, u), \qquad r \geqslant r_0$$

and shall obtain a contradiction if K is sufficiently small. Let r be a large positive number satisfying the conclusions (4.4.4) of Lemma 4.5. Then we deduce from Theorem 4.2 and Lemma 4.4 that we have for $|x| = r$

$$u(x) \leqslant 4^{m+q}(q + 2)K\left\{qr^q \int_1^r \frac{B(\rho)\,d\rho}{\rho^{q+1}} + (q + 1)r^{q+1} \int_r^\infty \frac{B(\rho)\,d\rho}{\rho^{q+2}} + o(r^{\lambda-\varepsilon})\right\}.$$

We choose x so that $u(x) = B(r)$ and use Lemma 4.5. This gives

$$\begin{aligned} B(r) &\leqslant 4^{m+q}(q + 2)K\, B(r)\left\{qr^q \int_1^r \left(\frac{\rho}{r}\right)^{\lambda-\varepsilon} \frac{d\rho}{\rho^{q+1}} \right. \\ &\qquad \left. + (q + 1)r^{q+1} \int_r^\infty \left(\frac{\rho}{r}\right)^{\lambda+\varepsilon} \frac{d\rho}{\rho^{q+2}}\right\} + o(r^{\lambda-\varepsilon}) \\ &= 4^{m+q}(q + 2)K\, B(r)\left\{\frac{q}{\lambda - \varepsilon - q} + \frac{q + 1}{q + 1 - \lambda - \varepsilon}\right\} + o(r^{\lambda-\varepsilon}). \end{aligned}$$

Also in view of (4.4.4) we have for a suitable fixed r_0

$$\frac{B(r)}{r^{\lambda-\varepsilon}} \geqslant \frac{B(r_0)}{r_0^{\lambda-\varepsilon}} = C_1,$$

say. Thus we may divide by $B(r)$, let $r \to \infty$ and obtain

$$1 \leqslant 4^{m+q}(q + 2)K\left\{\frac{q}{\lambda - \varepsilon - q} + \frac{q + 1}{q + 1 - \lambda - \varepsilon}\right\}.$$

Since ε may be chosen as small as we please, we may set $\varepsilon = 0$. This gives

$$K \geqslant \frac{(q + 1 - \lambda)(\lambda - q)}{\lambda(q + 2)4^{m+q}}.$$

Thus

$$\overline{\lim_{r\to\infty}} \frac{N(r)}{T(r)} \geqslant \overline{\lim_{r\to\infty}} \frac{N(r)}{B(r)} \geqslant \frac{(q + 1 - \lambda)(\lambda - q)}{\lambda(q + 2)4^{m+q}}.$$

This proves Theorem 4.5.

We note that $A(m, \lambda) \to 1$, as λ approaches any positive integer. However $A(m, \lambda)$ remains bounded above by a constant less than one as $\lambda \to 0$. In fact if $\lambda = 0$, we can carry out the above analysis as before and obtain $A(m, 0) = 1 - \frac{1}{2}4^{-m}$. We can improve the above analysis considerably and obtain some sharp estimates for functions of order less than one and this we proceed now to do.

4.5. FUNCTIONS OF ORDER LESS THAN ONE

We start off by proving a sharp form of the estimate (4.2.4). This is

LEMMA 4.6. *Suppose that $u(x)$ is s.h., at most of order one convergence class in R^m and finite at the origin. Then we have for all x*

$$u(x) = u(0) + \int_{|\xi|<\infty} K_0(x, \xi)\, d\mu e_\xi. \tag{4.5.1}$$

Thus we have for $|x| = \rho$

$$u(x) \leqslant u(0) + (m-1)\rho \int_0^\infty \frac{t^{m-2} N(t)\, dt}{(\rho + t)^m}. \tag{4.5.2}$$

Equality holds if $u(\xi)$ is harmonic except on the ray $\xi = -\lambda x$, $0 \leqslant \lambda < \infty$.

We have by Theorem 4.2 in this case

$$u(x) = \int_{|\xi|<1} K(x-\xi)\, d\mu e_\xi + \int_{|\xi|\geqslant 1} K_0(x, \xi)\, d\mu e_\xi + \text{constant}. \tag{4.5.3}$$

Next since $u(0)$ is finite, it follows from (3.9.4) and (3.9.5) that

$$N(1, u) = \int_{D(0,1)} g(0, \xi)\, d\mu e_\xi < \infty,$$

where

$$g(0, \xi) = \log \frac{1}{|\xi|}, \qquad m = 2,$$

$$g(0, \xi) = |\xi|^{2-m} - 1, \qquad m > 2.$$

Since

$$K_0(x, \xi) = K(x - \xi) + g(0, \xi) + \varepsilon,$$

where $\varepsilon = 0$, if $m = 2$ and $\varepsilon = 1$, if $m > 2$, we deduce that (4.5.3) may be written in this case as

$$u(x) = \int K_0(x, \xi)\, d\mu e_\xi + \text{constant}.$$

Since $K_0(0, \xi) = 0$, the constant must be $u(0)$ and this gives (4.5.1).

We now note that for $|x| = \rho$, $|\xi| = t$

$$K_0(x, \xi) = \log\left|\frac{x-\xi}{t}\right| \leqslant \log\left(1 + \frac{\rho}{t}\right), \qquad m = 2 \tag{4.5.4}$$

$$K_0(x, \xi) = |\xi|^{2-m} - |x-\xi|^{2-m} \leqslant t^{2-m} - (\rho + t)^{2-m}, \qquad m > 2. \tag{4.5.5}$$

Suppose first that $m = 2$. Then we deduce that

$$u(x) - u(0) \leqslant \int_1^\infty \log\left(1 + \frac{\rho}{t}\right) dn(t)$$

$$= \lim_{\substack{\varepsilon \to 0 \\ R \to \infty}} \left\{\rho \int_\varepsilon^R \frac{n(t)\, dt}{t(t+\rho)} + \left[n(t) \log\left(1 + \frac{\rho}{t}\right)\right]_\varepsilon^R\right\}.$$

The term correspondlng to $t = R$ goes out, since $u(x)$ has at most order one minimal type, so that

$$\frac{n(R)}{R} \to 0, \quad \text{as} \quad R \to \infty.$$

Similarly since $u(0)$ and so $N(r)$ is finite for every r

$$n(t) \log(1/t) \to 0 \quad \text{as} \quad t \to 0.$$

Thus

$$u(x) - u(0) \leqslant \rho \int_0^\infty \frac{n(t)\, dt}{t(t+\rho)} = \rho \int_0^\infty \frac{dN(t)}{(t+\rho)} = \rho \int_0^\infty \frac{N(t)\, dt}{(t+\rho)^2},$$

by a similar argument. Equality holds if equality holds in (4.5.4) for all t for which the measure μ is non-zero, i.e. if all the measure is concentrated at points of the form $-\lambda x$, where λ is positive. This proves Lemma 4.6 if $m = 2$. If $m > 2$ we deduce similarly

$$u(x) - u(0) \leqslant \int_0^\infty \{t^{2-m} - (\rho + t)^{2-m}\}\, dn(t)$$

$$= (m-2) \int_0^\infty \{t^{1-m} - (\rho + t)^{1-m}\} n(t)\, dt$$

$$= \int_0^\infty \left\{1 - \left(\frac{t}{\rho + t}\right)^{m-1}\right\} dN(t)$$

$$= (m-1)\rho \int_0^\infty \frac{t^{m-2} N(t)\, dt}{(\rho + t)^m}.$$

Equality holds under the same conditions as before. This completes the proof of Lemma 4.6.

4.5.1. A sharp inequality connecting $N(r)$ and $B(r)$

We can now prove†

THEOREM 4.6. *Suppose that $u(x)$ is s.h. of order λ less than one in R^m and not bounded above in R^m and that $u(0)$ is finite. Then*

$$\varlimsup_{r\to\infty} \frac{N(r)}{B(r)} \geqslant \frac{(m-2)!\sin\pi\lambda}{\pi\lambda(\lambda+1)\dots(\lambda+m-2)} = \frac{\Gamma(m-1)\,\Gamma(\lambda)\sin\pi\lambda}{\pi\Gamma(m+\lambda-1)}, \tag{4.5.6}$$

where the right-hand side must be interpreted to be one if $\lambda = 0$. Equality holds if $u(x)$ is harmonic except on a ray through the origin and $N(r) = r^\lambda$, $0 < r < \infty$, if $\lambda > 0$, (and for all functions $u(x)$ if $\lambda = 0$).

We apply Lemma 4.5, with $\phi_1(r) = r^{\lambda-\varepsilon}$, $\phi_2(r) = r^{\lambda+\varepsilon}$, $\phi(r) = B(r)$, and suppose that ρ is chosen so that (4.4.4) holds, with r, ρ interchanged. Suppose now that

$$N(r) \leqslant CB(r), \qquad r > r_0.$$

We choose x so that $|x| = \rho$, $u(x) = B(\rho)$ and deduce from (4.5.2) that

$$\begin{aligned}
B(\rho) &\leqslant u(0) + (m-1)\rho\int_0^\infty \frac{t^{m-2}N(t)\,dt}{(\rho+t)^m} \\
&\leqslant (m-1)\rho\, C\int_0^\infty \frac{t^{m-2}B(t)\,dt}{(\rho+t)^m} + O(1) \\
&\leqslant O(1) + (m-1)\rho C\, B(\rho)\left\{\int_0^\rho \frac{t^{m-2}(t/\rho)^{\lambda-\varepsilon}\,dt}{(\rho+t)^m} + \int_\rho^\infty \frac{t^{m-2}(t/\rho)^{\lambda+\varepsilon}\,dt}{(\rho+t)^m}\right\} \\
&= O(1) + (m-1)C\, B(\rho)\left\{\int_0^1 \frac{x^{m-2+\lambda-\varepsilon}\,dx}{(1+x)^m} + \int_1^\infty \frac{x^{m-2+\lambda+\varepsilon}\,dx}{(1+x)^m}\right\}.
\end{aligned}$$

We divide by $B(\rho)$, which tends to infinity with ρ. Thus

$$1 \leqslant (m-1)C\left\{\int_0^1 \frac{x^{m-2+\lambda-\varepsilon}\,dx}{(1+x)^m} + \int_1^\infty \frac{x^{m-2+\lambda+\varepsilon}\,dx}{(1+x)^m}\right\}.$$

The integrals are clearly continuous functions of ε, when ε is small, and so we may set $\varepsilon = 0$ and obtain

$$1 \leqslant (m-1)C\int_0^\infty \frac{x^{m-2+\lambda}\,dx}{(1+x)^m}.$$

† The case when $u(z) = \log|f(z)|$, and f is entire, is due to Valiron [1935].

Setting $x = t/(1-t)$, we obtain

$$\int_0^\infty \frac{x^{m-2+\lambda}}{(1+x)^m}\,dx = \int_0^1 t^{m-2+\lambda}(1-t)^{-\lambda}\,dt = \frac{\Gamma(m-1+\lambda)\,\Gamma(1-\lambda)}{\Gamma(m)}$$

$$= \frac{\lambda(\lambda+1)\ldots(\lambda+m-2)\,\Gamma(\lambda)\,\Gamma(1-\lambda)}{(m-1)!}$$

$$= \frac{\pi\lambda(\lambda+1)\ldots(\lambda+m-2)}{(m-1)!\,\sin\pi\lambda}.$$

Thus

$$C \geqslant \frac{(m-2)!\,\sin\pi\lambda}{\pi\lambda(\lambda+1)\ldots(\lambda+m-2)}, \qquad \lambda > 0; \qquad C \geqslant 1, \qquad \lambda = 0.$$

This yields (4.5.6). Since

$$N(r) \leqslant T(r) + O(1) \leqslant B(r) + O(1),$$

we deduce that equality must hold in (4.5.6) if $\lambda = 0$. Also if $\lambda > 0$, we set

$$N(r) = r^\lambda, \quad \text{i.e.} \quad n(r) = \frac{\lambda}{d_m} r^{\lambda+m-2},$$

and note that equality holds in (4.5.2) provided that $x_1 = \rho$ and $u(x)$ is harmonic outside the nonpositive x_1 axis. Thus in this case

$$B(\rho) = (m-1)\rho \int_0^\infty \frac{t^{m-2+\lambda}\,dt}{(\rho+t)^m} = \rho^\lambda / C(\lambda, m),$$

where

$$C(\lambda, m) = \frac{\Gamma(m-1)\,\Gamma(\lambda)\,\sin\pi\lambda}{\pi\,\Gamma(\lambda+m-1)}, \tag{4.5.7}$$

as required.

4.5.2.

We can deduce immediately a sharpened form of Theorem 4.5, if $\lambda < 1$. This is

THEOREM 4.7. *If $u(x)$ is s.h. of order $\lambda < 1$ in R^m, then*

$$\delta(u) \leqslant 1 - C(\lambda, m),$$

where $C(\lambda, m)$ is given by (4.5.7) *In particular if $\lambda = 0$, $\delta(u) = 0$.*

If $\varepsilon > 0$, we have by Theorem 4.6 for some arbitrarily large r

$$N(r) > (C(\lambda, m) - \varepsilon)B(r) \geqslant (C(\lambda, m) - \varepsilon)T(r),$$

and this yields Theorem 4.7. If $\lambda = 0$, $C(\lambda, m) = 1$, so that $\delta(u) = 0$. For $\lambda > 0$, Theorem 4.7 is not sharp. However for $0 < \lambda < 1$

$$C(\lambda, m) \geqslant \frac{\sin \pi\lambda}{\pi\lambda} \frac{(m-2)!}{(m-1)!} = \frac{1}{(m-1)} \frac{\sin \pi\lambda}{\pi\lambda} > \frac{1-\lambda}{4^m},$$

so that the inequality of Theorem 4.7 is much sharper than that of Theorem 4.5.

We also obtain for small λ an improvement on Theorem 4.3. This is

THEOREM 4.8. *If* $0 < \lambda < 1$ *then*

$$\varliminf_{r \to \infty} \frac{B(r, u)}{T(r, u)} \leqslant \frac{1}{C(\lambda, m)}.$$

For

$$\frac{B(r, u)}{T(r, u)} \leqslant \frac{B(r, u)}{N(r, u)},$$

so that Theorem 4.8 follows from Theorem 4.6.

The inequality of Theorem 4.8 is sharp at any rate for $m = 2$, $0 \leqslant \lambda \leqslant \frac{1}{2}$. For in this case

$$\frac{1}{C(\lambda, m)} = \frac{\pi\lambda}{\sin \pi\lambda}.$$

If we set† $u(z) = r^{\lambda} \cos \lambda\theta$, where $z = re^{i\theta}$, $|\theta| \leqslant \pi$, then for $\lambda \leqslant \frac{1}{2}$

$$T(r, u) = \frac{1}{\pi} \int_0^{\pi} r^{\lambda} \cos \lambda\theta \, d\theta = \frac{r^{\lambda} \sin \pi\lambda}{\lambda\pi} = N(r, u),$$

since

$$u(0) = m(r, u) = 0$$

in this case.

We shall prove in the next section that for $0 \leqslant \lambda \leqslant \frac{1}{2}$, and $m = 2$, we have $\delta(u) = 0$, so that Theorem 4.7 still does not give a best possible result. However, if $m > 2$, $\delta(u)$ can be positive for arbitrarily small λ. This is shown by the

† The function $u(z)$ is s.h., since if $\pi - \delta < \theta < \pi + \delta$, $z = \mathrm{r}e^{i\theta}$, $u(z) = \max\{r^{\lambda} \cos \lambda\theta, r^{\lambda} \cos(\lambda(2\pi - \theta))\}$, when δ is small.

extremal functions in Theorem 4.6. We set

$$\rho^2 = |x|^2 = \sum_{\nu=1}^{m} x_\nu^2, \qquad x_1 = \rho \cos\theta,$$

and suppose that $N(r) = r^\lambda$ and that $u(x)$ is harmonic except on the negative x_1 axis. Then, if j denotes the vector $(1, 0, 0, \ldots, 0)$,

$$u(x) = \int_0^\infty \{t^{2-m} - |x + jt|^{2-m}\}\, dn(t)$$

$$= \frac{\lambda(\lambda + m - 2)}{(m - 2)} \int_0^\infty \{t^{2-m} - (t^2 + \rho^2 + 2t\rho\cos\theta)^{1-(m/2)}\} t^{\lambda+m-3}\, dt.$$

We set $t = \rho y$ and deduce that

$$u(x) = \frac{\lambda(\lambda + m - 2)}{m - 2} \rho^\lambda I(\lambda, m, \theta)$$

where

$$\text{(4.5.8)} \quad I(\lambda, m, \theta) = \int_0^\infty \{y^{2-m} - (1 + y^2 + 2y\cos\theta)^{1-(m/2)}\} y^{\lambda+m-3}\, dy.$$

The function $I(\lambda, m, \theta)$ is essentially an associated Legendre function, and does not appear to be elementary if m is odd. We can express $I(\lambda, m, \theta)$ as a meromorphic function of λ for all λ by means of the contour integral

$$I(\lambda, m, \theta) = \frac{1}{(e^{2\pi i\lambda} - 1)} \int_\Gamma \frac{z^{\lambda+m-3}\, dz}{(1 + z^2 + 2z\cos\theta)^{(m/2)-1}},$$

where Γ is a Jordan contour surrounding the two points

$$z = -\cos\theta \mp i\sin\theta$$

and not meeting the positive real axis or zero. For when Γ becomes a large circle and a small circle centred on the origin joined by segments just above and just below the positive real axis the integral approximates to the right-hand side of (4.5.5), when $0 < \lambda < 1$.

Thus when m is even we can evaluate $I(\lambda, m, \theta)$ by means of residues. We obtain for instance

$$I(\lambda, 4, \theta) = \frac{\pi}{\sin \pi\lambda} \frac{\sin(\lambda + 1)\theta}{\sin\theta},$$

and also the recursion relation

$$\frac{\partial}{\partial\theta} I(\lambda, m, \theta) = (m - 2)\sin\theta\, I(\lambda - 1, m + 2, \theta),$$

which yields for instance

$$I(\lambda, 6, \theta) = \frac{\pi}{2 \sin \pi\lambda} \frac{\cos \theta \sin (\lambda + 2)\theta - (\lambda + 2) \sin \theta \cos (\lambda + 2)\theta}{\sin^3 \theta}$$

We also note that for $0 < \lambda < 1$, $m \geqslant 3$, $I(\lambda, m, \theta)$ has the following properties

(i) $I(\theta)$ is a steadily decreasing function of θ for $0 \leqslant \theta < \pi$.
(ii) $I(\theta) > 0$ for $0 \leqslant \theta \leqslant \pi/2$.
(iii) $I(\theta) \to -\infty$ as $\theta \to \pi$.

These all follow easily from (4.5.8). To deduce the last we note that as $\theta \to \pi$

$$I(\theta) \to \int_0^\infty \{y^{2-m} - |1 - y|^{2-m}\} y^{\lambda+m-3}\, dy = -\infty.$$

4.5.3. The sharp bound for $\delta(u)$; statement of results

We can with rather more effort obtain a best possible form of Theorem 4.7. We define $I(\lambda, m, \theta)$ by (4.5.8), when $0 < \lambda < 1$ and m is an integer such that $m \geqslant 3$, while

$$I(\lambda, 2, \theta) = \cos \lambda\theta.$$

Then if $\rho = |x|$,

$$u_0(x) = \rho^\lambda\, I(\lambda, m, \theta) \tag{4.5.9}$$

is s.h. in R^m and harmonic for $|\theta| < \pi$. Our result can now be stated as follows.

THEOREM 4.9. *If $u(x)$ is s.h. of order λ in R^m, where $0 < \lambda < 1$, then*†

$$\delta(u) \leqslant \delta(u_0) \tag{4.5.10}$$

where $u_0(x)$ is given by (4.5.9)

By Theorem 4.9 the problem of the exact upper bound for $\delta(u)$ is reduced to the evaluation of $\delta(u_0)$. This appears to be possible in terms of elementary functions only if m is even. We have in particular†

† For the case when $u(z) = \log|f(z)|$, where $f(z)$ is entire, this result is a special case of a theorem of Edrei and Fuchs [1960].

THEOREM 4.10. *If* $m = 2$, *and* $u_0(x)$ *is given by* (4.5.9) *then*

$$\delta(u_0) = 0, \quad 0 \leqslant \lambda \leqslant \tfrac{1}{2}; \tag{4.5.11}$$

$$\delta(u_0) = 1 - \sin(\pi\lambda), \quad \tfrac{1}{2} \leqslant \lambda < 1.$$

If $m = 4$,

$$\delta(u_0) = 1 - \frac{\sin \pi\lambda}{(\lambda + 1)\sin(\pi/(\lambda + 1))}, \qquad 0 < \lambda < 1. \tag{4.5.12}$$

To prove our results we need a number of lemmas.

LEMMA 4.7. *If*

$$|x| = \left(\sum_{\nu=1}^{m} x_\nu^2\right)^{1/2} = R,$$

we define $\rho \geqslant 0$, *and* θ, *such that* $0 \leqslant \theta \leqslant \pi$, *by the equations* $x_1 = R\cos\theta$, $\rho = R\sin\theta$. *Then if* $\sigma(\theta_0)$ *denotes the* $(m - 1)$*-dimensional area of the spherical cap* $|\theta| < \theta_0$ *on the hypersphere* $|x| = R$, *we have*

$$\sigma(\theta_0) = c_{m-1} R^{m-1} \int_0^{\theta_0} (\sin\theta)^{m-2}\, d\theta$$

where $c_m = 2\pi^{m/2}/\Gamma(\tfrac{1}{2}m)$ *as usual.*

The proof is an elementary exercise in the calculus of several variables and we omit it.

We have next

LEMMA 4.8. *Suppose that* $f(\theta)$ *is a real continuous decreasing function of* θ *for* $0 < \theta < \pi$, *such that*

$$\int_0^{\varepsilon} (\sin\theta)^{m-2} f(\theta)\, d\theta < +\infty$$

for some, and so every, positive ε. *Then if* E *is any measurable set on the surface of the sphere of radius* R *in* R^m *and* E_0 *is the spherical cap having the same area as* E, *we have*

$$\int_E f(\theta)\, d\sigma \leqslant \int_{E_0} f(\theta)\, d\sigma.$$

Suppose first that $f(\theta) \geqslant 0$, $0 < \theta < \theta_0$, $f(\theta) = 0$, $\theta_0 < \theta < \pi$, and that E_0 is the spherical cap $\theta < \theta_0$. Then Lemma 4.5 is obvious since in this case, if S denotes the whole sphere, we have

$$\int_E f(\theta)\,\mathrm{d}\sigma \leqslant \int_S f(\theta)\,\mathrm{d}\sigma = \int_{E_0} f(\theta)\,\mathrm{d}\sigma.$$

Next suppose only that $f(\theta)$ satisfies the hypotheses of Lemma 4.8 and in addition $f(\theta_0) = 0$, where E_0 is the spherical cap $0 < \theta < \theta_0$. Let

$$g(\theta) = f(\theta), \qquad \theta < \theta_0, \qquad g(\theta) = 0, \qquad \theta > \theta_0.$$

Then we can apply the previous argument to $g(\theta)$ and deduce that

$$\int_E f(\theta)\,\mathrm{d}\sigma \leqslant \int_E g(\theta)\,\mathrm{d}\sigma \leqslant \int_{E_0} g(\theta)\,\mathrm{d}\sigma = \int_{E_0} f(\theta)\,\mathrm{d}\sigma,$$

so that Lemma 4.8 still holds. Finally in the general case we apply the result we have just proved with $f(\theta) - f(\theta_0)$ instead of $f(\theta)$. This gives

$$\int_E \{f(\theta) - f(\theta_0)\}\,\mathrm{d}\sigma \leqslant \int_{E_0} \{f(\theta) - f(\theta_0)\}\,\mathrm{d}\sigma,$$

i.e.

$$\int_E f(\theta)\,\mathrm{d}\sigma \leqslant \int_{E_0} f(\theta)\,\mathrm{d}\sigma + f(\theta_0)\left\{\int_E \mathrm{d}\sigma - \int_{E_0} \mathrm{d}\sigma\right\}.$$

This gives Lemma 4.8, since E and E_0 have equal measure.

We have now

Lemma 4.9. *Suppose that $u(x)$, $u_1(x)$ are* s.h. *of order less than one in R^m, that $u_1(x)$ is harmonic except on the negative x_1 axis and that*

$$N(r, u_1) = N(r, u), \qquad 0 < r < \infty, \qquad u(0) = u_1(0).$$

Then if E, E_0 are defined as in Lemma 4.8 we have

$$\int_E u(x)\,\mathrm{d}\sigma(x) \leqslant \int_{E_0} u_1(x)\,\mathrm{d}\sigma(x).$$

In view of Lemma 4.6 we have

$$u(x) = u(0) + \int_{|\xi| < \infty} K_0(x, \xi)\,\mathrm{d}\mu e_\xi,$$

where

$$K_0(x, \xi) = \log\left|\frac{x - \xi}{\xi}\right|, \qquad m = 2,$$

$$K_0(x, \xi) = |\xi|^{2-m} - |x - \xi|^{2-m}, \qquad m > 2.$$

Let ξ_1 denote the point $(-|\xi|, 0, 0, \ldots, 0)$. Then

$$u_1(x) = u(0) + \int_{|\xi|<\infty} K_0(x, \xi_1)\, d\mu e_\xi.$$

Now let E be a measurable set on $|x| = R$ and E_0 the spherical cap on $|x| = R$, having the same measure as E. Then

$$\int_E u(x)\, d\sigma(x) = \int_E u(0)\, d\sigma(x) + \int_{|\xi|<\infty} d\mu e_\xi \int_E K_0(x, \xi)\, d\sigma(x). \tag{4.5.13}$$

The inversion of the double integral is justified since for all x on E, we have

$$K_0(x, \xi) \leqslant K_0(-x_1, \xi_1)$$

and

$$\int_E K_0(-x_1, \xi_1)\, d\sigma(x)\, d\mu(\xi) \leqslant mc_m R^m \int_0^\infty \frac{t^{m-2} N(t)\, dt}{(R+t)^m} < \infty,$$

by Lemma 4.6.

Next we note that we may make an isometric (orthogonal) transformation or rotation of coordinates in R^m, $y = T(x)$, which sends ξ to ξ_1. Also such a transformation does not affect the measure $d\sigma(x)$, so that E is transformed into a set E_1, whose measure is equal to that of E_0.

Thus, since $K_0(x, \xi)$ depends on $|\xi|$ and $|x - \xi|$ only

$$\int_E K_0(x, \xi)\, d\sigma(x) = \int_{E_1} K_0(y, \xi_1)\, d\sigma(y).$$

Now the function $K_0(y, \xi_1)$ can be written as

$$K_0(y, \xi_1) = f_m(\theta),$$

where, if $|\xi| = t$, $|y| = R$, $y_1 = R\cos\theta$, and y_1 is the first coordinate of y, we have

$$\begin{cases} f_2(\theta) = \frac{1}{2}\log\left\{\dfrac{R^2 + 2tR\cos\theta + t^2}{t^2}\right\} \\ f_m(\theta) = t^{2-m} - (R^2 + 2tR\cos\theta + t^2)^{1-(m/2)}, \qquad m > 2. \end{cases} \tag{4.5.14}$$

Thus, in view of Lemma 4.8, we have

$$\int_{E_1} K_0(y, \xi_1)\, d\sigma(y) \leqslant \int_{E_0} K_0(x, \xi_1)\, d\sigma(x).$$

Thus (4.5.10) gives

$$\int_E u(x)\,\mathrm{d}\sigma(x) \leqslant \int_E u(0)\,\mathrm{d}\sigma(x) + \int_{|\xi|<\infty} \mathrm{d}\mu e_\xi \int_{E_0} K_0(x, \xi_1)\,\mathrm{d}\sigma(x)$$

$$= \int_{E_0} u_1(x)\,\mathrm{d}\sigma(x).$$

This proves Lemma 4.9.

It will follow from Lemma 4.9 that we may confine our investigations to functions harmonic except on the negative x_1 axis, for the purpose of proving Theorem 4.9. We consider $x = (x_1, x_2, \ldots, x_m)$, write $|x| = R$ and define θ by

$$R\cos\theta = x_1.$$

Then

$$u(x) = u(0) + \int_0^\infty f_m(t, \theta, R)\,\mathrm{d}n(t),$$

where $n(t)$ is the Riesz mass in $|\xi| < t$, and so on the segment $(-t, 0)$ of the negative x_1 axis, and $f_m(t, \theta, R)$ is given by (4.5.14). On integrating by parts twice w.r.t. t we obtain

$$u(x) = u(0) + \int_0^\infty \left\{1 - \frac{(R\cos\theta + t)t^{m-1}}{(R^2 + 2tR\cos\theta + t^2)^{m/2}}\right\} \mathrm{d}N(t)$$

i.e.

$$u(x) = u(0) + \int_0^\infty \frac{P_m(t, R, \theta)N(t)\,\mathrm{d}t}{(R^2 + 2tR\cos\theta + t^2)^{(m/2)+1}}, \tag{4.5.15}$$

where

$$P_m(t, R, \theta) = (m-1)R^3t^{m-2}\cos\theta + R^2t^{m-1}(m + (m-2)\cos^2\theta) + Rt^m(m-1)\cos\theta. \tag{4.5.16}$$

We have next

Lemma 4.10. *If*

$$Q_m(t, R, \phi) = \int_0^\phi \frac{P_m(t, R, \theta)(\sin\theta)^{m-2}}{(R^2 + 2tR\cos\theta + t^2)^{(m/2)+1}}\,\mathrm{d}\theta,$$

then $Q_m(t, R, \phi) \geqslant 0$, *for all positive* t, R *and* $0 \leqslant \phi \leqslant \pi$.

We note that P_m is, for fixed t and R, a quadratic polynomial in $\cos\theta$, which is positive for $\cos\theta = 1$ and non-positive for $\cos\theta = -1$. Thus $P_m(t, R, \theta)$ has exactly one zero θ_0 such that $0 < \theta_0 \leqslant \pi$, and $P_m > 0$ for $0 \leqslant \theta < \theta_0$, $P_m < 0$ for $\theta_0 < \theta \leqslant \pi$. Thus $Q_m(t, R, \phi)$ increases with ϕ to a maximum at $\phi = \theta_0$ and then decreases again. Thus to prove the Lemma it is sufficient to prove the result for $\phi = \pi$.

In view of Lemma 4.7 we may write

$$Q_m(t, R, \pi) = \frac{1}{c_{m-1}R^{m-1}} \int_{S(0,R)} \frac{P_m(t, R, \theta)}{(R^2 + 2tR\cos\theta + t^2)^{(m/2)+1}} \, d\sigma(x),$$

where x is a point on $S(0, R)$ and θ is defined as before. Also if $m \geqslant 2$

$$f(t, x) = \frac{P_m(t, R, \theta)}{(R^2 + 2tR\cos\theta + t^2)^{(m/2)+1}} = \frac{1}{d_m}\frac{\partial}{\partial t} t^{m-1} \frac{\partial}{\partial t} f_m(t, \theta, R),$$

where $d_2 = 2$, $d_m = m - 2$, if $m \geqslant 3$.

We deduce that for $t \neq R$,

$$Q_m(t, R, \pi) = \frac{1}{d_m c_{m-1} R^{m-1}} \frac{\partial}{\partial t} t^{m-1} \frac{\partial}{\partial t} \int_{S(0,R)} f_m(t, \theta, R) \, d\sigma(x).$$

Since $f_m(t, \theta, R)$ is s.h., the right-hand side can be evaluated by the Poisson–Jensen formula, Theorem 3.14. We deduce that

$$\frac{1}{c_m R^{m-1}} \int_{S(0,R)} f_m(t, \theta, R) \, d\sigma(x) = \begin{cases} 0, & R < t; \\ g(t, R), & R > t; \end{cases}$$

where

$$g(t, R) = \log\frac{R}{t}, \qquad m = 2;$$

$$g(t, R) = t^{2-m} - R^{2-m}, \qquad m > 2.$$

This yields

$$Q_m(t, R, \pi) = 0, \qquad R \neq t.$$

Also for $t = R$ we see that

$$P_m(t, R, \theta) = R^{m+1}(1 + \cos\theta)(m + (m-2)\cos\theta) \geqslant 0 \text{ for all } \theta,$$

so that $Q_m(R, R, \phi) > 0$, $0 < \phi < \pi$, and $Q_m(R, R, \pi) = +\infty$. This completes the proof of Lemma 4.10.

4.5.4. Proof of Theorem 4.9

We now proceed to apply the technique of the Pólya peaks to the function $u_1(x)$ constructed in Lemma 4.9. In view of Lemma 4.5, we may choose a sequence of values of $r = r_n$ going to infinity such that

$$(4.5.17)\qquad \begin{cases} N(t) \leqslant N(r)\left(\dfrac{t}{r}\right)^{\lambda-\varepsilon}, & t_0 \leqslant t \leqslant r. \\ \\ N(t) \leqslant N(r)\left(\dfrac{t}{r}\right)^{\lambda+\varepsilon}, & t \geqslant r. \end{cases}$$

Let E be a set on $|x| = r$, where $u(x) > 0$, and let E_0 be the spherical cap on $|x| = r$ which is equimeasurable with E. Then by Lemma 4.9 we have

$$\int_E u(x)\,d\sigma(x) \leqslant \int_{E_0} u_1(x)\,d\sigma(x).$$

If E_0 is given by $|\theta| < \phi$, the right-hand side may, by (4.5.15), be written as

$$c_{m-1}r^{m-1}\int_0^{\phi} (\sin\theta)^{m-2}\,d\theta\left\{u(0) + \int_0^{\infty} \frac{P_m(t, r, \theta)N(t)\,dt}{(r^2 + 2tr\cos\theta + t^2)^{(m/2)+1}}\right\}.$$

We assume $|\phi| < \pi$. Then the double integral is absolutely convergent, since $N(t)$ has order less than 1, and so we may invert the order of integration and deduce that

$$\int_{E_0} u_1(x)\,d\sigma(x) = c_{m-1}r^{m-1}\int_0^{\infty} Q_m(t, r, \phi)N(t)\,dt + u(0)\sigma(E_0).$$

We take for E a set on $|x| = r$, where $u(x) > 0$, and such that $\sigma(E_0) < c_m r^{m-1}$, and

$$(4.5.18)\qquad T(r, u) - 1 < \frac{1}{c_m r^{m-1}}\int_E u(x)\,d\sigma(x) \leqslant \frac{1}{c_m r^{m-1}}\int_{E_0} u_1(x)\,d\sigma(x)$$

$$= \frac{c_{m-1}}{c_m}\int_0^{\infty} Q_m(t, r, \phi)N(t)\,dt + \max(u(0), 0).$$

We now use (4.5.17) and deduce, since $Q_m(t, R, \phi) \geqslant 0$

$$T(r, u) \leqslant \frac{c_{m-1}}{c_m} N(r)\left\{\int_0^r (t/r)^{\lambda-\varepsilon}Q_m(t, r, \phi)\,dt + \int_r^{\infty} (t/r)^{\lambda+\varepsilon}Q_m(t, r, \phi)\,dt\right\} + O(1)$$

$$= \frac{c_{m-1}}{c_m} N(r)\left\{\int_0^1 s^{\lambda-\varepsilon}Q_m(s, 1, \phi)\,ds + \int_1^{\infty} s^{\lambda+\varepsilon}Q_m(s, 1, \phi)\,ds\right\} + O(1).$$

In the range $\frac{1}{2} \leqslant s \leqslant 2$, $s^{\lambda-\varepsilon} \to s^{\lambda}$ uniformly as $\varepsilon \to 0$, and the integrals

$$\int_0^{\frac{1}{2}} s^{\lambda-\varepsilon}\,Q_m(s, 1, \phi)\,ds \quad\text{and}\quad \int_2^{\infty} s^{\lambda+\varepsilon}\,Q_m(s, 1, \phi)\,ds$$

converge uniformly in a range $|\varepsilon| < \varepsilon_0$, $0 \leqslant \phi \leqslant \pi$, if $\lambda < 1$. Hence given $\eta > 0$, we can choose ε_0 so small that for $0 \leqslant \phi \leqslant \pi$, $0 < \varepsilon < \varepsilon_0$

$$\int_0^1 s^{\lambda-\varepsilon} Q_m(s, 1, \phi)\, ds + \int_1^\infty s^{\lambda+\varepsilon} Q_m(s, 1, \phi)\, ds$$

$$< (1 + \eta) \int_0^\infty s^{\lambda} Q_m(s, 1, \phi)\, ds + \eta.$$

Thus, given $\eta > 0$, we have for a sequence $r = r_n$ which tends to ∞ with n and a corresponding sequence $\phi = \phi_n$

$$T(r, u) < (1 + 2\eta)\frac{c_{m-1}}{c_m} N(r)\left\{\int_0^\infty s^{\lambda} Q(s, 1, \phi)\, ds + \eta\right\}$$

$$\leqslant (1 + 2\eta)\frac{c_{m-1}}{c_m} N(r)\left\{\int_0^\infty s^{\lambda} Q(s, 1, \phi_0)\, ds + \eta\right\},$$

where ϕ_0 is so chosen as to maximize the integral on the right-hand side. Since η is any positive number, we deduce that

$$\varliminf_{r\to\infty} \frac{T(r, u)}{N(r, u)} \leqslant \frac{c_{m-1}}{c_m} \int_0^\infty s^{\lambda} Q(s, 1, \phi_0)\, ds. \tag{4.5.19}$$

We now note that if $u_0(x)$ is so chosen as to have all its mass on the negative x_1 axis and if the counting function $N(r)$ is given by $N(r) = r^{\lambda}$, then $u_0(x)$ is the function defined by (4.5.9). For this function we have for any spherical cap E_0 given by $|\phi| < \phi_1$, $|x| = r$, where $\phi_1 < \pi$,

$$\frac{1}{c_m r^{m-1}} \int_{E_0} u_0(x)\, d\sigma(x) = \frac{c_{m-1}}{c_m} \int_0^\infty Q_m(t, r, \phi_1) N(t)\, dt$$

$$= \frac{c_{m-1}}{c_m} r^{\lambda} \int_0^1 s^{\lambda} Q_m(s, 1, \phi_1)\, ds.$$

Thus, since $u_0(x) > 0$ in an angle $|\phi| < \phi_0$, where ϕ_0 is independent of r, we have

$$\frac{T(r, u_0)}{N(r, u_0)} = \max_{0\leqslant\phi\leqslant\pi} \frac{c_{m-1}}{c_m} \int_0^\infty s^{\lambda} Q(s, 1, \phi)\, ds = \frac{c_{m-1}}{c_m} \int_0^\infty s^{\lambda} Q(s, 1, \phi_0)\, ds$$

for all positive r, so that by (4.5.16) we have

$$1 - \delta(u) = \varlimsup_{r\to\infty} \frac{N(r, u)}{T(r, u)} \geqslant \left\{\frac{c_{m-1}}{c_m} \int_0^\infty s^{\lambda} Q(s, 1, \phi_0)\, ds\right\}^{-1} = 1 - \delta(u_0).$$

This completes the proof of Theorem 4.9.

4.5.5. Proof of Theorem 4.10

To prove Theorem 4.10 we return to (4.5.9). If $m = 2$ we have $u_0(x) = cr^\lambda \cos \lambda\theta$, so that if $\lambda \leqslant \frac{1}{2}$,

$$T(r, u_0) = \frac{cr^\lambda}{\pi} \int_0^\pi \cos \lambda\theta \, d\theta = \frac{cr^{\lambda 2}}{\pi} \frac{\sin \pi\lambda}{\lambda},$$

$$m(r, u_0) = 0,$$

and so

$$\delta(u_0) = 0.$$

Thus for functions $u(x)$ of order $\lambda \leqslant \frac{1}{2}$, we have, when $m = 2$, $\delta(u) = 0$, which is (4.5.11).

If $\frac{1}{2} < \lambda < 1$,

$$T(r, u_0) = \frac{cr^\lambda}{\pi} \int_0^{\pi/(2\lambda)} \cos \lambda\theta \, d\theta = \frac{cr^\lambda}{\lambda\pi},$$

$$m(r, u_0) = \frac{-cr^\lambda}{\pi} \int_{\pi/(2\lambda)}^{\pi} \cos \lambda\theta \, d\theta = \frac{cr^{\lambda 2}(1 - \sin \pi\lambda)}{\lambda\pi},$$

$$\delta(u_0) = \frac{m(r, u_0)}{T(r, u_0)} = 1 - \sin \pi\lambda.$$

This completes the proof of 4.5.11.

Next if $m = 4$ we have for $|x| = r$

$$u_0(x) = \frac{cr^{\lambda 2}\sin (\lambda + 1)\theta}{\sin \theta}$$

In view of Lemma 4.7 we deduce that

$$T(r, u_0) = c'r^\lambda \int_0^{\pi/(\lambda+1)} \sin (\lambda + 1)\theta \sin \theta \, d\theta = c'r^\lambda \frac{(\lambda + 1)}{\lambda(\lambda + 2)} \sin \frac{\pi}{\lambda + 1},$$

where c, c' are constants. Also

$$\begin{aligned} m(r, u_0) &= -c'r^\lambda \int_{\pi/(\lambda+1)}^{\pi} \sin (\lambda + 1)\theta \sin \theta \, d\theta \\ &= \frac{c'r^\lambda}{\lambda(\lambda + 2)} \left[(\lambda + 1) \sin \frac{\pi}{\lambda + 1} - \sin (\pi\lambda) \right]. \end{aligned}$$

Thus

$$\delta(u_0) = \frac{m(r, u_0)}{T(r, u_0)} = 1 - \frac{\sin \pi\lambda}{(\lambda + 1) \sin \{\pi/(\lambda + 1)\}}.$$

This gives (4.5.12) and completes the proof of Theorem 4.10.

We note that

$$\delta(u_0) \sim \frac{\pi^2\lambda^3}{3},$$

as $\lambda \to 0$, whereas Theorem 4.7 only gives

$$\delta(u_0) = O(\lambda).$$

4.6. TRACTS AND ASYMPTOTIC VALUES

Suppose that $f(z)$ is an entire function having k ($\geqslant 2$) distinct asymptotic values a_ν, $\nu = 1$ to k. This means that there exist k curves Γ_ν going from 0 to ∞ in the z plane, such that

$$f(z) \to a_\nu, \quad \text{as} \quad z \to \infty \quad \text{along} \quad \Gamma_\nu.$$

Since the a_ν are distinct, the Γ_ν do not intersect near ∞ and so we can without loss of generality assume them to be non-intersecting in the z plane. We may assume the Γ_ν to be ordered anticlockwise around the origin, and set $\Gamma_{k+1} = \Gamma_1$. Then, for $\nu = 1$ to k, Γ_ν and $\Gamma_{\nu+1}$ bound a simply connected domain D_ν and $D_\mu \cap D_\nu = \phi$ for $\mu \neq \nu$. Also, since $a_\mu \neq a_{\mu+1}$, $f(z)$ is unbounded† in each domain D_μ, while $f(z)$ is bounded on each of the curves Γ_μ.

From these facts, Ahlfors [1930] deduced his celebrated proof of Denjoy's conjecture, namely that the lower order of $f(z)$ is at least $\frac{1}{2}k$. In order to generalize this result to subharmonic functions, Heins [1959] showed that it is necessary to use only the fact that

$$u(z) = \log|f(z)|$$

is bounded above on the curves Γ_μ, but unbounded in each domain D_μ. Thus if M is large the set

$$E_M = \{z \mid u(z) \geqslant M\}$$

has points in each domain D_μ but not on any of the Γ_μ, so that E_M has at least k components. For a function $u(z)$ s.h. in R^m we make only this assumption, namely that E_M has $k \geqslant 2$ distinct components and see what we can deduce from this about the lower order of $u(z)$. The resulting lower bound which we shall obtain, while less sharp than the $\frac{1}{2}k$ of Ahlfors, will, for fixed $m \geqslant 2$, have the correct order $k^{1/(m-1)}$ as a function of k as $k \to \infty$. We shall return to the case $m = 2$ in the second volume. The work in this section is largely due to M. N. M. Talpur [1975, 1976].

† See e.g. Titchmarsh [1939, p. 179].

4.6.1. Preliminary results

Suppose that $u(x)$ is s.h. in a neighbourhood of a compact set E in R^m. For each $M \geqslant 0$ the set

$$E_M = \{x \mid x \in E \text{ and } u(x) \geqslant M\}$$

is a closed subset of the compact set E and consequently E_M is compact. We need to investigate more closely the structure of E_M and so recall some notions from point set topology.

We recall that a set E is *connected* if E does not permit a *partition* into sets E_1, E_2, such that

(a) $E_1 \cup E_2 = E, \quad E_1 \cap E_2 = \phi$;
(b) $E_j \neq \phi, \quad j = 1, 2$; and
(c) E_j is closed in E, $j = 1, 2$.

If E is closed and not connected and (E_1, E_2) is a partition of E, then the last condition ensures that E_1 and E_2 are closed.

We can now define an equivalence relation among points $x \in E$. We say that $x_1 \sim x_2$ if there exists a connected subset $E(x_1, x_2)$ of E containing x_1 and x_2. The relation is clearly symmetric and reflexive since points are connected. It is also transitive, since the union of two connected sets $E(x_1, x_2) \cup E(x_2, x_3)$, each of which contains x_2 is connected. Thus under the relation $\sim$, the points of E separate into disjoint subsets of E, which are called *components*. They are maximal connected subsets of E. For a component E' containing the point x_1 is the union of all connected subsets of E which contain x_1 and so is connected. Also E' is by definition the largest connected subset containing x_1.

A compact connected set containing at least two points is called a *continuum*. We need the following:

LEMMA 4.11. *The components of a compact set in R^m are continua or points.*

Let E_0 be a component of the compact set E, containing the point x_0. If E_0 reduces to the point x_0 there is nothing to prove. Suppose then that E_0 contains at least two points. It is enough to prove that E_0 is closed in E, since then E_0 is closed in R^m. Since E_0 is also bounded, as a subset of E, E_0 is then compact and so a continuum.

To prove that E_0 is closed, we show that $\bar{E}_0 = E_0$, where $\bar{E}_0$ is the closure of E_0. For this it is sufficient to show that $\bar{E}_0$ is connected, since in this case $\bar{E}_0$ lies in E_0, the union of all connected subsets of E, which contain x_0.

Suppose contrary to this that $\bar{E}_0$ has a partition (F_1, F_2) such that $x_1 \in F_1$ and $x_2 \in F_2$. Then since $\bar{E}_0$ is closed, F_1 and F_2 are closed and so compact and so they lie at a positive distance δ from each other. Let E_1, E_2 be the subsets of E_0, which lie in F_1 and F_2 respectively. Then E_1, E_2 form a partition of E_0.

In fact, (a) is obvious. Next $x_1 \in E_1$ is a point or limit point of E_0. In the first case E_1 is not empty. In the second $D(x_1, \delta)$ contains a point ξ_1 of E_0, and this point, being distant less than δ from $x_1 \in F_1$, cannot belong to F_2. Thus $\xi_1 \in F_1 \cap E_0 = E_1$, so that E_1 is not empty. Similarly E_2 is not empty. Thus (b) holds.

Finally $E_j = E_0 \cap F_j$, and since F_j is closed E_j is closed in E_0, for $j = 1,2$. Thus (c) holds and $\{E_1, E_2\}$ gives a partition of E_0, contrary to hypothesis. This completes the proof of Lemma 4.9.

We can now prove a result which, although only a tool in our theory, has independent interest. This is:

THEOREM 4.11. *Suppose that $u(x)$ is* s.h. *in the neighbourhood of a compact set E in R^m. For any real K let E_K be the subset of E in which $u(x) \geqslant K$, and let $C(K)$ be a component of E_K. Let*

$$v(x) = u(x), \qquad x \in C(K), \qquad v(x) = K, \qquad \textit{elsewhere}.$$

Then $v(x)$ is s.h. *in the interior of E.*

Suppose first that $u(x)$ is continuous. Then E_K is closed and so compact and hence so is $C(K)$ by Lemma 4.11. Suppose that x_0 is a point of E. Then, if $v(x_0) > K$, $x_0 \in C(K)$ and so $u(x_0) > K$. Since $u(x)$ is continuous, we deduce that $u(x) > K$ in a neighbourhood $D(x_0, r)$ of x_0. Thus $D(x_0, r) \subset C(K)$ and so $v(x) = u(x)$ in $D(x_0, r)$. Thus $v(x)$ is s.h. at x_0. Again if $v(x_0) = K$, suppose first that x_0 is an exterior point of $C(K)$. Then a neighbourhood $D(x_0, r)$ lies outside $C(K)$ and so $v(x) = K$ in $D(x_0, r)$. Thus again $v(x)$ is s.h. at x_0. A similar conclusion holds if x_0 is an interior point of $C(K)$ and $u(x_0) = K$.

Suppose finally that x_0 is a boundary point of $C(K)$. Then $u(x_0) = K$. For, by continuity, if $u(x) < K$ or $u(x) > K$ at $x = x_0$, the same inequality holds in a neighbourhood of x_0, so that x_0 would be either exterior or interior to $C(K)$. Also since $u(x)$ is continuous at x_0 and $v(x) = K = u(x_0)$ or $v(x) = u(x)$, we deduce that

$$v(x) \to K = u(x_0), \quad \text{as} \quad x \to x_0.$$

Thus $v(x)$ is continuous at x_0. Finally the mean value-inequality is obvious at x_0, since we have everywhere

$$v(x) \geqslant K = v(x_0).$$

This proves Theorem 4.11 if $u(x)$ is continuous.

The general result is less obvious but can be proved by means of Theorem 3.8. We assume that $u(x)$ is s.h. in a neighbourhood of E. Then by Theorem

3.8 there exists a sequence $u_n(x)$ of functions, s.h. and continuous in a (smaller) neighbourhood of E, and such that $u_n(x)$ decreases strictly with n and

$$u_n(x) \to u(x) \quad \text{as} \quad n \to \infty.$$

Let x_0 be a fixed point of $C(K)$, let $C_n(K)$ be the component of

$$\{x \,|\, u_n(x) \geqslant K \text{ and } x \in E\}$$

which contains x_0 and set

$$v_n(x) = u_n(x), \qquad x \in C_n(K)$$
$$v_n(x) = K, \qquad \text{elsewhere.}$$

Then by what we have just seen $v_n(x)$ is s.h. in the interior of E. We note that $C_{n+1}(K) \subset C_n(K)$, since $u_{n+1}(x) \leqslant u_n(x)$. Thus if x is outside $C_{n+1}(K)$ we have

$$K = v_{n+1}(x) \leqslant v_n(x),$$

while if $x \in C_{n+1}(K)$,

$$K \leqslant v_{n+1}(x) = u_{n+1}(x) < u_n(x) = v_n(x).$$

Thus the sequence $v_n(x)$ is a decreasing sequence of s.h. functions in E and so

$$v_0(x) = \lim_{n\to\infty} v_n(x)$$

is s.h. in (a neighbourhood of) E. In fact $v_0(x)$ is u.s.c. by Theorem 1.3, $v_0(x) < +\infty$, since $v_n(x) < +\infty$, and finally since $v_n(x) \downarrow v_0(x)$ we have

$$\int_{S(x_0,r)} v_0(x)\,d\sigma(x) = \lim_{n\to\infty} \int_{S(x_0,r)} v_n(x)\,d\sigma(x)$$

$$\geqslant \lim_{n\to\infty} c_m r^{m-1} v_n(x_0) = c_m r^{m-1} v_0(x_0).$$

Thus $v_0(x)$ satisfies the conditions of Section 2.1 and so $v_0(x)$ is s.h.

Finally we note that $v_0(x)$ is the function $v(x)$ defined in Theorem 4.11. In fact let

$$C_0(K) = \bigcap_{n=1}^{\infty} C_n(K).$$

Then $C_0(K)$ is the countable intersection of the nested compact connected sets $C_n(K)$ and so $C_0(K)$ is compact and connected,† i.e. $C_0(K)$ is a point or

† See e.g. Newman [1951].

a continuum. Also in $C_0(K)$ we have for each n

$$v_n(x) = u_n(x) \geqslant K,$$

so that $v_0(x) = \lim u_n(x) = u(x) \geqslant K$.

Since $C_n(K)$ contains x_0 for each n, $C_0(K)$ contains x_0 and since $C_0(K)$ is a connected set containing x_0 in which $u(x) \geqslant K$, we have $C_0(K) \subset C(K)$. On the other hand we have in $C(K)$

$$K \leqslant u(x) \leqslant u_n(x),$$

so that $C(K) \subset C_n(K)$ for each n. Thus $C(K) \subset C_0(K)$, i.e. $C(K) = C_0(K)$.

We have seen that $v_0(x) = v(x)$ in $C(K)$. If x is outside $C(K)$ then x is outside $C_n(K)$ for some n and so for all $n > n_0$. Thus

$$v_0(x) = \lim_{n\to\infty} v_n(x) = \lim_{n\to\infty} K = K = v(x).$$

Thus $v(x) = v_0(x)$ in E, and since $v_0(x)$ is s.h. in E, we have proved Theorem 4.11.

4.6.2.

We now suppose that $u(x)$ is s.h. in $|x| < r_0$, where $0 < r_0 \leqslant \infty$ and that the set $u(x) \geqslant 0$ contains at least $k \geqslant 2$ components in $C(0, r)$ for $r_1 \leqslant r < r_0$. We wish to deduce from this a lower bound for the growth of $u(x)$. Let $C_1(r)$, $C_2(r), \ldots, C_k(r)$ be distinct components of $u(x) \geqslant 0$ in $C(0, r)$ and set

$$v_\nu(x) = u(x) \quad \text{in} \quad C_\nu(r)$$

$$v_\nu(x) = 0 \quad \text{elsewhere.}$$

Then, by Theorem 4.11, $v_\nu(x)$ is s.h. in $D(0, r)$. Also

$$v_\nu(x) \geqslant 0, \quad \text{in} \quad C(0, r) \tag{4.6.1}$$

$$v_\mu(x) \cdot v_\nu(x) = 0, \quad \text{for} \quad \mu \neq \nu. \tag{4.6.2}$$

We also assume that $v_\nu(x)$ is not identically zero in $|x| \leqslant r_1$ so that

$$B_\nu(r_1) = \sup_{|x|=r_1} v_\nu(x) > 0, \qquad \nu = 1 \text{ to } k. \tag{4.6.3}$$

We shall see that (4.6.1) to (4.6.3) are sufficient in themselves to give us a good deal of information about the growth of the functions $v_\nu(x)$.

Let $e_\nu(r)$ be the subset of $|x| = r$, in which $v_\nu(x) > 0$, and let

$$\theta_\nu(r) = \frac{1}{c_m r^{m-1}} \int_{e_\nu(r)} d\sigma(x),$$

where $d\sigma(x)$ denotes surface area on $S(0, r)$. Then the sets $e_\nu(r)$ are disjoint by (4.6.2) so that

$$\sum_{\nu=1}^{k} \theta_\nu(r) \leqslant 1. \tag{4.6.4}$$

Next if

$$T_\nu(r) = \frac{1}{c_m r^{m-1}} \int_{e_\nu(r)} v_\nu(x)\, d\sigma(x)$$

is the characteristic of $v_\nu(x)$ and

$$B_\nu(r) = \sup_{x \in e_\nu(r)} v_\nu(x), \tag{4.6.5}$$

is the maximum of $v_\nu(x)$ on $|x| = r$, we deduce at once that

$$T_\nu(r) \leqslant \theta_\nu(r) B_\nu(r).$$

On the other hand we have from Theorem 3.19 for $0 < \rho < r$

$$B_\nu(\rho) \leqslant \frac{r^{m-2}(r+\rho)}{(r-\rho)^{m-1}} T_\nu(r) \leqslant \frac{\theta_\nu(r) r^{m-2}(r+\rho)}{(r-\rho)^{m-1}} B_\nu(r). \tag{4.6.6}$$

In view of (4.6.6) we see that

$$\begin{aligned} \prod_{\nu=1}^{k} B_\nu(\rho) &\leqslant \left[\frac{r^{m-2}(r+\rho)}{(r-\rho)^{m-1}}\right]^k \prod_{\nu=1}^{k} \{\theta_\nu(r) B_\nu(r)\} \\ &\leqslant \left[\frac{1}{k}\frac{r^{m-2}(r+\rho)}{(r-\rho)^{m-1}}\right]^k \prod_{\nu=1}^{k} B_\nu(r), \end{aligned} \tag{4.6.7}$$

by the geometric–arithmetic mean Theorem and (4.6.4).

We can now prove our first result.

THEOREM 4.12. *Suppose that the* $v_\nu(x)$ *satisfy* (4.6.1) *to* (4.6.3) *in* $D(0, r)$, *that* $B_\nu(r)$ *is defined by* (4.6.5) *and*

$$B_0(r) = \left(\prod_{\nu=1}^{k} B_\nu(r)\right)^{1/k}$$

Then for $r_1 \leqslant r < r_0$ *we have*

$$B_0(r) \geqslant \frac{2}{3}\left(\frac{r}{r_1}\right)^{c_1} B_0(r_1), \tag{4.6.8}$$

where $c_1(k, m)$ *depends on* k *and* m *only and we may take*

$$c_1 = \log\left(\frac{3}{2}\right) \Big/ \log \frac{1 + [3/(2k)]^{1/(m-1)}}{1 - [3/(2k)]^{1/(m-1)}}. \tag{4.6.9}$$

We note that as $k \to \infty$ for fixed m

$$c_1(k,m) \sim \frac{1}{2}\log\frac{3}{2}\left(\frac{2k}{3}\right)^{1/(m-1)}.$$

so that c_1 has the right order of magnitude as a function of k, at least when $m = 2$. We shall show by an example at the end of the chapter, that this remains true for $m > 2$.

We set

$$\phi = [3/(2k)]^{1/(m-1)}, \qquad \lambda = \frac{1+\phi}{1-\phi}, \; r_\mu = r_1\lambda^{\mu-1}, \qquad B_0(r_\mu) = B_\mu.$$

Then setting $r = r_1\lambda^{\mu-1}$, $\rho = r_1\lambda^{\mu-2}$, we deduce from (4.6.7) if $r < r_0$

$$B_\mu \geqslant \frac{k(\lambda-1)^{m-1}}{\lambda^{m-2}(\lambda+1)}B_{\mu-1} \geqslant k\left(\frac{\lambda-1}{\lambda+1}\right)^{m-1}B_{\mu-1} = k\phi^{m-1}B_{\mu-1} = \frac{3}{2}B_{\mu-1}.$$

Thus we deduce by induction that for $\mu \geqslant 1$

$$B_\mu \geqslant \left(\frac{3}{2}\right)^{\mu-1}B_1.$$

We choose for μ the largest integer such that $r_\mu \leqslant r$. Then

$$B_0(r) \geqslant B_0(r_\mu) \geqslant \left(\frac{3}{2}\right)^{\mu-1}B_1 = \frac{2B_1}{3}\left(\frac{3}{2}\right)^{\mu} = \frac{2B_1}{3}\left(\frac{r_{\mu+1}}{r_1}\right)^{c_1} \geqslant \frac{2}{3}B_0(r_1)\left(\frac{r}{r_1}\right)^{c_1},$$

where $c_1 = \log(\frac{3}{2})/\log\lambda$, as required.

4.6.3. Components $C(K)$ in domains

Theorem 4.12 enables us to extend Theorem 4.11 to functions s.h. in domains, i.e. open connected sets. We need the following

LEMMA 4.12. *Suppose that $u(x)$ is* s.h. *in a neighbourhood of the closed ball $C(x_1, r)$ and further that $u(x_1) > K > -\infty$. Let C_1 be that component of the set $u(x) \geqslant K$ in $C(x_1, r)$, which contains x_1. Then given K_1, such that $K < K_1 \leqslant u(x_1)$, there exists $\delta > 0$, such that $u(x) < K_1$, at all points of $D(x_1, \delta)$ not in C_1.*

Suppose that x_2 is a point of the open ball $D(x_1, r)$, such that x_2 is not in C_1 and $u(x_2) \geqslant K_1$. Let C_2 be the component of $u(x) \geqslant K$ in $C(x_1, r)$ which contains x_2. Since x_2 is not on C_1, C_1 and C_2 are disjoint. For $\nu = 1, 2$ we now set

$$u_\nu(x) = u(x), \qquad x \in C_j,$$

$$u_\nu(x) = K, \text{ elsewhere in } D(x_0, r).$$

In view of Theorem 4.11 the functions $u_\nu(x)$ are s.h. in $D(x_0, r)$.

We now define

$$v_\nu(x) = u_\nu(x_1 + x) - K, \quad |x| \leqslant r,$$

and note that the $v_\nu(x)$ satisfy the conditions (4.6.1), (4.6.2) and for $r_1 = |x_2 - x_1|$, we have for $\nu = 1, 2$

$$B_\nu(r_1) = \sup_{|x| \leqslant r_1} v_\nu(x) \geqslant K_1 - K = B_0 > 0.$$

Hence in view of Theorem 4.12, we deduce for $r_1 < \rho < r$

$$B_0(\rho) \geqslant \frac{2}{3}\left(\frac{\rho}{r_1}\right)^{c_1} B_0.$$

with the notation of that theorem. Here

$$B_0(\rho) \leqslant \sup_{|x - x_1| \leqslant r} \{u(x) - K\} = B_1$$

say. Thus we deduce that

$$B_1 \geqslant \frac{2}{3}\left(\frac{\rho}{r_1}\right)^{c_1} B_0, \qquad r_1 \geqslant \rho\left(\frac{2B_0}{3B_1}\right)^{1/c_1}.$$

Making $\rho \to r$, we obtain Lemma 4.12 with

$$\delta = r\left(\frac{2B_0}{3B_1}\right)^{1/c_1} = r\left(\frac{2(K_1 - K)}{3B_1}\right)^{1/c_1}.$$

Suppose now that $u(x)$ is s.h. in a domain D in R^m, possibly the whole space. For $n = 1, 2, \ldots$, let E_n be compact subsets of D, such that E_n lies in the interior of E_{n+1} and

$$\bigcup_{n=1}^{\infty} E_n = D.$$

If x_0 is any point of D, then $x_0 \in E_n$ for $n \geqslant n_0$ say. If

$$u(x_0) \geqslant K > -\infty, \text{ and } n \geqslant n_0$$

we define $C_n(x_0, K)$ to be the component of $u(x) \geqslant K$ in E_n which contains x_0.

Evidently $C_n(x_0, K)$ expands with increasing n. We set

$$C = C(x_0, K, D) = \bigcup_{n=n_0}^{\infty} C_n(x_0, K),$$

and call C a limit-component, or sometimes simply component of $u(x) \geqslant K$

in D. It must be pointed out however that this terminology is in some ways an abuse of language. For C, although connected, is not in general a maximal connected set in D. Nor is C in general closed in D. But any two points x_1, x_2 of C belong to the continuum $C_n(x_0, K)$ for sufficiently large n, and so they are joined to x_0 and each other by a continuum on which $u(x) \geqslant K$.

Conversely if γ is any continuum containing x_0 and on which $u(x) \geqslant K$, then γ lies in $\bigcup_{n=1}^{\infty} D_n$, where D_n is the interior of E_n, and so, by the Heine–Borel Theorem γ lies in $\bigcup_{n=1}^{N} D_n$ for some N and so in D_N. Thus γ lies in $C_N(x_0, K)$ and so in C. Thus C is the union of all continua γ, which lie in D and contain x_0 and on which $u(x) \geqslant K$. In particular we see that C is independent of the particular exhausting sequence E_n. We can now prove

THEOREM 4.13. *If $u(x)$ is* s.h. *in a domain D, and $C = C(x_0, K, D)$ is a limit-component in D, and if*

$$v(x) = u(x) \text{ in } C$$

$$v(x) = K \text{ elsewhere,}$$

then $v(x)$ is s.h. *in D.*

Let x_1 be any point of D. We proceed to prove that $v(x)$ is s.h. at x_1, i.e. that $v(x)$ satisfies the conditions (i), (ii) and (iii) of Section 2.1. We distinguish 3 cases.

(a) Suppose first that x_1 is exterior to C, so that some neighbourhood N of x_1 does not meet C. Then $v(x) = K$ in N and so $v(x)$ is constant and so s.h. near x_1.

(b) Suppose next that x_1 is a point of C. Then

$$v(x_1) = u(x_1) \geqslant K.$$

Since $u(x)$ is s.h. and so u.s.c., given $\varepsilon > 0$ we can find a neighbourhood $N(\varepsilon)$ of x_1 in which

$$u(x) < u(x_1) + \varepsilon,$$

and so

$$v(x) \leqslant \max(u(x), K) < u(x_1) + \varepsilon = v(x_1) + \varepsilon.$$

Thus $v(x)$ is finite and u.s.c. at x_1. It remains to prove that the mean value inequality is satisfied. To see this let E_1 be a compact subset of D, containing x_1 in its interior D_1 and let C_1 be the component of $u(x) \geqslant K$ in E_1, which contains x_1. Set

$$v_1(x) = u(x), \quad x \in C_1$$
$$= K, \text{ elsewhere in } E_1.$$

By Theorem 4.11, $v_1(x)$ is s.h. in D_1. On the other hand C_1 is a continuum on which $u(x) \geqslant K$ and so C_1 is contained in C. Thus

$$v_1(x) \leqslant v(x) \text{ in } E_1, \text{ while } v_1(x_1) = v(x_1) = u(x_1).$$

Thus for all sufficiently small r, we have

$$v(x_1) = v_1(x_1) \leqslant \frac{1}{c_m r^{m-1}} \int_{S(x_1, r)} v_1(x)\, d\sigma(x) \leqslant \frac{1}{c_m r^{m-1}} \int_{S(x_1, r)} v(x)\, d\sigma(x).$$

Thus the mean-value inequality is also satisfied at x_1, and so $v(x)$ is s.h. at x_1.

(c) Suppose finally that x_1 is a limit-point of C, which does not belong to C. Then $v(x_1) = K$, and $v(x) \geqslant K$ everywhere so that (i) and (iii) of Section (2.1) are clearly satisfied. It remains to prove that $v(x)$ is u.s.c. at x_1. Suppose first that $u(x_1) \leqslant K$. Then since $u(x)$ is u.s.c. at x_1 we can find a neighbourhood N of x_1, in which

$$u(x) < K + \varepsilon,$$

and hence

$$v(x) \leqslant \max(u(x), K) < K + \varepsilon = v(x_1) + \varepsilon.$$

Thus in this case $v(x)$ is u.s.c. at x_1.

Lastly suppose that

$$u(x_1) > K.$$

Let F_1 be a closed ball $|x - x_1| \leqslant r_1$ contained in D and let C_1 be the component of $u(x) \geqslant K$ in F_1, which contains x_1. Then since x_1 is not in C, C_1 cannot meet C. Also in view of Lemma 4.12, given ε, such that $0 < \varepsilon < u(x_1) - K$, we can find $\delta > 0$ such that C_1 contains all points of $|x - x_1| \leqslant \delta$ where $u(x) \geqslant K + \varepsilon$. These points do not therefore belong to C and $v(x) = K$ at all such points x. At all other points of $|x - x_1| \leqslant \delta$, we have

$$v(x) \leqslant \max(u(x), K) < K + \varepsilon.$$

Thus $v(x) \leqslant K + \varepsilon = v(x_1) + \varepsilon$ in $|x - x_1| \leqslant \delta$, and so $v(x)$ is u.s.c. at x_1. This completes the proof of Theorem 4.13.

By means of Theorem 4.13 we can obtain useful information about the structure of components. We shall call a component C of $u(x) \geqslant K$ in a domain D *thin* if $u(x) \equiv K$ on C. Otherwise we shall call C *thick*. Rather surprisingly† there exist s.h. functions with continuum many distinct thin components. We defer examples of this to the second volume. On the other hand thick components always have positive m-dimensional measure and so their total number is at most countable. These results are contained in†

THEOREM 4.14. *Suppose that C is a limit component of $u(x) \geqslant K$ in a domain D. Then C goes to the boundary of D, i.e. C does not lie in any compact subset of D. Further, if C is thick, then C has positive m-dimensional volume.*

We suppose that $u(x)$ is not constant in D, since otherwise Theorem 4.14 is trivial. Suppose first that C is thick, and set $v(x) = u(x)$ in C, $v(x) = K$ elsewhere in D, so that $v(x)$ is s.h. in D by Theorem 4.13.

Since C is thick, C contains a point x_0, where $u(x_0) > K$. Thus if $v(x)$ is constant in D, $v(x) = v(x_0) = u(x_0) > K$. Thus C contains the whole of D and $u(x)$ is constant in D. Suppose next that $v(x)$ is not constant. Then $v(x)$ cannot attain its upper bound M at an interior point of D and so there exists a sequence x_n of points in D, which approach the boundary of D, such that

$$v(x_n) \to M > K.$$

Thus for sufficiently large n, we have $v(x_n) > K$, so that $x_n \in C$ and so C goes to the boundary of D.

Next suppose that x_0 is a point of C where $u(x_0) > K$. For a small positive δ let G_1 and C_1 be the subsets of $D(x_0, \delta)$ where $v(x) = K$ and $v(x) > K$ respectively. Suppose now that C has measure zero. Then so does C_1 for all small δ and

$$\iint_{D(x_0,\,\delta)} (v(x) - v(x_0))\, dx = \iint_{G_1} (v(x) - v(x_0))\, dx = (K - v(x_0))m < 0$$

where m is the measure of $D(x_0, \delta)$. This contradicts the fact that $v(x)$ is s.h. (in view of Theorem 2.12 for instance). Thus C has positive measure.

It remains to prove that even if C is thin, C goes to the boundary of D. To do this we argue as follows. Let E_0 be a compact subset of D containing x_0, let F_0 be the frontier of E_0 and let C be the component of $u(x) \geqslant K$ in E_0, containing x_0. We wish to show that C meets F_0. Let C_n be the component of $u(x) \geqslant K - (1/n)$, in E_0, which contains x_0. Then C_n is thick, since $u(x_0) \geqslant K$, and so by what we proved above C_n meets F_0. Let

$$\Gamma = \bigcap_{n=1}^{\infty} C_n$$

† Talpur [1975].

Then Γ is a countable intersection of nested continua, since $C_{n+1} \subset C_n$, and so Γ is a non-empty connected compact set. Also the sets $C_n \cap F_0$ are compact and not empty and hence so is their intersection $\Gamma \cap F_0$. Thus Γ contains the point x_0 and meets the frontier of E_0. Also we have clearly

$$u(x) \geqslant K \text{ on } \Gamma,$$

so that $\Gamma \subset C$. Thus C meets the frontier of F_0 and so C goes to the boundary of D. This completes the proof of Theorem 4.14.

We proceed to prove

THEOREM 4.15. *Suppose that $u(x)$ is* s.h. *and not constant in R^m, that $C = C(K)$ is a thick component of $u(x) \geqslant K$ there and that $u(x) \leqslant M$ on C, where M is finite. Then $m \geqslant 3$ and either*

(a) *C contains all points of R^m in which $u(x) > K$, so that in particular $u(x) < M$ in R^m, or*

(b) *there exists at least one thick component C_1, disjoint from C. Then $u(x)$ is unbounded above and in fact has infinite lower order on any such component C_1. More precisely*† *if*

$$B_1(R) = \sup_{|x| \leqslant R,\, x \in C_1} u(x)$$

then

$$\log B_1(R)/\log R \to \infty, \quad \text{as} \quad R \to \infty.$$

COROLLARY. *If $u(x)$ is* s.h. *in R^m there exists K_0 such that $u(x)$ is unbounded above on all thick components $C(K)$ for $K > K_0$.*

If C contains all points where $u(x) > K$, we have case (a) and there is nothing further to prove. Suppose contrary to this that there exists a point x_1 such that $u(x_1) > K$ and x_1 is not on C. Then the component of $u(x) \geqslant K$ which contains x_1 is disjoint from C. Thus there exists at least one thick component disjoint from C. We suppose that C_1 is such a component of $u(x) \geqslant K_1$ say, and proceed to show that $u(x)$ has infinite lower order on C_1. For this purpose we set

$$v_1(x) = u(x) - K_1, \quad x \in C_1, \qquad v_1(x) = 0 \quad \text{elsewhere}.$$

$$v_2(x) = u(x) - K, \quad x \in C, \qquad v_2(x) = 0 \quad \text{elsewhere}.$$

Then $v_1(x)$, $v_2(x)$ are s.h. by Theorem 4.13. It is enough to show that $v_1(x)$ has infinite lower order.

† More generally we define the lower order, order and type class of $u(x)$ on an unbounded set C_1 by the corresponding quantities for $B_1(R)$.

The functions $v_1(x)$ and $v_2(x)$ satisfy (4.6.1) and (4.6.2). We define $\theta_\nu(r)$ as in Section 4.6.2, $B_\nu(r)$ by (4.6.5) and use (4.6.6). Thus we have for $0 < \rho < r < \infty$

$$B_\nu(\rho) < \frac{\theta_\nu(r) r^{m-2}(r+\rho)}{(r-\rho)^{m-1}} B_\nu(r). \tag{4.6.10}$$

Since the components C and C_1 are thick, $B_\nu(\rho) > 0$ for large ρ. Also by hypothesis

$$B_2(\rho) \to M - K, \text{ as } \rho \to \infty, \text{ where } 0 < M - K < \infty.$$

Thus given $\varepsilon > 0$, we have for sufficiently large ρ and $r > \rho$

$$\theta_2(r) > \frac{(1-\varepsilon)(r-\rho)^{m-1}}{r^{m-2}(r+\rho)}.$$

Letting $r \to \infty$ for fixed ρ, we deduce that

$$\theta_2(r) > 1 - 2\varepsilon, \; r > r_0,$$

i.e.

$$\theta_2(r) \to 1, \quad \text{as} \quad r \to \infty.$$

In view of (4.6.4) we now deduce that

$$\theta_1(r) \leqslant 1 - \theta_2(r) \to 0, \text{ as } r \to \infty.$$

Choose now r_0 so that $B_1(r_0) > 0$, and set

$$r_\mu = r_0 2^\mu, \theta_\mu = \theta_1(r_\mu).$$

Then (4.6.10) with $\rho = r_\mu$, $r = r_{\mu+1}$, gives

$$B_1(r_\mu) < \theta_\mu \,.\, 3 \,.\, 2^{m-2} B_1(r_{\mu+1}),$$

so that

$$\frac{B_1(r_{\mu+1})}{B_1(r_\mu)} \to \infty, \quad \text{as} \quad \mu \to \infty.$$

Thus

$$\frac{\log B_1(r_{\mu+1}) - \log B_1(r_\mu)}{\log(r_{\mu+1}) - \log r_\mu} \to \infty,$$

and so by addition we deduce that

$$\frac{\log B_1(r_\mu)}{\log r_\mu} \to \infty.$$

If

$$r_\mu < r < r_{\mu+1}, \; B_1(r) \geqslant B_1(r_\mu)$$

and

$$\log r \leqslant \log r_\mu + \log 2,$$

so that

$$\frac{\log B_1(r)}{\log r} \to \infty.$$

This gives (b).

Also if $m = 2$, $v_2(x)$ is bounded above and so $v_2(x) = M = \text{const.}$ by Theorem 2.14. Since C is thick, $M > K$, so that $u(x) \equiv M$, contrary to hypothesis. Thus $m \geqslant 3$. In the Corollary, the conclusion is obvious if $u(x)$ is unbounded above on every thick component. If there is a thick component $C(K_1)$ on which $u(x)$ is bounded above by K_0, say, and $C_1(K)$ is a thick component for $K > K_0$, then $C(K_1)$ and $C_1(K)$ are disjoint and case (b) holds, so that $u(x)$ is unbounded on $C_1(K)$. This proves the Corollary and completes the proofs of Theorem 4.15 and its Corollary.

4.6.4. Tracts and growth

Suppose that $u(x)$ is s.h. in R^m and bounded above there. Then it follows from Theorem 3.21 that

$$u(x) \to M, \text{ as } x \to \infty \tag{4.6.11}$$

on almost all fixed straight lines, where M is the least upper bound of $u(x)$ in R^m. Suppose next that $u(x)$ is bounded above by M on a thick component $C(K)$. Then we can apply this result to the function $v_2(x)$ in the last section. Thus on almost all straight lines Γ

$$v_2(x) \to M, \text{ as } x \to \infty \text{ on } \Gamma$$

i.e.

$$M - \varepsilon < v_2(x) < M, \qquad x \in \Gamma, \; |x| > r_0(\varepsilon).$$

If ε is so small that $M - \varepsilon > K$, we deduce that $v_2(x) = u(x)$ so that (4.6.11) still holds. Suppose next that $u(x)$ is unbounded in R^m. Then for every K there exists at least one component $C(K)$, and by the corollary to Theorem 4.15 if $K_2 > K_1 > K_0$, then $u(x)$ is unbounded on every thick component $C(K_1)$ so that each thick component $C(K_1)$ contains at least one thick component $C(K_2)$.

Thus if $N(K)$ denotes the number (possibly infinite) of distinct thick components $C(K)$, then $N(K)$ is nondecreasing with increasing K for $K > K_0$, in the sense that if $N(K) = \infty$ for some value K_1 then $N(K) = \infty$ for $K > K_1$. Thus in every case

$$N_0 = \lim_{K\to\infty} N(K)$$

exists and $0 \leqslant N_0 \leqslant \infty$. Also $N_0 = 0$ if and only if $u(x)$ is bounded above in R^m, which can happen for a non-constant $u(x)$ only if $m \geqslant 3$.

We define N_0 to be the *number of tracts* of $u(x)$. If N_0 is finite, then $N(K) = N_0$ for $K > K_0'$ say. In this case if $K_2 > K_1 > K_0'$ each component $C(K_1)$ contains at least one and so exactly one component $C(K_2)$. All those points in $C(K_1)$ where $u(x) > K_2$ belong to the same thick component $C(K_2)$. If $u(z) = \log|f(z)|$, where $f(z)$ is a non-constant entire function, then N_0 is not less than the number of distinct finite asymptotic values in the classical sense. We can now prove†

THEOREM 4.16. *Suppose that $u(x)$ is s.h. in R^m and of lower order λ. Then if $u(x)$ has N_0 traits, where $N_0 \geqslant 2$, we have*

$$\lambda \geqslant c_1(N_0, m),$$

where $c_1(k, m)$ is the quantity defined in Theorem 4.12. In particular if $N_0 \geqslant 2$, then $\lambda > 0$ and if $N_0 = \infty$, then $\lambda = \infty$.

If $\lambda = \infty$, there is nothing to prove. Thus we may assume that λ is finite. We take N to be a positive integer such that $2 \leqslant N \leqslant N_0$. Then $u(x)$ is unbounded on every component $C(K)$ for $K \geqslant K_0$ say. Also since $N(K)$ is a positive integer or ∞, we may choose K so large that $N(K) \geqslant N$. Thus there exist N mutually disjoint thick components $C_\nu(K)$ for $\nu = 1$ to N.

We now set

$$v_\nu(x) = u(x) - K, \qquad x \in C_\nu(K)$$

$$v_\nu(x) = 0, \qquad \text{elsewhere.}$$

Then the $v_\nu(x)$ satisfy (4.6.1) and (4.6.2) and they are s.h. in R^m by Theorem 4.13. Also $v_\nu(x)$ is not identically zero since the components $C_\nu(K)$ are thick. We may thus choose r_1 so large that $B_0(r_1) > 0$ in Theorem 4.12. Thus we may apply that Theorem and deduce that

$$B_0(r) \geqslant \tfrac{2}{3}B_0(r_1)\left(\frac{r}{r_1}\right)^{c_1}, r > r_1.$$

† For the sharpest known results in this direction see Friedland and Hayman [1976].

Since

$$B_0(r) \leqslant \sup_{|x| \leqslant r} (u(x) - K),$$

we deduce that the lower order of $u(x)$ is at least c_1. If $N_0 = \infty$, we can choose N as large as we please and since $c_1(N, m) \to \infty$ as $N \to \infty$, we deduce that in this case the lower order of $u(x)$ is infinite.

4.6.5. Iversen's Theorem

A classical Theorem of Iversen [1915–1916] states that if $f(z)$ is an entire function then there exists a path Γ going to ∞ such that

$$f(z) \to \infty, \text{ as } z \to \infty \text{ on } \Gamma.$$

We proceed to establish an analogue of this result for subharmonic functions. Serious problems arise from the fact that these functions are not in general continuous. This forces us in the first instance to replace the path Γ by a continuum going to ∞. This concept we now define.

We say that Γ is *a continuum going to* ∞ in R^m if

$$\Gamma = \bigcup_{n=1}^{\infty} \gamma_n$$

where the γ_n are continua such that

$$\gamma_n \cap \gamma_{n+1} \neq \varnothing, \qquad n = 1, 2 \ldots$$

and, given any compact set E, there exists $n_0 = n_0(E)$ such that

$$\gamma_n \cap E = \varnothing, \qquad n > n_0.$$

We shall call Γ an *asymptotic continuum* and say that

$$u(x) \to a, \qquad \text{as } x \to \infty \text{ on } \Gamma,$$

if either a is finite and, given $\varepsilon > 0$, $\exists R_0(\varepsilon)$, such that

$$|u(x) - a| < \varepsilon, \qquad \text{if } x \in \Gamma \text{ and } |x| > R_0(\varepsilon),$$

or if $a = +\infty$ and given $K > 0$, $\exists R_0(K)$, such that

$$u(x) > K, \text{ if } x \in \Gamma \text{ and } |x| > R_0(\varepsilon).$$

Then we have the following result of Talpur [1976]

THEOREM 4.17. *If $u(x)$ is s.h. in R^m, and C is a thick component of $u(x) \geqslant K_1$ for some K_1, then there is an asymptotic continuum $\Gamma \subset C$ such that*

$$u(x) \to M, \qquad \textit{as } x \to \infty \textit{ on } \Gamma,$$

where M is the least upper bound of $u(x)$ on C. In particular we may always choose M to be the upper bound of $u(x)$ in R^m.

Suppose first that M is finite. Then the conclusion holds for almost all fixed straight lines Γ going through the origin, in view of (4.6.11).

Suppose next that $M = +\infty$. In view of Theorem 4.15, Corollary, there exists K_0, such that $u(x)$ is unbounded on every thick component $C(K)$ for $K > K_0$. Take $K_2 = \max(K_0, K_1)$, set $M_n = K_2 + n$, and choose a point x_1 in $C(K_1)$, such that $u(x_1) > M_1$. Since $M = +\infty$, x_1 exists. If x_n has already been defined such that $u(x_n) > M_n$, we define C_n to be the component of $u(x) \geqslant M_n$, which contains x_n. Then $u(x)$ is unbounded on C_n, and so we can find x_{n+1} on C_n such that $u(x_{n+1}) > M_{n+1}$. Also $x_1 \in C(K_1)$ and hence so does C_1 and so x_2. Since $x_{n+1} \in C_n$ and $M_{n+1} > M_n$, $C_{n+1} \subset C_n$. Thus

$$C_n \subset C_{n-1} \subset \ldots \subset C_1 \subset C(K_1).$$

Also since x_n and x_{n+1} lie in C_n, there exists a continuum γ_n containing x_n and x_{n+1}, such that $\gamma_n \subset C_n$. We now set

$$\Gamma = \bigcup_{n=1}^{\infty} \gamma_n.$$

Then Γ lies in C_1 and so in $C(K)$ as required. Also

$$u(x) \geqslant M_n \quad \text{on} \quad \gamma_n.$$

If E is any compact set, then $u(x) < M_{n_0}$ on E for some n_0, so that γ_n does not meet E for $n \geqslant n_0$. Thus Γ is a continuum going to ∞ and

$$u(x) \to \infty \quad \text{as} \quad x \to \infty \quad \text{on} \quad \Gamma.$$

If $u(x)$ is bounded by M in R^m and M is finite, then by Theorem 4.15, we may take $K_1 < M$ and then M is the upper bound of $u(x)$ on the unique thick component of $u(x) \geqslant K_1$. If $M = +\infty$ then if we choose $K_1 > K_0$, where K_0 is the quantity of Theorem 4.15, Corollary, then there exists a thick component $C(K_1)$ and $u(x)$ is unbounded above on $C(K_1)$. This completes the proof of Theorem 4.17.

It is easy to deduce that if $u(x)$ is continuous, we can replace Γ by a path. We have

THEOREM 4.18. *If $u(x)$ is continuous in Theorem 4.17, then Γ may be chosen to be a sectionally polygonal path, in other words the γ_n can be chosen to be straight line segments in the definition of Γ.*

In fact in this case $u(x)$ is uniformly continuous on each γ_n. We suppose that $u(x) \geqslant K_n > K_{n-1}$ on γ_n and chose $\varepsilon = K_n - K_{n-1}$. Then by the uniform continuity we can find δ_n such that

$$u(x) > K_{n-1}, \qquad x \in \gamma_n(\delta_n),$$

where $\gamma_n(\delta_n)$ is the δ_n-neighbourhood of γ_n, i.e. the set of all points distant less than δ_n from some point of γ_n. Evidently $\gamma_n(\delta_n)$ is open. Also if ξ_n, ξ'_n are points of $\gamma_n(\delta_n)$ they can be joined to points x_n, x'_n in γ_n by straight line-segments in $\gamma_n(\delta_n)$. Thus ξ_n, ξ'_n can be joined by a continuum in $\gamma_n(\delta_n)$, so that $\gamma_n(\delta_n)$ is a connected open set, or domain. Now any two points in a domain D can be joined by a broken line or polygonal path in D. For if D_1 is the set of points which can be so joined to a point x_1 in D and D_2 is the remainder, then D_1 and D_2 are both open in D and form a partition of D unless D_2 is empty.

Thus if x_n denotes any point in $\gamma_n \cap \gamma_{n-1}$, we can join x_n to x_{n+1} by a broken line in $\gamma_n(\delta_n)$ on which

$$u(x) \geqslant K_{n-1}.$$

Since $K_{n-1} \to M$, we deduce as before that Γ is a continuum going to infinity which is now a countable union of straight line segments. This proves Theorem 4.18.

4.6.6. Construction of an asymptotic path

The problem of finding a path instead of an asymptotic continuum for general subharmonic functions is rather more difficult. Talpur [1975] proved that in the plane a sectionally polygonal path can indeed be found. More recently Fuglede [1975] showed that a path can also be found in higher dimensions. His proof depends on a deep theorem about Brownian motion by Nguyen-Xuan-Loc and T. Watanabe [1972]. This latter result shows that if x_1, x_2 lie in a connected set G, in which a subharmonic function is positive then x_1, x_2 can be joined by a path (essentially a Brownian motion) in G. From this result it follows at once that the continua γ_n can be replaced by paths in the definition of an asymptotic continuum, so that we get in effect an asymptotic path Γ.

Recently Carleson [1974] has shown by a subtle direct argument that a sectionally polygonal asymptotic path always exists. The proof does not use the arcwise connectedness of connected sets in which $u(x) > 0$.

We have not space here for all the above results, some of which lie very deep. Instead we shall prove a somewhat weaker arcwise connectedness theorem, which yields the existence of an asymptotic path, though not a polygonal path. Our result is

THEOREM 4.19. *Suppose that $u(x)$ is s.h. in a neighbourhood N of a continuum F and that $u(x) \geqslant K$ in F. Then if x_1, x_2 are two points of F, there exists a path joining x_1 to x_2 in N, on which $u(x) \geqslant K - 1$.*

COROLLARY. *The continua γ_n in the construction of Γ for Theorem 4.17 can be taken to be paths. Thus Γ becomes an asymptotic path, i.e. the continuous image of the positive real axis.*

To deduce the Corollary from Theorem 4.19, it is sufficient to suppose that $M = +\infty$, since otherwise Γ may be taken to be a straight line. Let x_n be the points in the proof of Theorem 4.17. Then by the construction x_n, x_{n+1} can be joined by a continuum on which $u(x) \geqslant K_2 + n$, where K_2 is a constant. Thus, in view of Theorem 4.19, x_n, x_{n+1} can be joined by a path on which $u(x) \geqslant K_2 + n - 1$. We can write this path as

$$x = \alpha(t), \qquad n - 1 \leqslant t \leqslant n,$$

where $\alpha(t)$ is continuous and $\alpha(n-1) = x_n$, $\alpha(n) = x_{n+1}$. Thus $\alpha(t)$ is defined for $t \geqslant 0$ and is continuous and further

$$u\{\alpha(t)\} \geqslant K_2 + t - 1$$

for all t. Thus the path Γ given by $x = \alpha(t)$ for $t \geqslant 0$ is the required asymptotic path.

We proceed to prove Theorem 4.19. Our argument will be based on Lemma 4.12. However we now require a somewhat more precise statement of the result obtained there. This is

LEMMA 4.13. *Suppose that $u(x)$ is s.h. in $C(x_1, r)$, that $u(x_1) \geqslant K + h$ and that $u(x) \leqslant K + B$ in $C(x_1, r)$, where h, B are positive constants and K is real. Then if $u(x_2) \geqslant K + h$, and $|x_2 - x_1| \leqslant \delta$, where*

$$\delta = r\left(\frac{2h}{3B}\right)^{\alpha},$$

and $\alpha = 1/c_1$ is a positive constant depending only on m, then x_1, x_2 can be joined by a continuum in $C(x_1, r)$ on which $u(x) \geqslant K$.

We now proceed to our construction. The neighbourhood N of F will contain the set F' of all points distant at most r from F for some positive r. Since F' is compact $u(x) \leqslant K - 1 + B$ in F', where B is a positive quantity.

With this definition of B, we define

$$r_n = r\,2^{-n}$$

$$\delta_n = r_n\left(\frac{2^{1-n}}{3B}\right)^{\alpha}$$

where α is the constant of Lemma 4.13. We define positive integers $j_1, j_2, \ldots, j_n, \ldots$ and points ξ_j^n, for $0 \leqslant j \leqslant j_1 j_2 \ldots j_n$ as follows. In the first instance, since F is connected, we can find a chain of points $\xi_j^1, 0 \leqslant j \leqslant j_1$ in F, such that

$$\xi_0^1 = x_1, \qquad \xi_{j_1}^1 = x_2$$

and

$$|\xi_{j+1}^1 - \xi_j^1| < \delta_1.$$

In view of Lemma 4.13, it follows that ξ_j^1 and ξ_{j+1}^1 can be joined by a continuum γ'_j, lying in $C(\xi_j, r_1)$ and such that $u(x) \geqslant K - 1 + \frac{1}{2}$, on γ'_j. Since γ'_j is connected we can find a finite number of points on γ'_j forming a chain from ξ_j^1 to ξ_{j+1}^1 and having mutual distance less than δ_2. By repeating points if necessary we may assume that there is the same number, $j_2 + 1$, of points in each such chain including ξ_j^1 and ξ_{j+1}^1. We can number all these points ξ_j^2, $0 \leqslant j \leqslant j_1 j_2$. If p is an integer, $0 \leqslant p \leqslant j_1$, we have

$$\xi_{pj_2}^2 = \xi_p^1.$$

If $pj_2 \leqslant j \leqslant (p+1)j_2$, then ξ_j^2 lies on a continuum γ'_p, joining ξ_p^1 to ξ_{p+1}^1, lying in $C(\xi_p^1, r_2)$ and on which $u(x) \geqslant K - 1 + \frac{1}{2}$.

We suppose that ξ_p^n, has been defined for $0 \leqslant p \leqslant j_1 j_2 \ldots j_n$, such that

$$|\xi_{p+1}^n - \xi_p^n| < \delta_n, \qquad 0 \leqslant p \leqslant j_1 j_2 \ldots j_n - 1,$$

and such that

$$u(\xi_p^n) \geqslant K - 1 + 2^{1-n}.$$

Then ξ_p^n, ξ_{p+1}^n can be joined by a continuum in $C(\xi_p^n, r_n)$ on which $u(x) \geqslant K - 1 + 2^{-n}$, and on this continuum we define the points ξ_j^{n+1}, $pj_{n+1} \leqslant j \leqslant (p+1)j_{n+1}$, such that

$$\xi_{pj_{n+1}}^{n+1} = \xi_p^n, \qquad \xi_{(p+1)j_{n+1}}^{n+1} = \xi_{p+1}^n \tag{4.6.12}$$

and

$$|\xi_{j+1}^{n+1} - \xi_j^{n+1}| < \delta_{n+1}. \tag{4.6.13}$$

Thus the points ξ_j^n are defined for all n.

Let Q be the set of all rational numbers t of the form

$$t = \frac{p}{j_1 j_2 \ldots j_n},$$

for some n, where p is an integer, $0 \leqslant p \leqslant j_1 j_2 \ldots j_n$. Then we define a function $\alpha(t)$ on these rationals by

$$\alpha(t) = \xi_p^n.$$

The definition is unique in view of (4.6.12). Next we note that $\alpha(t)$ is uniformly continuous in Q. In fact suppose that t, t' are two numbers in Q, such that

$$|t - t'| < \frac{1}{j_1 j_2 \cdots j_n}. \tag{4.6.14}$$

Then t, t' either belong to the same or adjacent intervals of the form

$$\frac{p}{j_1 j_2 \cdots j_n} \leqslant t \leqslant \frac{p+1}{j_1 j_2 \cdots j_n}. \tag{4.6.15}$$

Consider the points t in the interval (4.6.15). If $t = q/(j_1 j_2 \cdots j_{n+1})$, then $\alpha(t)$ is a point in the continuum in $C(\xi_p^n, r_n)$ joining ξ_p^n and ξ_{p+1}^n. Thus if $t_0 = p/(j_1 j_2 \cdots j_n)$, we have

$$|\alpha(t) - \alpha(t_0)| \leqslant r_n.$$

If t lies in (4.6.15) and is of the form

$$t = \frac{q}{j_1 j_2 \cdots j_{n+k}}$$

we deduce inductively that

$$|\alpha(t) - \alpha(t_0)| \leqslant r_n + r_{n+1} + \cdots r_{n+k} \leqslant 2r_n.$$

Thus this inequality holds for all $t \in Q$ which satisfy (4.6.15). If t, t' satisfy (4.6.14) they lie in the same or adjacent intervals of the type (4.6.15) and we deduce that

$$|\alpha(t) - \alpha(t')| \leqslant 4r_n. \tag{4.6.16}$$

Thus $\alpha(t)$ is uniformly continuous in Q and, since Q is dense in $[0, 1]$, $\alpha(t)$ can be uniquely extended as a continuous function in $[0, 1]$. For any t in $[0, 1]$ we need only find a sequence $t_n \in Q$, such that $t_n \to t$. In view of the uniform continuity $\alpha(t_n)$ also converges to a limit which we define to be $\alpha(t)$. Clearly this definition is independent of the approximating sequence t_n. Further if t, t' are any numbers in $[0, 1]$ satisfying (4.6.14), it follows that (4.6.16) still holds, so that $\alpha(t)$ is continuous. Thus $x = \alpha(t)$ is a path γ joining x_1, x_2. Also points of γ are distant at most $r = 2r_1$ from F, so that γ lies in N.

It remains to show that $u(x) \geqslant K - 1$ on γ. To see this, suppose that $z = \alpha(t)$ is a point on γ and let t_n be the largest number of the form

$$t_n = \frac{p}{j_1 j_2 \cdots j_n},$$

such that p is an integer and $t_n \leqslant t$. Let $z_n = \alpha(t_n)$. Then by our construction z_n

and z_{n+1} lie on a continuum γ_n in $C(z_n, r_n)$ on which $u(x) \geqslant K - 1 + 2^{-n}$. Let

$$\Gamma_N = \bigcup_{n=1}^{N} \gamma_n.$$

Then we deduce that Γ_N lies in $C(z_1, r)$, since for ξ on γ_n we have

$$|\xi - z_1| \leqslant \sum_{\nu=1}^{n-1} |z_\nu - z_{\nu+1}| + r_n \leqslant \sum_{\nu=1}^{n} r_\nu = r_1 \sum_{\nu=1}^{n} 2^{1-\nu} < 2r_1 = r.$$

Thus Γ_N lies entirely on the component C, of $u(x) \geqslant K - 1$ in $C(z_1, r)$, which contains z_1. In particular all the points z_n lie on C. Since C is closed and $z_n \to z$ as $n \to \infty$, we deduce that z lies on C, i.e. $u(z) \geqslant K - 1$. This completes the proof of Theorem 4.19.

It is sometimes useful to have a Jordan arc joining distinct points x_1, x_2, i.e. a path without selfintersections. The possibility of doing this derives from†

THEOREM 4.20. *If γ is a path with distinct endpoints x_1, x_2 lying on a set E, then there exists a Jordan arc on E, with endpoints x_1, x_2.*

We write I_0 for the closed interval $[0, 1]$ and let γ be given by

$$x = \alpha(t), \qquad t \in I_0, \qquad \alpha(0) = x_1, \qquad \alpha(1) = x_2.$$

If for $t \neq t'$ we always have $\alpha(t) \neq \alpha(t')$ then γ is a Jordan arc. If not let (t_1, t_1') be an interval I_1, such that $t_1' - t_1$ is maximal and $\alpha(t_1') = \alpha(t_1)$. If $I_1, I_2, \ldots, I_{n-1}$ have been defined we choose (t_n, t_n') to be an interval I_n of maximal length in

$$I_0 - \bigcup_{\nu=1}^{n-1} I_\nu,$$

such that $\alpha(t_n) = \alpha(t_n')$, provided such a pair of distinct points t_n, t_n' exists. The process either terminates when no new interval I_n can be defined or continues indefinitely. In either case we write $J = I_0 - \cup I_\nu$. Then

(i) J is a perfect set, i.e. a closed set without isolated points, apart possibly from the points 0, 1.

(ii) If t, t' are distinct points of J, which are not endpoints of the same interval I_ν, then $\alpha(t) \neq \alpha(t')$.

Clearly J is closed since $\cup I_\nu$ is open. If J had an isolated point $t_0 \neq 0, 1$, then there would be intervals I_μ, I_ν on each side of t_0. This contradicts our construction which took the intervals I_μ, I_ν to be of maximal length at each stage, since $I_\mu \cup t_0 \cup I_\nu$ would be a larger interval than I_μ or I_ν. This proves (i).

† See e.g. Kerékjártó [1923, p. 103].

Also (ii) follows since if $\alpha(t) = \alpha(t')$, then at some stage $|t - t'|$ would be larger than the length of I_ν, which leads to a contradiction.

We next construct a function $\beta(t)$ which is continuous and non-decreasing in $[0, 1]$, and is such that $\beta(0) = 0$, $\beta(1) = 1$ and $\beta(t)$ is constant precisely in the intervals I_ν. If one of the intervals I_ν has 0 as an endpoint we set $\beta(t) = 0$ in the closure of this interval. Similarly if I_ν has 1 as an endpoint we set $\beta(t) = 1$ in the closure of I_ν. If there are any other intervals I_ν, we choose one of largest length, relabel it $I_{1/2}$ and set $\beta(t) = \frac{1}{2}$ in $I_{1/2}$. Next if there are intervals I_ν to the left of $I_{1/2}$ we choose one of largest length, redefine it to be $I_{1/4}$ and set $\beta(t) = \frac{1}{4}$ there. Similarly $I_{3/4}$ is defined and $\beta(t) = \frac{3}{4}$ there. Suppose at any stage that points or intervals I_ν have been formed where $\beta(t) = p/2^q$ and $(p + 1)/2^q$ respectively. If between these there is an interval I_ν, we set $\beta(t) = (2p + 1)/2^{q+1}$ in the largest such interval. Otherwise, since J does not have isolated points, there must be a whole interval J_0 of J between the points where $\beta(t) = p/2^q$ and those where $\beta(t) = (p + 1)/2^q$. We define $\beta(t)$ as a linear function in J_0 and in particular we define $\beta(t) = (2p + 1)/2^{q+1}$ at the midpoint of J_0.

Since at each stage we have defined $\beta(t)$ in the largest interval I_ν where it was not previously defined, it is clear that $\beta(t)$ is defined and constant in each of the intervals I_ν and at their endpoints. In complete intervals of J between I_μ and I_ν, $\beta(t)$ is continuous and strictly increasing. At other points t_0 we see that t_0 is a limit of intervals where $\beta(t)$ is defined and so, since $\beta(t)$ is non-decreasing, where it is defined, we obtain a value for $\beta(t_0)$. Since $\beta(t)$ assumes every value of the form $p/2^q$, $\beta(t)$ cannot have any jumps and so is continuous. From our construction $\beta(t)$ is only constant on intervals I_ν, and these correspond to (some) values of the form $p/2^q$.

For any τ, such that $0 \leqslant \tau \leqslant 1$, let $P = \beta^{-1}(\tau)$, i.e. P is the set of points t where $\beta(t) = \tau$. If P is a single point t, then t is not an endpoint if any of the intervals I_ν, and we set $h(\tau) = \alpha(t)$. If P is an interval, given by $t_1 \leqslant t \leqslant t_1'$, then (t_1, t_1') is an interval I_ν and so $\alpha(t_1) = \alpha(t_1')$. We set the common value equal to $h(\tau)$. We see that $h(\tau)$ is still continuous, but assumes different values at different points, i.e. $x = h(\tau)$ defines a Jordan arc from x_1 to x_2. Also the values assumed by $h(\tau)$ form a subset of those assumed by $\alpha(t)$. This proves Theorem 4.20. We note in particular that the path in Theorem 4.19 may always be assumed to be a Jordan arc.

4.6.7. Growth on asymptotic paths

It is natural to ask how quickly $u(x)$ must grow to infinity on some asymptotic path. If we confine ourselves to upper growth we can give a fairly complete answer to this problem.†

† When $u(z) = \log|f(z)|$, and f is entire, the next two results were proved by Hayman [1960].

THEOREM 4.21. *Suppose that $u(x)$ is unbounded above in R^m and has finite lower order λ, or more generally that the number N_0 of tracts of $u(x)$ is finite. Then given any sequence x_n, such that*

$$u(x_n) \to \infty,$$

there exists an asymptotic path Γ containing an infinite subsequence of the points x_n. If $N_0 = 1$, we may choose Γ to contain all but a finite number of the points x_n.

COROLLARY. *If $N_0 < \infty$, there exists an asymptotic path Γ, on which $u(x)$ has the same order and type-class as in R^m.*

By Theorem 4.16, N_0 is finite if λ is finite. Suppose then that N_0 is finite. Then there exists K_0, such that if $K \geqslant K_0$, there exist exactly N_0 thick components $C_\nu(K)$, $\nu = 1$ to N_0 of $u(x) \geqslant K$. Also each such component contains exactly one component $C_\nu(K')$, if $K' > K$. If x_n is the sequence of Theorem 4.21, then all but a finite number of the x_n lie in

$$\bigcup_{\nu=1}^{N} C_\nu(K_0).$$

Thus at least one of the components, $C_1(K_0)$ say, contains infinitely many of the points x_n.

Suppose that y_n is an infinite subsequence of the x_n, lying in $C_1 = C_1(K_0)$ and suppose further that

$$u(y_n) > u(y_{n-1}),$$

and that

$$u(y_1) > K_0.$$

Choose K_n so that $u(y_n) < K_n < u(y_{n+1})$, $n \geqslant 1$. Then the component $C_1(K_n)$ contains all points in C_1 on which $u(x) > K_n$ and in particular y_{n+1} and y_{n+2}. Hence we can find a continuum in $C_1(K_n)$ which contains y_{n+1} and y_{n+2}, and hence, by Theorem 4.19, a path γ_n, on which $u(x) \geqslant K_n - 1$ from y_{n+1} to y_{n+2}. Then

$$\Gamma = \bigcup_{n=1}^{\infty} \gamma_n$$

is the required asymptotic path. For clearly Γ contains all the y_n. Also outside the continuum

$$\bigcup_{\nu=1}^{n-1} \gamma_\nu,$$

we have

$$u(x) \geqslant K_{n-1} \text{ on } \Gamma,$$

so that

$$u(x) \to +\infty, \quad \text{as} \quad x \to \infty \text{ on } \Gamma.$$

Also since on γ_n we have $u(x) \geqslant K_n - 1$, γ_n lies outside a fixed compact set if n is large enough.

If $N_0 = 1$, there is only one thick component $C(K)$ for $K \geqslant K_0$ which necessarily contains all the points x_n, for which $u(x_n) > K$. We may assume without loss in generality that $u(x_n)$ is non-decreasing with n. Let n_0 be the first index such that

$$u(x_{n_0}) > K_0.$$

Let n_p be the first index such that $u(x_{n_p}) > u(x_{n_{p-1}})$ and choose K_p such that

$$u(x_{n_p}) < K_p < u(x_{n_{p+1}}).$$

Then the component $C(K_p)$ contains all the x_ν for $\nu \geqslant n_{p+1}$, and we can find a continuum γ_p in $C(K_p)$ which contains all the x_ν for $n_{p+1} \leqslant \nu \leqslant n_{p+2}$. For we can find such a continuum $\gamma_{p,\nu}$ containing $x_{n_{p+1}}$ and x_ν and set

$$\gamma_p = \bigcup_{\nu = n_{p+1}}^{n_{p+2}} \gamma_{p,\nu}$$

Now we use Theorem 4.19 to find a path γ'_ν from x_ν to $x_{\nu+1}$, $n_{p+1} \leqslant \nu < n_{p+2}$, on which $u(x) \geqslant K_p - 1$. Then

$$\Gamma = \bigcup_{p=1}^{\infty} \gamma_p'$$

is the required path, which contains all the x_ν for $\nu \geqslant n_1$. This completes the proof of Theorem 4.21.

It remains to prove the corollary. We need merely choose for x_n a sequence on which the growth is maximal. If $u(x)$ has order ρ we choose x_n so that

$$\frac{\log u(x_n)}{\log |x_n|} \to \rho.$$

If ρ is finite and $u(x)$ has at least type T, we choose x_n so that

$$\varliminf_{n \to \infty} \frac{u(x_n)}{|x_n|^\rho} \geqslant T.$$

Then if y_n is the subsequence of x_n which lies on Γ we have

$$B_\Gamma(|y_n|, u) \geqslant u(y_n),$$

and the Corollary follows.

We next consider the case of infinite lower order. Here the results weaken. We have

THEOREM 4.22. *If $u(x)$ has infinite order then given any $\rho > 0$, there exists an asymptotic path $\Gamma = \Gamma(\rho)$ on which $u(x)$ has order at least ρ.*

Thus there exists an asymptotic path on which $u(x)$ has an arbitrarily large finite order. However we shall see that $u(x)$ need not have infinite order on any asymptotic path.

We proceed to prove Theorem 4.22. If the number N_0 of tracts is finite, it follows from Theorem 4.21, corollary, that $u(x)$ has infinite order on some asymptotic path. Thus we may assume that $N_0 = \infty$.

We choose a positive integer N so large that $c_1(N, m) > \rho$, where $c_1(N, m)$ is the quantity of Theorem 4.12. Since $N_0 = \infty$, we can find N mutually disjoint components $C_\nu = C_\nu(K_0)$, $\nu = 1$ to N, if K_0 is sufficiently large. We prove first that at least one of the components C_ν has the property that $u(x)$ has order greater than ρ on every thick sub-component C'_ν contained in C_ν.

Suppose contrary to this that for $\nu = 1$ to N, there exists a thick component $C'_\nu = C_\nu(K_\nu)$, contained in C_ν on which $u(x)$ has order at most ρ. We set

$$\begin{aligned} v_\nu(x) &= u(x) - K_\nu, \qquad x \in C'_\nu \\ &= 0, \qquad \text{elsewhere.} \end{aligned}$$

Then the functions $v_\nu(x)$ satisfy the hypotheses of Theorem 4.12 in $D(0, r)$, when r is sufficiently large. Thus for all large r we have with the notation of Theorem 4.12.

$$B_0(r) \geqslant Ar^{c_1},$$

where A is a constant, and so for at least one ν

$$B_\nu(r) \geqslant Ar^{c_1}, \tag{4.6.17}$$

where

$$B_\nu(r) = K_\nu + \sup_{x \in C'_\nu, |x| = r} u(x).$$

Thus for at least one ν the inequality (4.6.17) holds for some arbitrarily large r. This contradicts the assumption that the order of $u(x)$ on C'_ν is at most $\rho < c_1(N, m)$.

Suppose now that $C = C(K_0)$ is a component such that $u(x)$ has order greater than ρ on every thick sub-component of C and in particular on the whole of C. We define $K_n = K_0 + n$, and choose for x_1 any point of C, such that

$$u(x_1) > K_1 + |x_1|^\rho.$$

If x_n has already been defined, we choose C_n to be the component of $u(x) \geqslant K_n$, containing x_n, and find x_{n+1} on C_n, such that

$$u(x_{n+1}) > K_{n+1} + |x_{n+1}|^\rho. \tag{4.6.18}$$

We prove by induction that

$$C_n \subset C_{n-1} \subset \ldots \subset C_1 \subset C_0.$$

Thus $u(x)$ has order greater than ρ on C_n, so that x_{n+1} exists. Further x_n and x_{n+1} lie in a continuum γ_n in C_n, on which

$$u(x) \geqslant K_n,$$

and so there is a path γ_n' joining x_n to x_{n+1} on which $u(x) \geqslant K_n - 1$. Thus

$$\Gamma = \bigcup_{n=1}^{\infty} \gamma_n'$$

is an asymptotic path containing the sequence x_n. In view of (4.6.18), $u(x)$ has order at least ρ on Γ. This proves Theorem 4.22.

4.6.8. Three examples

We conclude the chapter by three examples to show that the conclusions of the previous section cannot be significantly sharpened.

Examples

1. The function $u(x) = |x_1|$, where x is $(x_1, x_2, \ldots x_m)$, is s.h. of order 1 in R^m, when $m \geqslant 2$. If $K > 0$, the set $u(x) > K$ splits into the two components $x_1 > K$ and $x_1 < -K$. Thus there are two tracts. If y_n is the sequence $\sim((-1)^n n, 0, \ldots 0,)$, then

$$u(y_n) = n \to \infty.$$

But any continuum joining y_n to $y_{n'}$ where n is even and n' is odd must cross the hyperplane $x_1 = 0$. Thus if

$$\Gamma = \bigcup_{n=1}^{\infty} \gamma_n$$

is an asymptotic continuum, all the γ_n from a certain point onwards must lie in the same half-space $x_1 > 0$ or $x_1 < 0$, and so Γ cannot contain infinitely many y_n for both odd n and even n. Thus the condition $N_0 = 1$ is essential for the last part of Theorem 4.21. It can be shown (Friedland and Hayman [1976]), that if $N_0 = 2$, then the lower order is at least one.

2. Let η_n be a sequence of positive numbers such that

$$\sum_{n=1}^{\infty} \eta_n = 2,$$

let ε_n be a sequence of positive numbers such that

$$\varepsilon_n r^{1/\eta_n} \to 0, \qquad \text{as } n \to \infty$$

for any fixed positive r, and set

$$s_n = \sum_{\nu=1}^{n} \eta_\nu, \qquad s_0 = 0.$$

Then for $z = re^{i\theta}$,

$$\pi s_{n-1} \leqslant \theta \leqslant \pi s_n, \qquad n = 1 \text{ to } \infty$$

we define

$$u(z) = \varepsilon_n r^{1/\eta_n} \sin\left(\frac{\theta - \pi s_{n-1}}{\eta_n}\right).$$

Then $u(z)$ is s.h. in the z-plane. This is obvious for z not on the lines $\theta = \pi s_n$, since except on these lines $u(z)$ is harmonic. Since $u(z) = 0$ on all these lines and $u(z) \geqslant 0$ elsewhere, the mean-value inequality holds also on the lines.

Finally $u(z)$ is continuous everywhere. This is obvious except on the positive real axis. Clearly as $\theta \to 0$ from above we have

$$u(z) = \varepsilon_1 r^{1/\eta_1} \sin(\theta/\eta_1) \to 0$$

if $|z|$ remains bounded. If $\theta \to 2\pi$, while $r \leqslant R$,

$$u(z) = O\{\varepsilon_n R^{1/\eta_n}\} \to 0$$

also. Thus $u(z)$ is continuous on the positive axis.

Evidently $u(z)$ has infinite order, since for any fixed n

$$\varliminf_{r\to\infty} \frac{B(r, u)}{r^{1/\eta_n}} \geqslant \varepsilon_n > 0,$$

and $\eta_n \to 0$. By a suitable choice of the η_n and ε_n it is not difficult to see that we can make $B(r, u)$ tend to infinity as rapidly as we please (by making the η_n tend to zero extremely rapidly and then choosing $\varepsilon_n = n^{-1/\eta_n}$) or as slowly as we please, (by choosing the ε_n very small) subject to having infinite lower order (the reader should verify this).

On the other hand $u(z) = 0$ on all the lines $\arg z = \pi s_n$. Thus if

$$\Gamma = \bigcup_{\nu=1}^{\infty} \gamma_\nu$$

is an asymptotic continuum, then $u(z) > 0$ on γ_ν for $\nu \geqslant \nu_0$, and so γ_ν cannot meet any of these lines. If γ_{ν_0} contains a point z for which

$$\pi s_{n-1} < \arg z < \pi s_n$$

this inequality must hold on the whole of γ_{ν_0}, hence on γ_{ν_0+1}, and so on all the γ_ν for $\nu \geqslant \nu_0$. Thus Γ finally lies in the above angle and so the order of $u(z)$ on Γ is at most $1/\eta_n$. Thus $u(z)$ has finite order on any asymptotic continuum.

3. The quantity $c_1(k, m)$ of Theorem 4.12 has played a key role in the above theory. It is therefore of interest to obtain a lower bound for this quantity. We proceed to show that the estimate of Theorem 4.12 gives at least the right order of magnitude as $k \to \infty$ for fixed m.

For this purpose we recall the examples of Section 4.3.1. They prove that, given $\lambda > 2$, there exists a function $u(x)$, s.h. in R^m, and such that

(4.6.19) $$u(x) = 0, \qquad \pi/\lambda \leqslant \theta \leqslant \pi,$$

(4.6.20) $$u(x) = |x|^\alpha(1 + \cos \lambda\theta), \qquad 0 \leqslant \theta < \pi/\lambda,$$

where

(4.6.21) $$\alpha = 3\lambda m^{3/2}.$$

Suppose that $\xi = (\xi_1, \xi_2, \ldots, \xi_m)$ is any point on the unit sphere and for any $x \in R^m$, such that $x \neq 0$, let $x' = (x_1', x_2', \ldots, x_m') = x/|x|$ be the corresponding point on the unit sphere and define θ by

(4.6.22) $$\cos \theta = \sum_{\mu=1}^{m} x_\mu' \xi_\mu.$$

Then

$$|x' - \xi|^2 = 2(1 - \cos \theta).$$

Since subharmonicity is invariant under orthogonal transformations in R^m, we deduce that the function $u(x)$ defined by (4.6.19) to (4.6.21) with θ given by (4.6.22) is still subharmonic. In particular we note that the function $u(x)$ defined in this way is non-negative, not identically zero and vanishes for

$$|x' - \xi|^2 \geqslant 2\left(1 - \cos \frac{\pi}{\lambda}\right),$$

i.e. for

$$|x' - \xi| \geqslant 2\sin\frac{\pi}{2\lambda}.$$

We now note that given $\delta > 0$, we can certainly find N points $\xi_j, j = 1$ to N on the unit-sphere in R^m, such that

(4.6.23) $$|\xi_i - \xi_j| \geqslant \delta > 0, \qquad 1 \leqslant i < j \leqslant N$$

where

$$N > A_1\left(\frac{1}{\delta}\right)^{m-1},$$

and A_1 is a constant depending on m only. If $\delta > 1/\sqrt{m}$, we may take $N = 1$. Otherwise we choose for k the largest positive integer such that

$$k\delta \leqslant \frac{1}{\sqrt{m}},$$

and consider for the ξ_i all points of the form $(x_1, x_2, \ldots, x_m)$, where

$$x_j = v_j\delta, \qquad -k \leqslant v_j \leqslant k,$$

for $j = 1$ to $m - 1$, and

$$x_m = \sqrt{\left(1 - \sum_{j=1}^{m-1} x_j^2\right)}.$$

Since there are $(2k + 1)$ choices for each v_j we obtain in this way

$$N = (2k + 1)^{m-1} > (1/(\delta\sqrt{m}))^{m-1}$$

distinct points, whose mutual distance is at least δ.

We label these points ξ_1 to ξ_N and given $\alpha > 3m^{3/2}$, we define λ by (4.6.21), choose

$$\delta = 4\sin(\pi/2\lambda)$$

and construct the function $v_j(x)$ as $u(x)$ was constructed above with ξ_j instead of ξ. Then the sets $v_j(x) > 0$ are disjoint for different j, since the sets

$$|x' - \xi_j| < 2\sin\left(\frac{\pi}{2\lambda}\right) = \tfrac{1}{2}\delta$$

are disjoint from each other. Thus if

$$u(x) = \sum_{j=1}^{N} v_j(x)$$

the function $u(x)$ is s.h. in R^m and has order α, mean type and for $K > 0$ there are N components $u(x) \geqslant K$, where

$$N \geqslant \{4\sqrt{m} \sin (\pi/2\lambda)\}^{1-m} = (4\sqrt{m} \sin (1{\cdot}5\pi m^{3/2}/\alpha))^{1-m}$$
$$\geqslant \left(\frac{\alpha}{6\pi m^2}\right)^{m-1}.$$

In particular we see that for this value of N we must have

$$c_1(N, m) \leqslant \alpha < 6\pi m^2 N^{1/(m-1)}$$

in Theorems 4.12 and 4.16. More precise results have been obtained recently by Brannan, Fuchs, Hayman and Kuran [1975]. They show that the correct order for $c_1(N, m)$ within a factor which is an absolute constant is $\log N$, if $2 \leqslant N < 2^m$, and $mN^{1/(m-1)}$ if $N \geqslant 2^m$.

Chapter 5

Capacity and Null Sets

5.0. INTRODUCTION

We saw in Chapter 3 that a s.h. function can locally be expressed as the sum of a potential and a harmonic function. In this chapter we study the behaviour of the potentials which arise in this way and (when we can do this without extra trouble) slightly more general potentials. Our study leads us naturally to the notion of the capacity of sets in R^m, which is an extension of the notion of size in the sense of volume, length or area. Certain sets of capacity zero turn out to act as null sets from various points of view, such as the maximum principle for bounded harmonic functions and the problem of Dirichlet and we are led in this way to a much more general solution to the problem of Dirichlet, Green's function and the Poisson–Jensen formula. Finally in the last two sections we develop Choquet's results on capacitability, which show that for analytic sets and in particular for Borel sets inner and outer capacity are equal.

5.1. POTENTIALS AND α-CAPACITY

We define

$$K_\alpha(x) = -|x|^{-\alpha}, \qquad \alpha > 0$$

$$K_0(x) = \log |x|.$$

Let E be a compact set in R^m, μ a mass distribution in E, such that $\mu(E) = 1$.
Then for $x \in R^m$ we define the α-potential

$$V_\alpha(x) = \int_E K_\alpha(x - y)\, \mathrm{d}\mu(y).$$

Since $K_\alpha(x)$ is s.h. in R^m for $\alpha \leq m - 2$, it follows from Theorem 3.6(i) that in this case $V_\alpha(x)$ is s.h. in R^m. Also if $\alpha = m - 2$, $V_\alpha(x)$ is harmonic outside E. This is the most important case, and we write $K(x)$ for $K_\alpha(x)$ and $V(x)$ for $V_\alpha(x)$ if $\alpha = m - 2$. Also, if $\alpha > m - 2$, $V_\alpha(x)$ is super-harmonic outside E.

We note that in any case $K_\alpha(x)$ is u.s.c. and so $V_\alpha(x)$ is u.s.c. in R^m by Theorem 3.6. Also $V_\alpha(x)$ is continuous outside E for the same reason. We can prove a little more.

Theorem 5.1. (The Continuity Principle)†. *If x_0 is a point of E and $V_\alpha(x)$ is continuous at x_0 as a function on E only, then $V_\alpha(x)$ is also continuous at x_0 as a function in R^m.*

Since $V_\alpha(x)$ is continuous at x_0 as a function on E, we deduce that $V_\alpha(x_0) = V_0 > -\infty$. Also $V_\alpha(x)$ is in any case u.s.c. so that

$$\overline{\lim_{x \to x_0}}\ V_\alpha(x) \leqslant V_0$$

We proceed to prove that

$$\underline{\lim_{x \to x_0}}\ V_\alpha(x) \geqslant V_0, \tag{5.1.1}$$

given that this inequality holds as $x \to x_0$ along points of E.

Since V_0 is finite we deduce that the point x_0 cannot have a positive mass. Let ρ be a small positive number. Let D be the complement of E and let

$$D_\rho = D \cap D(x_0, \rho), \qquad E_\rho = E \cap C(x_0, \rho).$$

If x_0 is an isolated point of E, then x_0 carries no mass and so $V_\alpha(x)$ is continuous at x_0 in any case. Also if x_0 is an interior point of E then there is some neighbourhood $D(x_0, \rho_0)$ which lies entirely in E and Theorem 5.1 is trivial. Thus we may suppose that x_0 is a limit-point of both E and D.

Let $x \in D_\rho$ and let x_1 be a nearest point to x of E_ρ so that

$$|x_1 - x| \leqslant |x - a|$$

for any $a \in E_\rho$. Then

$$|x_1 - a| \leqslant |x - x_1| + |x - a| \leqslant 2|x - a| \tag{5.1.2}$$

for any $a \in E_\rho$.

We now suppose that, given $\varepsilon > 0$, ρ is chosen so small that

$$\int_{E\rho} - K_\alpha(x_0 - a)\, \mathrm{d}\mu(a) < \varepsilon.$$

This is possible since $V_\alpha(x_0)$ is finite. Next it is clear that

$$\int_{E - E\rho} K_\alpha(x - a)\, \mathrm{d}\mu(a)$$

† Evans [1933].

is continuous at x_0. Hence, since $V_\alpha(x_1)$ is continuous as a function of x_1 at x_0, for $x_1 \in E$, it follows that

$$\int_{E_\rho} K_\alpha(x_1 - a)\,\mathrm{d}\mu(a) = V_\alpha(x_1) - \int_{E - E_\rho} K_\alpha(x - a)\,\mathrm{d}\mu(a)$$

is continuous at x_0 as a function of x_1, for x_1 on E. Thus if $|x_1 - x_0| < \rho_1 < \rho$ we have

$$\int_{E_\rho} - K_\alpha(x_1 - a)\,\mathrm{d}\mu(a) < \int_{E_\rho} - K_\alpha(x_0 - a)\,\mathrm{d}\mu(a) + \varepsilon < 2\varepsilon.$$

Suppose now that $|x - x_0| < \rho_2 < \frac{1}{2}\rho_1$. Then $|x_1 - x_0| < \rho_1$ by (5.1.2).

Also if $\alpha = 0$, we have from (5.1.2)

$$- K_\alpha(x - a) \leqslant - K_\alpha(x_1 - a) + \log 2,$$

so that

$$\int_{E_\rho} - K_\alpha(x - a)\,\mathrm{d}\mu(a) \leqslant \int_{E_\rho} - K_\alpha(x_1 - a)\,\mathrm{d}\mu(a) + \mu(E_\rho)\log 2 < 3\varepsilon,$$

if we choose ρ so small that $\mu(E_\rho) < \varepsilon$. If $\alpha > 0$, we have from (5.1.2)

$$\int_{E_\rho} - K_\alpha(x - a)\,\mathrm{d}\mu(a) \leqslant 2^\alpha \int_{E_\rho} - K_\alpha(x_1 - a)\,\mathrm{d}\mu(a) < 2^{1+\alpha}\,\varepsilon.$$

Also since

$$\int_{E - E_\rho} K_\alpha(x - a)\,\mathrm{d}\mu(a)$$

is continuous at $x = x_0$, we clearly have for $|x - x_0| < \rho_2$, and so $|x_1 - x_0| < 2p_2$, if ρ_2 is sufficiently small,

$$\left| \int_{E - E_\rho} K_\alpha(x - a)\,\mathrm{d}\mu(a) - \int_{E - E_\rho} K_\alpha(x_1 - a)\,\mathrm{d}\mu(a) \right| < \varepsilon,$$

since both integrals approximate to

$$\int_{E - E_\rho} K_\alpha(x_0 - a)\,\mathrm{d}\mu(a).$$

Thus we deduce finally that

$$|V_\alpha(x) - V_\alpha(x_1)| \leqslant \left| \int_{E - E_\rho} \{K_\alpha(x - a) - K_\alpha(x_1 - a)\}\,\mathrm{d}\mu(a) \right|$$
$$+ \left| \int_{E_\rho} K_\alpha(x - a)\,\mathrm{d}\mu(a) \right| + \left| \int_{E_\rho} K_\alpha(x_1 - a)\,\mathrm{d}\mu(a) \right| < 6^{1+\alpha}\,\varepsilon.$$

Since $V_\alpha(x_1)$ is continuous at $x_1 = x_0$, we deduce (5.1.1). This proves Theorem 5.1.

By a similar method we can also prove a restricted minimum principle, which has important applications.

THEOREM 5.2. *Suppose that with the hypotheses of Theorem* 5.1 *we have*

$$V_\alpha(x) \geqslant \mu_0 \text{ on } E.$$

Then we have

$$V_\alpha(x) \geqslant \mu_0' \text{ in } R^m,$$

where

$$\mu_0' = 2^\alpha \mu_0, \text{ if } \alpha > 0, \qquad \mu_0' = \mu_0 - \log 2, \text{ if } \alpha = 0.$$

As before we define D to be the complement of E in R^m. We take for x any point in D and choose x_1 to be nearest point of E to x. Then we have for any $a \in E$

$$|x_1 - x| \leqslant |x - a|$$

and

$$|x_1 - a| \leqslant 2|x - a|$$

just as in (5.1.2). This gives for $\alpha > 0$

$$K_\alpha(x - a) \geqslant 2^\alpha K_\alpha(x_1 - a)$$

so that

$$V_\alpha(x) \geqslant 2^\alpha V_\alpha(x_1) \geqslant \mu_0'$$

as required. Also if $\alpha = 0$

$$K_0(x - a) \geqslant K_0(x_1 - a) - \log 2,$$

so that we obtain in this case

$$V_0(x) \geqslant V_0(x_1) - \log 2 \int_E d\mu \geqslant \mu_0 - \log 2 = \mu_0'$$

as before.

We now define

$$I_\alpha(\mu) = \int_E V_\alpha(x)\, d\mu(x) = \iint_{E \times E} K_\alpha(x - y)\, d\mu(x)\, d\mu(y).$$

Since $K_\alpha(x - y)$ is bounded above on E, the integral always exists. We also set

$$V_\alpha = \sup_\mu I_\alpha(\mu)$$

and call V_α the equilibrium value of E.

Suppose that d is the diameter of E, so that $|x - y| \leqslant d$ in E. Then

$$K_0(x - y) \leqslant \log d,$$

$$K_\alpha(x - y) \leqslant -d^{-\alpha}, \alpha > 0.$$

Thus since $\mu(E) = 1$, we deduce that

$$V_0 \leqslant \log d, \qquad V_\alpha \leqslant -d^{-\alpha}, \qquad \alpha > 0.$$

We now define the α-capacity $C_\alpha(E)$ (or $\mathrm{Cap}_\alpha(E)$) of E by

$$C_0(E) = e^{V_0}, \qquad \alpha = 0$$

$$C_\alpha(E) = (-V_\alpha)^{-1/\alpha}, \qquad \alpha > 0.$$

Since $-\infty \leqslant V_\alpha < +\infty$, we deduce that

$$0 \leqslant C_\alpha(E) < \infty$$

Also if $C_\alpha(E) = 0$, we see that $I_\alpha(\mu) = -\infty$ for every measure μ. This is the case whenever $\alpha \geqslant m$ and for certain sets E, when $\alpha < m$.

5.1.1. Weak convergence

Our next aim is to prove that the supremum V_α is attained for a suitable measure μ on E, which will be called the *equilibrium distribution*. For this we need a result on weak convergence of measures which has considerable importance in its own right.

THEOREM† 5.3. *Suppose that μ_n is a sequence of measures on a compact set E in R^m such that $\mu_n(E) \leqslant A$ for all n, where A is a constant. Then we can select a subsequence μ_{n_p} of measures which converge weakly to a limiting measure μ in E; i.e. for any continuous function $\phi(x)$ on E,*

$$\int_E \phi(x)\,\mathrm{d}\mu_{n_p} \to \int_E \phi(x)\,\mathrm{d}\mu, \quad \textit{as} \quad p \to \infty.$$

Suppose that E lies in the hypercube $|x_i| \leqslant M$, for $i = 1$ to m, where x_i are the coordinates of x. Consider all hypercubes C of the form

$$a_i \leqslant x_i < b_i, \qquad i = 1 \quad \text{to} \quad m, \tag{5.1.3}$$

† Frostman [1935].

where a_i, b_i are rational numbers such that $-M \leqslant a_i < b_i \leqslant M$. The totality of these hypercubes is countable and so they may be enumerated in a sequence C_k, $k = 1$ to ∞.

The measures $\mu_n(C_k)$ are bounded by A for fixed k and so we can find a subsequence $(n, 1)$ of the integers such that

$$\mu_{n,1}(C_1) \to l_1, \quad \text{as} \quad n \to \infty,$$

then a further subsequence $(n, 2)$ such that

$$\mu_{n,2}(C_2) \to l_2, \quad \text{as} \quad n \to \infty$$

and by a step by step process we find for every positive integer k a subsequence (n, k) such that

$$\mu_{n,k}(C_k) \to l_k, \quad \text{as} \quad n \to \infty.$$

Here (n, k) denotes the nth member of the kth subsequence.

We now consider the sequence (n, n) of positive integers and note that (n, n) belongs to the kth subsequence if $n > k$. Thus

$$\mu_{n,n}(C_k) \to l_k, \quad \text{as} \quad n \to \infty$$

for every positive integer k. We denote the integer (p, p) by n_p and deduce that $n_p \geqslant p$, so that

$$n_p \to \infty \quad \text{as} \quad p \to \infty$$

and also that $\mu_{n_p}(C_k)$ converges to a limit $\mu(C_k)$ as $p \to \infty$ for every hypercube C_k bounded by faces with rational coordinates.

We now proceed to construct a linear functional on continuous functions on E. Let $\phi(x)$ be such a function. Let Δ: $\{C_1, C_2, \ldots, C_t\}$ be any finite collection of hypercubes C which are disjoint and whose union contains E.

Let M_j, m_j be the least upper bound and greatest lower bound of $\phi(x)$ on C_j. We define the lower and upper sums

$$s(\Delta, \phi) = \sum_{j=1}^{t} m_j \mu(C_j), \qquad S(\Delta, \phi) = \sum_{j=1}^{t} M_j \mu(C_j).$$

We note that if C_j does not meet E, then $\mu_n(C_j) = 0$ for each n and so $\mu(C_j) = 0$. Thus the corresponding terms in the sums can be put equal to zero.

We say that a collection Δ' is a simple refinement of Δ, if Δ' can be obtained from Δ by successively splitting a hypercube C into two others C' and C''. Thus for instance if C is given by (5.1.3) C' and C'' could be given by the same inequalities for all values of i except $i = i_0$, say, and for $i = i_0$, we should have

$$a_{i_0} \leqslant x_{i_0} < c_{i_0} \text{ on } C' \text{ and } c_{i_0} \leqslant x_{i_0} < b_{i_0} \text{ on } C''.$$

Since

$$\mu_{n_p}(C') + \mu_{n_p}(C'') = \mu_{n_p}(C),$$

we deduce that $\mu(C') + \mu(C'') = \mu(C)$. Also if m', m'', m denote the lower bounds and M', M'', M the upper bounds of $\phi(x)$ on C', C'' and C respectively we see that $s(\Delta')$ differs from $s(\Delta)$ in that we replace $m_j\mu(C_j)$ by $m_j'\mu(C_j') + m_j''\mu(C_j'')$ which is not smaller. If C' or C'' do not meet E, then $s(\Delta') = s(\Delta)$. Thus refinement increases the lower sums and decreases the upper sums similarly.

If Δ, Δ' are any two finite collections containing E, it is easy to see that Δ, Δ' have a common refinement Δ'', obtained by taking all hypercubes whose faces are any of the faces of Δ' or Δ''.

Thus

$$s(\Delta, \phi) \leqslant s(\Delta'', \phi) \leqslant S(\Delta'', \phi) \leqslant S(\Delta', \phi),$$

so that any lower sum is not greater than any upper sum.

Let I be the least upper bound of all lower sums and J the greatest lower bound of all upper sums. Then

$$I \leqslant J.$$

On the other hand since $\phi(x)$ is continuous on E, $\phi(x)$ is uniformly continuous on E and so given $\varepsilon > 0$, we can find δ, such that if $x, x' \in E$ and $|x - x'| < \delta$, then $|\phi(x) - \phi(x')| \leqslant \varepsilon$. This will always be the case if x, x' belong to the same hypercube of diameter less than δ.

Thus if Δ is any collection of hypercubes C_i of diameters less than δ, we deduce that $M_i - m_i < \varepsilon$, and so

$$J - I \leqslant S(\Delta, \phi) - s(\Delta, \phi) \leqslant \sum_i (M_i - m_i)\, \mu(C_i) \leqslant \varepsilon \sum_i \mu(C_i) \leqslant \varepsilon A,$$

since the C_i are disjoint. Since ε is arbitrary, we deduce that $I = J$.

We now define

$$L(\phi) = I(\phi) = J(\phi)$$

and show that $L(\phi)$ is a functional, in the sense of Section 3.2. In fact (3.2.1) is obvious. Also if $a > 0$, we clearly have

$$s(\Delta, a\phi) = a\, s(\Delta, \phi), \text{ so that } L(a\phi) = a\, L(\phi).$$

Similarly if $a < 0$ the conclusion follows from

$$s(\Delta, a\phi) = aS(\Delta, \phi).$$

Finally if, f, g are continuous we clearly have

$$s(\Delta, f) + s(\Delta, g) \leqslant s(\Delta, f + g) \leqslant S(\Delta, f + g) \leqslant S(\Delta, f) + S(\Delta, g).$$

If Δ is a collection of sufficiently small hypercubes the first and last members differ little from $L(f) + L(g)$, so that

$$L(f + g) = L(f) + L(g).$$

Thus (3.2.2) also holds and $L(f)$ is a positive linear functional on E. Thus $L(f)$ can, in view of Theorem 3.4, be written as an integral with respect to a measure μ on E, so that for any continuous function f on E

$$L(f) = \int_E f \, d\mu.$$

It remains to show that for any continuous function f on E we have

$$\int f \, d\mu_{n_p} \to L(f) = \int f \, d\mu, \quad \text{as} \quad p \to \infty.$$

To see this we take a collection Δ of hypercubes C_i.

Then if m_i, M_i are the lower and upper bounds of f on C_i, we have

$$\varliminf_{p \to \infty} \int f \, d\mu_{n_p} \geqslant \varliminf_{p \to \infty} \sum m_i \mu_{n_p}(C_i) = \sum_i m_i \mu(C_i) = s(\Delta, f).$$

Similarly

$$\varlimsup_{p \to \infty} \int f \, d\mu_{n_p} \leqslant S(\Delta, f),$$

Since this is true for every collection Δ, we deduce that

$$I \leqslant \varliminf \int f \, d\mu_{n_p} \leqslant \varlimsup \int f \, d\mu_{n_p} \leqslant J,$$

hence

$$\lim_{p \to \infty} \int f d\mu_{n_p} = L(f)$$

as required. This completes the proof of Theorem 5.3.

5.2. CONDUCTOR POTENTIALS AND CAPACITY†

We can now prove

† The results in this section are mainly due to Frostman [1935].

THEOREM 5.4. *Suppose that E is a compact set in R^m and that $V_\alpha(E) > -\infty$. Then there exists a unit mass distribution μ on E such that*

$$I_\alpha(\mu) = V_\alpha(E).$$

Let μ_n be a sequence of unit measures in E such that

$$I_\alpha(\mu_n) \to V_\alpha(E).$$

In view of Theorem 5.3 we may suppose, by taking a subsequence if necessary that μ_n converges weakly to a measure μ on E. By taking $f = 1$ on E, we deduce that

$$\mu(E) = \int f \, \mathrm{d}\mu = \lim_{n \to \infty} \int f \, \mathrm{d}\mu_n = \lim \mu_n(E) = 1.$$

Thus μ is a unit measure.

We now prove the following useful

LEMMA 5.1. *If the sequence of measures μ_n on a fixed compact set F converges weakly to μ and $f(x)$ is* u.s.c. *on F then*

$$\overline{\lim_{n \to \infty}} \int_F f(x) \, \mathrm{d}\mu_n(x) \leqslant \int_F f(x) \, \mathrm{d}\mu(x).$$

Since $f(x)$ is u.s.c. there exists a sequence $f_m(x)$ of continuous functions such that $f_m(x)$ decreases with increasing m and $f_m(x) \to f(x)$ as $m \to \infty$.

Thus

$$\int_F f_m(x) \, \mathrm{d}\mu(x) \to \int_F f(x) \, \mathrm{d}\mu(x), \quad \text{as} \quad m \to \infty$$

and, given $\varepsilon > 0$, we can find m, such that

$$\int_F f_m(x) \, \mathrm{d}\mu(x) < \int_F f(x) \, \mathrm{d}\mu(x) + \varepsilon.$$

Thus

$$\overline{\lim_{n \to \infty}} \int_F f \, \mathrm{d}\mu_n \leqslant \lim_{n \to \infty} \int_F f_m \, \mathrm{d}\mu_n = \int_F f_m \, \mathrm{d}\mu < \int_F f \, \mathrm{d}\mu + \varepsilon.$$

This proves Lemma 5.1. We apply the result to the function $K_\alpha(x - y)$ on $E \times E$ and deduce that $I_\alpha(\mu) \geqslant \overline{\lim}\, I_\alpha(\mu_n) = V_\alpha(E)$. The opposite inequality follows from the definition, and Theorem 5.4 is proved.

The measure μ is called an *equilibrium distribution* of E and

$$U_\alpha(y) = \int_E K_\alpha(x - y) \, \mathrm{d}\mu(x)$$

is called the *conductor potential* of E.

It is convenient at this stage to define the capacity of sets which are not necessarily compact. This we do as follows. If E is any set the (inner) capacity $C_\alpha(E)$ is the least upper bound of $C_\alpha(F)$ for all compact sets F contained in E. We also define the outer† capacity

$$C_\alpha^*(E) = \inf C_\alpha(G)$$

for all open sets G containing E.

It is clear that if E_1, E_2 are any sets, such that $E_1 \subset E_2$, then $C_\alpha(E_1) \leqslant C_\alpha(E_2)$. Hence if E is any bounded set we always have $C_\alpha(E) \leqslant C_\alpha^*(E)$. If equality holds here, we say that E is *capacitable*.

We shall prove in Section 5.8 that all Borel sets are capacitable. For the time being we must distinguish between capacity and outer capacity.

We note that if E_1, E_2 are any sets such that $E_1 \subset E_2$ it is immediate that

$$C_\alpha^*(E_1) \leqslant C_\alpha^*(E_2).$$

Also it follows from the definition that if G is an open set then

$$C_\alpha^*(G) = C_\alpha(G),$$

so that open sets are capacitable. The next theorem implies the corresponding result for compact sets.

THEOREM 5.5. *If E is a compact set and G_n a sequence of bounded open sets, such that $\bar{G}_{n+1} \subset G_n$, $i = 1, 2, \ldots, n = 1, 2, \ldots$ and $\bigcap_{n=1}^{\infty} G_n = E$. Then*

$$C_\alpha(G_n) \to C_\alpha(E), \quad \text{as} \quad n \to \infty. \tag{5.2.1}$$

Thus compact sets are capacitable.

It follows from the monotonic property of capacity that if $C_n = \mathrm{Cap}_\alpha(G_n)$, then $C_n \geqslant C_{n+1}$, $n = 1, 2, \ldots$. Also for each fixed n

$$\mathrm{Cap}_\alpha(E) \leqslant C_n.$$

Thus

$$C = \lim_{n\to\infty} C_n \text{ exists}$$

and

$$\mathrm{Cap}_\alpha(E) \leqslant C.$$

Let μ_n be a unit mass distribution over the compact set $\bar{G}_n$ such that

† Brelot [1939b].

$$V_n = I_\alpha(\mu_n) = \int_{\bar{G}_n \times \bar{G}_n} K_\alpha(x - y)\, d\mu_n(x)\, d\mu_n(y)$$

is maximal. Then in view of the selection Theorem 5.3 we can find a subsequence μ_{n_p} converging weakly to a unit measure μ.

The limit measure μ is distributed on $\bar{G}_{n_p}$ for every p and so it is distributed on E. By taking a subsequence if necessary we now assume without loss of generality that μ_n converges weakly to μ. We apply Lemma 5.1 and deduce that

$$\overline{\lim_{n\to\infty}}\ V_n = \overline{\lim_{n\to\infty}} \int_{\bar{G}_n \times \bar{G}_n} K_\alpha(x - y)\, d\mu_n(x)\, d\mu_n(y) \leqslant V$$

$$= \int_{E \times E} K_\alpha(x - y)\, d\mu(x)\, d\mu(y).$$

In view of the definition of capacity this implies that

$$C = \lim_{n\to\infty} \operatorname{Cap}_\alpha(\bar{G}_n) \leqslant \operatorname{Cap}_\alpha(E).$$

This proves (5.2.1).

Next if E is any compact set we may take for G_n the set of points distant less that $1/n$ from E. Then the conditions of Theorem 5.5 are satisfied and so we have (5.2.1). In particular given $\varepsilon > 0$, we can find an open set G_n containing E, such that

$$C_\alpha(G_n) < C_\alpha(E) + \varepsilon.$$

Thus

$$C_\alpha^*(E) < C_\alpha(E) + \varepsilon,$$

and so

$$C_\alpha^*(E) \leqslant C_\alpha(E).$$

Thus $C_\alpha^*(E) = C_\alpha(E)$ so that E is capacitable.

The property of Theorem 5.5 is usually called the u.s.c. property of capacity. It plays an important role in many considerations including the general theory of capacitability to which we shall return in Section 5.8.

5.2.1. The nature of the conductor potential

We now investigate more closely the nature of the conductor potential for a general compact set E. Before starting and proving the fundamental theorem we need

THEOREM 5.6. *Suppose that E is a compact set and ν a positive mass distribution on E such that* $0 < \nu(E) < \infty$ *and*

$$I(\nu) = \int_E K_\alpha(x - y)\, d\nu(x)\, d\nu(y) > -\infty,$$

(so that in particular $\mathrm{Cap}_\alpha(E) > 0$*). Then if* E_1 *is a compact subset of E such that* $\mathrm{Cap}_\alpha(E_1) = 0$, *or a countable union of such sets, we have* $\nu(E_1) = 0$.

We have, on E, $|x - y| \leqslant d$, for some positive d, so that

$$K_\alpha(x - y) \leqslant 0, \qquad \alpha > 0$$

$$K_\alpha(x - y) \leqslant \log^+ d, \qquad \alpha = 0.$$

Thus if we set $d_1 = \log^+ d$, we see that for $x, y \in E$

$$K(x - y) = K_\alpha(x - y) - d_1 \leqslant 0.$$

Also

$$\int_E K(x - y)\, d\nu(x)\, d\nu(y) > -\infty,$$

and hence if E_1 is a compact subset of E we have

$$\int_{E_1} K(x - y)\, d\nu(x)\, d\nu(y) \geqslant \int_E K(x - y)\, d\nu(x)\, d\nu(y) > -\infty.$$

Thus

$$\int_{E_1} K_\alpha(x - y)\, d\nu(x)\, d\nu(y) > -\infty.$$

If $\nu(E_1) > 0$, this implies that $\mathrm{Cap}(E_1) > 0$, contrary to hypothesis.

If $E_1 = \bigcup_{n=1}^{\infty} F_n$, where $\mathrm{Cap}_\alpha(F_n) = 0$, then $\nu(F_n) = 0$ for each n and since measure is additive it follows that $\nu(E_1) = 0$ as before.

A set E will be called† an α-polar set if $E \subset \bigcup_{n=1}^{\infty} F_n$, where the F_n are compact sets, such that $\mathrm{Cap}_\alpha(F_n) = 0$. We shall see later that this implies that $\mathrm{Cap}_\alpha^*(E) = 0$, but we do not need this result for the time being. If a conclusion holds outside an α-polar set, we say it holds nearly everywhere (α) or n.e.α.

† The name polar was originally introduced by Brelot [1941] for a rather wider class of sets.

Let E be a compact set and μ a positive mass distribution in E. We shall say that a point x_0 belongs to the carrier E^* of μ, if for every positive r we have $\mu\{D(x_0, r)\} > 0$. Evidently E^* is closed, so that E^* is a compact subset of E. We have also

THEOREM 5.7. *Suppose that E is a compact set such that* $\mathrm{Cap}_\alpha(E) > 0$ *and let μ be a corresponding equilibrium distribution on E and E^* the carrier of μ. Then E^* is a compact subset of E such that*

$$\mathrm{Cap}_\alpha(E^*) = \mathrm{Cap}_\alpha(E).$$

We note first that if $E_0 = E - E^*$ is the part of E not in E^*, then $\mu(E_0) = 0$. In fact if x_0 is any point of E_0 then we can find an open ball D, with centre at x_0 and not meeting E^*, such that $\mu(D) = 0$. We can then find an open ball D_0, with centre at a point with rational coordinates and whose radius is also rational, such that $x_0 \in D_0 \subset D$, so that $\mu(D_0) = 0$. The set of all such balls D_0 is countable and so we can arrange them in a sequence D_i, so that

$$\mu(E_0) \leqslant \sum_i \mu(D_i) = 0.$$

Thus μ is a mass distribution of unit mass on E^* and

$$V_\alpha(E^*) \geqslant \int_{E^* \times E^*} K_\alpha(x-y)\,\mathrm{d}\mu(x)\,\mathrm{d}\mu(y) = \int_{E \times E} K_\alpha(x-y)\,\mathrm{d}\mu(x)\,\mathrm{d}\mu(y) = V_\alpha(E).$$

Hence

$$\mathrm{Cap}_\alpha(E^*) \geqslant \mathrm{Cap}_\alpha(E).$$

The converse inequality is obvious since $E^* \subset E$, so that Theorem 5.7 is proved.

The following result due to Frostman is called by Tsuji [1959, p. 60] not without reason the fundamental theorem on conductor potentials.

THEOREM 5.8.† *Let E be a compact set in R^m, such that* $\mathrm{Cap}_\alpha(E) > 0$, *and let $u(x)$ be an associated conductor potential on E, μ the corresponding equilibrium distribution, and E^* the carrier of μ so that*

$$I_\alpha(\mu) = \iint_{E^*} K_\alpha(x-y)\,\mathrm{d}\mu(x)\,\mathrm{d}\mu(y) = V > -\infty. \tag{5.2.3}$$

Then

$$u(x) = \int_{E^*} K_\alpha(x-y)\,\mathrm{d}\mu(y) \geqslant V \text{ on } E^* \tag{5.2.4}$$

† Frostman [1935].

and

(5.2.5) $$u(x) \leqslant V, \text{n.e.}\alpha \quad \text{on} \quad E,$$

so that in particular

(5.2.6) $$u(x) = V, \text{n.e.}\alpha. \quad \text{on} \quad E^*.$$

We also have

(5.2.7) $$u(x) \geqslant V' \quad \text{in} \quad R^m,$$

where

(5.2.8) $$V' = 2^\alpha V, \textit{ if } \alpha > 0, \qquad V' = V - \log 2, \textit{ if } \alpha = 0.$$

We prove first of all (5.2.5).

Let A_n be the set of $x \in E$, such that

$$u(x) \geqslant V + \frac{1}{n}.$$

Then it is enough to show that $\mathrm{Cap}_\alpha(A_n) = 0$ for, since $u(x)$ is u.s.c., A_n is compact. Also if A is the set of all points of E where (5.2.5) is false then $A = \bigcup_{n=1}^{\infty} A_n$.

Suppose then that our conclusion in false, and let E_1 be a compact subset of E, such that we have for some $\varepsilon > 0$

$$u(x) \geqslant V + 2\varepsilon, \qquad x \in E_1,$$

and $\mathrm{Cap}_\alpha(E_1) > 0$. We may take $E_1 = A_n$ for a suitable n, and then choose $\varepsilon = (3n)^{-1}$.

By Theorem 5.7 μ is an equilibrium distribution for E^* as well as E and

$$\int_{E^*} u(x)\, \mathrm{d}\mu = V.$$

In particular there exists $a_0 \in E^*$, such that $u(a_0) < V + \varepsilon$. Thus a_0 lies outside E_1 and since $u(x)$ is u.s.c. we have, in a suitable neighbourhood $D(a_0, r)$ of a_0,

$$u(x) < V + \varepsilon.$$

We choose r so small, that $D(a_0, r)$ is at a positive distance from E_1. Also since $a_0 \in E^*$ we have

$$\mu\{D(a_0, r)\} = m > 0.$$

Since $\mathrm{Cap}_\alpha(E_1) > 0$ there exists a positive mass distribution σ on E_1, such that

$$I_\alpha(\sigma) = \int_{E_1 \times E_1} K_\alpha(x - y)\, d\sigma(x)\, d\sigma(y) > -\infty, \qquad \sigma(E_1) = m > 0.$$

We now define a mass distribution σ_1 in E by

$$\sigma_1 = \sigma \text{ on } E_1, \qquad \sigma_1 = -\mu \text{ in } D(a_0, r), \qquad \sigma_1 = 0 \text{ elsewhere.}$$

Then, for $0 < \eta < 1$, $\mu_1 = \mu + \eta\sigma_1$ is a positive mass distribution in E and

$$\int_E d\mu_1 = 1.$$

Also

$$\delta(I) = I_\alpha(\mu_1) - I_\alpha(\mu) = \iint_E K_\alpha(x - y)\, d\mu_1(x)\, d\mu_1(y)$$

$$- \int_{E \times E} K_\alpha(x - y)\, d\mu(x)\, d\mu(y) = 2\eta \int_{E \times E} K_\alpha(x - y)\, d\mu(x)\, d\sigma_1(y)$$
$$+ \eta^2 \int_{E \times E} K_\alpha(x - y)\, d\sigma_1(x)\, d\sigma_1(y)$$

$$= 2\eta \int_E u(y)\, d\sigma_1(y) + \eta^2 I_\alpha(\sigma_1)$$
$$\geqslant 2\eta\left[(V + 2\varepsilon)m - (V + \varepsilon)m + \tfrac{1}{2}\eta I_\alpha(\sigma_1)\right]$$
$$= 2\eta(\varepsilon m + \tfrac{1}{2}\eta I_\alpha(\sigma_1)) > 0$$

if η is small enough. This contradicts the maximality property of $I_\alpha(\mu)$. Thus $\mathrm{Cap}_\alpha(A_n) = 0$ for all n, and (5.2.5) is proved.

We next prove (5.2.4). In fact suppose contrary to this that for some $x = x_0 \in E^*$ we have

$$u(x) < V - \varepsilon.$$

Then since $u(x)$ is u.s.c. the same inequality holds in some neighbourhood $D(x_0, r)$ of x_0, and since $x_0 \in E^*$, we have

$$\mu\{D(x_0, r)\} = m_0 > 0.$$

Let E_2 be the set of all points of E^*, such that $u(x) > V$. Then in view of (5.2.5) and Theorem 5.6 we have $\mu(E_2) = 0$. Let E_3 be the set of all points of

E^* in $D(x_0, r)$ and E_4 the remainder of E^*. Then

$$V = \int_E u(x)\,d\mu = \int_{E^*} u(x)\,d\mu = \sum_{j=2}^{4} \int_{E_j} u(x)\,d\mu$$

$$= \int_{E_3} u(x)\,d\mu + \int_{E_4} u(x)\,d\mu \leqslant m_0(V - \varepsilon) + (1 - m_0)\,V = V - m_0\,\varepsilon.$$

This gives a contradiction and proves that

$$u(x) \geqslant V \qquad \text{on } E^*.$$

In view of (5.2.5) we deduce that

$$u(x) = V \text{ n.e.}\alpha \quad \text{on } E^*.$$

Finally we note that, in view of Theorem 5.2, and since μ is distributed over E^*, (5.2.4) implies (5.2.7). This completes the proof of Theorem 5.8.

5.3. POLAR SETS

We proceed to investigate in what sets a potential can be $-\infty$. Our positive result is contained in

THEOREM 5.9. *Suppose that v is a positive mass distribution over a compact set E,*

$$v(x) = \int_E K_\alpha(x - y)\,d\nu(y)$$

is the associated potential and E_1 is a compact set in R^m such that $v(x) = -\infty$ on E_1.

Then

$$\mathrm{Cap}_\alpha(E_1) = 0.$$

Suppose contrary to this that

$$\mathrm{Cap}_\alpha(E_1) > 0.$$

Let $u(x)$ be an associated conductor potential on E_1, μ the corresponding equilibrium distribution and E_1^* the carrier of μ. Then by Theorem 5.8 we know that

$$u(x) = \int_{E_1^*} K_\alpha(x - y)\,d\mu(y) \geqslant V' > -\infty \text{ in } R^m.$$

Also by Fubini's Theorem we have, since $K_\alpha(x - y)$ is bounded above,

$$\int_{E_1^*} v(x)\,d\mu(x) = \int_{E_1^*} d\mu(x) \int_E K_\alpha(x - y)\,d\nu(y) = \int_E u(y)\,d\nu(y).$$

Here the left-hand side is $-\infty$, while the right-hand side is at least $-V'$ and so finite. This contradiction proves Theorem 5.9.

As a corollary we deduce at once

THEOREM 5.10. *Suppose that $u(x)$ is* s.h. *and not identically $-\infty$ in the neighbourhood of a compact set E in R^m and that $u(x) = -\infty$ on E. Then*

$$\mathrm{Cap}_{m-2}(E) = 0.$$

Let F be a compact set containing E in its interior and such that $u(x)$ is s.h. in a neighbourhood of F. Then it follows from Riesz' Representation Theorem 3.9 that we have in the interior of F

$$u(x) = \int_F K_{m-2}(x-y)\,\mathrm{d}\mu(y) + h(x),$$

where $h(x)$ is harmonic in the interior of F and so bounded on E. Thus we deduce that

$$\int_F K_{m-2}(x-y)\,\mathrm{d}\mu(y) = -\infty, \qquad x \in E.$$

Now it follows from Theorem 5.9 that $\mathrm{Cap}_{m-2}(E) = 0$.

We proceed to prove a converse to Theorem 5.10.

THEOREM† 5.11. *Suppose that E is a compact set in R^m, such that $\mathrm{Cap}_{m-2}(E) = 0$, or more generally an $(m-2)$—polar set. Then $\exists u(x)$,* s.h. *in R^m and finite at a preassigned x_0 in $R^m - E$, such that $u(x) = -\infty$ in E. If $m > 2$, we may assume in addition that $u(x) < 0$ in R^m. If E is compact $u(x)$ is finite outside E.*

We suppose first that E is compact and let $D(0, R)$ be an open ball containing E. Let E_n be a sequence of compact sets such that E_{n+1} lies in the interior of E_n,

$$E_1 \subset D(0, R) \quad \text{and} \quad \bigcap_{n=1}^{\infty} E_n = E,$$

and further such that all the E_n are unions of a finite number of closed balls. Let G_n be that complementary domain to E_n in $D(0, R)$, which contains $S(0, R)$ as part of its boundary. Then G_n is admissible for the problem of Dirichlet and so we can construct a function $\omega_n(x)$, such that

† Evans [1936], see also Selberg [1937].

$\omega_n(x)$ is continuous in $C(0, R)$, $\omega_n(x)$ is harmonic in G_n,
$\omega_n(x) = 0$ on $S(0, R)$,
$\omega_n(x) = -1$ at all other points and in particular on E_n.

It is evident that $\omega_n(x)$ is s.h. in $D(0, R)$ and so it follows from the Poisson–Jensen formula (3.7.3) that we have for $y \in D(0, R)$, $|y| < r < R$

$$\omega_n(y) = \frac{1}{c_m} \int_{S(0,r)} \omega_n(\eta) \frac{r^2 - |y|^2}{r|\eta - y|^{m-1}} \, d\sigma(\eta) - \int_{E_n} g(x, y, r) \, d\mu_n(x),$$

where $g(x, y, r)$ is the Green's function in $D(0, r)$ and $d\sigma(\eta)$ is the superficial measure on $S(0, r)$. Letting $r \to R$, we deduce that

$$\omega_n(y) = -\int_{E_n} g(x, y, R) \, d\mu_n(x),$$

and μ_n is the Riesz measure in E_n.

We now note that by Theorem 1.10 we have

$$|g(x, y, R) + K_{m-2}(x - y)| \leqslant C \tag{5.3.1}$$

for x, y on E_1, where C is a constant. Thus we deduce that

$$\left| \int_{E_n} K_{m-2}(x - y) \, d\mu_n(x) - \omega_n(y) \right| \leqslant C\mu_n(E_n).$$

If we write $\mu_n(E_n) = M_n$, and $\nu_n = \mu_n/M_n$, then ν_n is a unit mass distribution on E_n and we have on E_n

$$\int_{E_n} K_{m-2}(x - y) \, d\nu_n(y) \geqslant -C - M_n^{-1},$$

since $\omega_n(y) \geqslant -1$ there.

We now deduce that $M_n \to 0$, as $n \to \infty$. For otherwise, by choosing a subsequence if necessary, we could assume that M_n^{-1} is bounded and that ν_n converges weakly to a unit measure ν that is necessarily distributed on E and in view of Lemma 5.1 we have on E

$$\int_E K_{m-2}(x - y) \, d\nu(y) \geqslant -C - \overline{\lim} \, M_n^{-1}.$$

This contradicts the hypothesis that $\mathrm{Cap}_{m-2} E = 0$.

By choosing a subsequence if necessary, we may thus suppose that $M_n \leqslant 2^{-n}$, and with this hypothesis we set

$$\omega(y) = \sum_1^\infty \omega_n(y) = -\int g(x, y, R) \, d\mu(x),$$

where for any Borel set e in $|x| < R$, we have

(5.3.2.) $$\mu(e) = \sum_1^\infty \mu_n(e) \leqslant \sum_1^\infty 2^{-n} \leqslant 1.$$

Thus μ is a Borel measure. Finally we set

$$u(x) = \int_{E_1} K_{m-2}(x - y)\,\mathrm{d}\mu(y).$$

It follows from (5.3.1) and (5.3.2) that we have in E_1

(5.3.3) $$|u(y) - \omega(y)| \leqslant C.$$

Evidently $\omega(y)$ and so $u(y)$ is $-\infty$ in E. Suppose next that y is a point in $D(0, R)$ but outside E. Then y is outside E_n for $n \geqslant n_0$, and so y is at a positive distance δ from E_n for $n > n_0$. We write

$$\omega(y) = \sum_{n=1}^{n_0-1} \omega_n(y) + \sum_{n=n_0}^{\infty} \omega_n(y) = \Sigma_1(y) + \Sigma_2(y)$$

say. Then $\Sigma_1(y) \geqslant -n_0$. Also for $x \in E_{n_0}$, we have

$$g(x, y, R) \leqslant C(\delta),$$

and so

$$-\Sigma_2(y) \leqslant \sum_{n=n_0}^{\infty} \int_{E_n} C(\delta)\,\mathrm{d}\mu_n(x) \leqslant C(\delta).$$

Thus $\omega(y)$ is finite at all points of $D(0, R)$ outside E and hence so is $u(y)$ by (5.3.3). Thus $\omega(y)$ is s.h. in R^m, being a potential, and $\omega(y)$ is $-\infty$ precisely in E. For since μ is distributed on E_1, $u(x)$ is evidently harmonic and so finite outside E_1 and so certainly outside $D(0, R)$. This completes the proof of Theorem 5.11 when E is compact.

Suppose finally that $E \subset \bigcup_{n=1}^{\infty} F_n$, and that E is an $(m-2)$-polar set, so that the F_n are compact sets of $(m-2)$-capacity zero. Then since there exists a s.h. function in R^m which is $-\infty$ on F_n but not elsewhere, F_n must have m dimensional measure zero by Theorem 2.6. Hence the same is true of E. Thus $\exists$ a point x_0 outside E. We now construct a function $u_n(x)$ with the following properties:

(5.3.4) $u_n(x)$ is s.h. in R^m, $u_n(x) = -\infty$ precisely in F_n;

(5.3.5) $$|u_n(x_0)| \leqslant \frac{1}{n^2};$$

(5.3.6) $$u_n(x) < \frac{1}{n^2}, \qquad |x| < n.$$

The construction of (5.3.4) is made as above. We then achieve (5.3.5) and (5.3.6) by multiplying $u_n(x)$ by a sufficiently small positive constant if necessary. We now set

$$u(x) = \sum_1^\infty u_n(x)$$

and assert that $u(x)$ has the required properties.

In fact write

$$S_N(x) = \sum_{n=1}^N u_n(x).$$

Then if

$$\sigma_N(x) = \sum_1^N \left(u_n(x) - \frac{1}{n^2} \right) = S_N(x) - \sum_1^N \frac{1}{n^2}$$

it follows that $\sigma_N(x)$ decreases with N for large N in $|x| < R$ for any fixed R and so

$$u(x) = \lim_{N \to \infty} \sigma_N(x) + \sum_1^\infty \frac{1}{n^2}$$

is s.h. in R^m. Also $\sigma_N(x) = -\infty$ in F_n for $N > n$, and so $u(x) = -\infty$ in F_n and so in E. Also, in view of (5.3.5) $u(x_0)$ is finite, so that $u(x)$ is not identically $-\infty$. If $m > 2$ we can replace (5.3.6) by $u_n(x) < 0$ in R^m, since $K_{m-2}(x) < 0$ in R^m in this case we deduce that $u(x) < 0$ in R^m. This completes the proof of Theorem 5.11.

5.4. CAPACITY AND HAUSDORFF MEASURE

The $(m - 2)$-polar or simply polar sets play a considerable role in potential theory. We have seen in the last section that among F_σ sets the polar sets are precisely those, on which s.h. functions can be $-\infty$. This makes it of interest to study these sets from the point of view of their size. It follows from Theorem 2.6 that if $u(x)$ is s.h. and $u(x) \not\equiv -\infty$, then the set where $u(x) = -\infty$ has m-dimensional measure zero. In particular this property holds for polar sets. However a good deal more than this is true. In order to prove more precise results we proceed to introduce and develop the notion of Hausdorff measure.† A hypercube of side d is a set of points $x = \{x_1, x_2, \ldots, x_m)$, whose coordinates satisfy

$$a_i < x_i < a_i + d, \qquad i = 1 \text{ to } m,$$

† Hausdorff [1918].

where the a_i are real numbers. We also allow any of the $<$ signs to be replaced by $\leqslant$. If strict inequality holds throughout, the hypercube is open, if $\leqslant$ stands in all the inequalities the hypercube is said to be closed.

Suppose now that $h(t)$ is a positive increasing function of t for $0 < t \leqslant t_0$, such that $h(t) \to 0$, as $t \to 0$. Let e be a bounded set and ε a positive number. We cover e by at most countably many hypercubes I_ν of sides d_ν less than ε and set

$$H_\varepsilon(e) = \inf \sum_\nu h(d_\nu)$$

where the infimum is taken over all such coverings.

Then $H_\varepsilon(e)$ is always finite and is a non-increasing function of ε. Thus

$$h^*(e) = \lim_{\varepsilon \to 0} H_\varepsilon(e)$$

always exists and $0 \leqslant h^*(e) \leqslant \infty$. The quantity of $h^*(e)$ is called the *Hausdorff measure* of e with respect to the function h.

Instead of hypercubes of sides d_i we might have chosen balls of radius d_i or general sets (or convex sets) of diameters d_i for our coverings. The corresponding quantities $h^*(e)$ are bounded by constant multiples of each other. In most cases we are only interested in whether $h^*(e)$ is zero, positive and finite, or infinite, so that all these methods give equivalent results. The definition in terms of hypercubes has advantages because of the stacking properties of hypercubes. We note that if $h(t) = t^m$ in R^m, then $h^*(e)$ is just m-dimensional Lebesgue measure.

If $\alpha > 0$ and $h(t) = t^\alpha$, then $h^*(e)$ is called α-dimensional measure, if $h(t) = (\log 1/t)^{-1}$, then $h^*(e)$ is logarithmic measure. The following result is almost obvious.

Lemma 5.2. *If $h_1(t)/h_2(t) \to 0$ as $t \to 0$, and $h_2^*(e) < \infty$, then $h_1^*(e) = 0$.*

In fact, given $\eta > 0$, we have $h_1(t) < \eta h_2(t)$, $t < \rho_0(\eta)$. This yields

$$H_{1\varepsilon}(e) \leqslant \eta H_{2\varepsilon}(e), \qquad \varepsilon < \rho_0(\eta),$$

so that

$$h_1^*(e) \leqslant \eta h_2^*(e).$$

Since η is any positive number, Lemma 5.2 follows.

We now denote α-dimensional measure by $l_\alpha(e)$ and logarithmic measure by $l_0(e)$. Our next result gives rise to an important definition.

LEMMA 5.3. *If e is any bounded set in R^m, then there exists a constant α_0, such that $0 \leqslant \alpha_0 \leqslant m$, and*

$$l_\alpha(e) = 0, \qquad \alpha > \alpha_0, \qquad l_\alpha(e) = \infty, \qquad \alpha < \alpha_0,$$

The quantity α_0 is called the *Hausdorff dimension* of e. We note that $l_m(e) < \infty$. In fact since e is bounded, e can be included in a hypercube I_0 of side $d < \infty$. We can write I_0 as the union of N^m hypercubes I_ν, $\nu = 1$ to N^m, each of side $d_\nu = d/N$.

Clearly
$$\Sigma(d_\nu)^m = N^m(d/N)^m = d^m.$$

Thus if we choose $d/N < \varepsilon$, we deduce that if $h(t) = t^m$, then

$$H_\varepsilon(e) \leqslant d^m.$$

Hence

$$h^*(e) \leqslant d^m, \qquad \text{i.e.} \quad l_m(e) \leqslant d^m < \infty.$$

It follows from Lemma 5.2. that $l_\alpha(e) = 0$, $\alpha > m$. We now define α_0 to be the lower bound of all $\alpha > 0$, such that $l_\alpha(e) = 0$. Then $\alpha_0 \leqslant m$. If $\alpha_1 > \alpha_0$, it follows from Lemma 5.2. that $l_{\alpha_1}(e) = 0$, since otherwise $l_\alpha(e) = \infty$ for $\alpha < \alpha_1$. Also if $\alpha < \alpha_0$, we can find α_2, such that $\alpha < \alpha_2 < \alpha_0$, and $l_{\alpha_2}(e) > 0$. Thus by Lemma 5.2. $l_\alpha(e) = \infty$. This proves Lemma 5.3.

The next result is useful later.

THEOREM 5.12. *Suppose that e is a set in R^m, and for fixed x_0 let E be the set of positive r, for which e meets the hypersphere $S(x_0, r)$. Then*

$$l_1(E) \leqslant l_1(e)\sqrt{m}.$$

Thus if $l_1(e) = 0$, the complement of E is everywhere dense on the positive axis. In this case e is totally disconnected.

Let I be a closed hypercube of side d. Then the diameter of I is $d\sqrt{m}$. Thus the set of r for which I meets the spheres $S(x_0, r)$ is a closed interval of length at most $d\sqrt{m}$. Suppose now that I_ν is an arbitrary set of hypercubes covering e.

Then I_ν meets $S(x_0, r)$ for an interval i_ν of values of r having length at most $d_\nu\sqrt{m}$, where d_ν is the side of i_ν.

Our hypotheses imply that the union of the intervals i_ν covers E. For each $\varepsilon > 0$, we can assume that $d_\nu < \varepsilon$ and $\Sigma d_\nu < l_1(e) + \varepsilon$. Thus also $\Sigma d_\nu\sqrt{m} < \sqrt{m}(l_1(e) + \varepsilon)$, and since ε can be chosen as small as we please we deduce that $l_1(E) \leqslant l_1(e)\sqrt{m}$.

If $l_1(e) = 0$, we deduce that $l_1(E) = 0$, so that E cannot contain an interval $[a, b]$, where $b > a$. Thus the complement of E is everywhere dense. If e contains any connected subset e_0, with two distinct points x_1, x_2, then e_0

must meet $S(x_1, r)$ for $0 < r < |x_2 - x_1|$, since otherwise the subsets of e_0, for which $|x - x_1| < r$, and $|x - x_1| > r$, would form a partition of e_0. Thus if $x_0 \neq x_1$, E contains an interval, and this gives a contradiction.

Hence e_0 can contain at most one point and all the connected subsets of e reduce to points, so that e is totally disconnected. This completes the proof of Theorem 5.12.

Our next aim is to describe the capacity of sets in terms of Hausdorff measure. For this, the following lemma due to Frostman [1935] is useful.

LEMMA 5.4. *Let E be a compact set in R^m, such that $h^*(E) > 0$. Then there exists a measure μ on E, such that $0 < \mu(E) < \infty$, and such that for any point a we have*

(5.4.1) $$\mu[D(a, r)] < Ch(r), \qquad 0 < r \leqslant 1,$$

where C is a constant.

Clearly having found a measure with some constant C, we may always consider μ/C instead of μ, so that we may take $C = 1$. Alternatively we can consider $\mu/\mu(E)$ instead of μ, so that we may assume that $\mu(E) = 1$.

We assume without loss of generality, that $h(t)$ is defined and increasing for all $t > 0$.

Since E is compact, we may cover E by a hypercube Q_0 of sides $2\delta_0$, given by $-\delta_0 \leqslant x_\nu < \delta_0$, $\nu = 1$ to m. We also remark the following. If E is covered by a finite number of hypercubes of sides d_i, then

(5.4.2) $$\sum_i h(d_i) > \eta > 0,$$

where η is independent of the covering. For if we could make $\Sigma h(d_i)$ arbitrarily small, then d_i would also have to be small for each i and we should deduce that $H_\varepsilon(E) = 0$ for each $\varepsilon > 0$ so that $h^*(E) = 0$, contrary to hypothesis. We now divide Q_0 into $2^{m(n+1)}$ equal hypercubes of sides $\delta_n = \delta_0 2^{-n}$ $(n = 0, 1, 2 \ldots)$. Let

$$Q_n^j = \left\{x \,\middle|\, \frac{k_\nu \delta_0}{2^n} \leqslant x_\nu < \frac{k_\nu + 1}{2^n}\,\delta_0,\ \nu = 1 \text{ to } m\right\}$$

where all the k_ν run from -2^n to $2^n - 1$. We note that Q_0, like the hypercubes Q_n^j, is semi-open. Also if $j \neq j'$, Q_n^j and $Q_n^{j'}$ are disjoint. We consider only such Q_n^j which contain points of E and denote this set by $\mathscr{M}_n$. We now define a mass distribution on the hypercubes $\mathscr{M}_n$ as follows. The distribution is obtained by a series of steps.

At the first stage we define the density of μ_n^1 at every point of a hypercube $Q_n^j \in \mathscr{M}_n$ to be constant and such that

$$\mu_n^1(Q_n^j) = h(\delta_n).$$

I

At the second stage we consider the measures of hypercubes $\mu_n^1(Q_{n-1}^j)$ when μ_n^1 is defined as above. If $\mu_n^1(Q_{n-1}^j) \leqslant h(\delta_{n-1})$, we define

$$\mu_n^2 = \mu_n^1 \quad \text{in} \quad Q_{n-1}^j.$$

Otherwise we define for sets $e \in Q_{n-1}^j$

$$\mu_n^2(e) = C_j \mu_n^1(e)$$

where the constant $C_j < 1$ is so chosen that

$$\mu_n^2(Q_{n-1}^j) = h(\delta_{n-1}), \quad \text{i.e,} \quad C_j = h(\delta_{n-1})/\mu_n^1(Q_{n-1}^j).$$

Having defined μ_n^2, we consider $\mu_n^2(Q_{n-2}^j)$ for all the hypercubes Q_{n-2}^j. We again define for sets $e \in Q_{n-2}^j$

$$\mu_n^3(e) = C_j \mu_n^2(e),$$

where

$$C_j = \inf\{1, h(\delta_{n-2})/\mu_n^2(Q_{n-2}^j)\}.$$

Continuing in this way we define the measures μ_n^k in Q_0 for $0 \leqslant k \leqslant n$, and finally we set $\mu_n^n = \mu_n$.

From the construction it is clear that every point $a \in E$ is contained in some hypercube Q_p^j, such that $\mu_n(Q_p^j) = h(\delta_p)$. If there are several such hypercubes we select that one for which p is least. The set of all hypercubes selected in this way we call special hypercubes and denote them by $Q^{(j)}$. It is evident from the construction that distinct special hypercubes are disjoint. Also the union of all the special hypercubes contains E, so that by (5.4.2)

$$\Sigma \mu_n(Q^{(j)}) = \Sigma h(\delta_{p(j)}) \geqslant \eta, \tag{5.4.3}$$

where $\delta_{p(j)}$ is the side of $Q^{(j)}$. On the other hand our construction gives

$$\Sigma \mu_n(Q^{(j)}) \leqslant \mu_n(Q_0) \leqslant h(\delta_0), \tag{5.4.4}$$

and more generally for any hypercube Q_p^j, whether or not it contains points of E we have from our construction

$$\mu_n(Q_p^i) \leqslant h(\delta_p). \tag{5.4.5}$$

We now apply Theorem 5.3 to the sequence of measures μ_n on the fixed hypercube $\bar{Q}_0$. This is legitimate in view of (5.4.4). Thus by Theorem 5.3 there exists a subsequence μ_{n_p} weakly converging to a measure μ on $\bar{Q}_0$. We proceed to show that μ has the required properties.

First let E_0 be any compact set and set $\phi(x) \equiv 1$. Then

$$\mu(E_0) \leqslant \int \phi(x)\,\mathrm{d}\mu = \lim_{p\to\infty} \int \phi(x)\,\mathrm{d}\mu_{n_p} = \lim_{p\to\infty} \mu_{n_p}(Q_0) \leqslant h(\delta_0)$$

by (5.4.4). Thus, if $r \geqslant \delta_0$, (5.4.1) holds in any case with $C = 1$.

Suppose that $r < \delta_0$, and that the integer n, is such that

$$\delta_n = \delta_0 2^{-n} \leqslant r < \delta_{n-1}.$$

We choose r' so that

$$r < r' < \delta_{n-1}$$

and define

$$\phi(x) = 1 \quad \text{for} \quad |x - a| \leqslant r,$$

$$\phi(x) = \sup\{1 - (|x - a| - r)/(r' - r), 0\},$$

otherwise. Then $\phi(x)$ is continuous and $\phi(x) = 1$ on $C(a, r)$, so that if $C = C(a, r)$, $C' = C(a, r')$ we have

$$\mu(C) \leqslant \int \phi(x)\,\mathrm{d}\mu = \lim_{p\to\infty} \int \phi(x)\,\mathrm{d}\mu_{n_p} \leqslant \overline{\lim_{p\to\infty}}\, \mu_{n_p}(C').$$

Since C' has diameter $2r' < 4\delta_n$, it is clear that C' can meet at most 5^m hypercubes of $\mathscr{M}_n$, each of which has measure at most $h(\delta_n) \leqslant h(r)$ by (5.4.5). Thus for each p

$$\mu_{n_p}(C') \leqslant 5^m h(r), \quad \text{so that} \quad \mu(C) \leqslant 5^m h(r),$$

which proves (5.4.1) with $C = 5^m$. Also if E' is a compact set disjoint from E, E' is at a positive distance 2δ from E. We define

$$\phi(x) = \sup(0, 1 - \mathrm{d}(x, E')/\delta),$$

where $d(x, E')$ denotes the distance from E'. Then if $\delta_n\sqrt{m} < \delta$, $\phi(x) = 0$ on every hypercube Q_n^j which meets E and so

$$\int \phi(x)\,\mathrm{d}\mu_n = 0, \qquad n > n_0.$$

Then $\mu(E') \leqslant \int \phi(x)\,\mathrm{d}\mu = 0$. Hence μ is distributed on E.

Finally we set $\phi(x) = 1$ on Q_0. Then (5.4.3) gives

$$\mu(Q_0) = \int \phi(x)\,\mathrm{d}\mu = \lim_{p\to\infty} \int \phi(x)\,\mathrm{d}\mu_{n_p} \geqslant \eta.$$

so that $\mu(Q_0) = \mu(E) > 0$. This completes the proof of Lemma 5.4.

5.4.1. The main comparison theorems

We can now proceed to compare Hausdorff measure and capacity.

THEOREM† 5.13. *Suppose that E is a compact set and $h^*(E) > 0$, where $h(t)$ is a function such that $\int_0^1 (h(t)\,\mathrm{d}t)/(t^{\alpha+1}) < \infty$. Then $C_\alpha(E) > 0$. Thus conversely if $C_\alpha(E) = 0$ then $h^*(E) = 0$.*

† Frostman [1935].

Set

$$I_0 = \int_0^1 \frac{h(t)\,dt}{t^{\alpha+1}},$$

and let $\mu(E)$ be a unit mass distribution on E satisfying the conditions of Lemma 5.4. We proceed to show that the corresponding potential

$$V_\alpha(x) = \int K_\alpha(x-\xi)\,d\mu e_\xi$$

is uniformly bounded. Suppose first that $\alpha = 0$. Let $n(t) = \mu(D(x,t))$. Then if R is greater than the diameter of $E \cup \{x\}$,

$$V_0(x) = \int_0^R \log t\, dn(t) = \lim_{\varepsilon\to 0}\left\{[n(t)\log t]_\varepsilon^R - \int_\varepsilon^R \frac{n(t)\,dt}{t}\right\}$$

$$= \lim_{\varepsilon\to 0}\left\{n(R)\log R + n(\varepsilon)\log\frac{1}{\varepsilon}\right\} - \int_0^R \frac{n(t)\,dt}{t} \geqslant -C\int_0^1 \frac{h(t)\,dt}{t} - \int_1^R \frac{dt}{t}$$

$$\geqslant -CI_0 - \log R,$$

in view of Lemma 5.4. Thus $V_0(x)$ is uniformly bounded below on E and $C_0(E) > 0$.

Similarly if $\alpha > 0$, we have

$$V_\alpha(x) = \int_0^R -t^{-\alpha}\,dn(t) = \lim_{\varepsilon\to 0}\left[\frac{-n(t)}{t^\alpha}\right]_\varepsilon^R - \alpha\int_\varepsilon^R \frac{n(t)\,dt}{t^{\alpha+1}}$$

$$\geqslant -\frac{n(R)}{R^\alpha} - \alpha\int_0^R \frac{n(t)\,dt}{t^{\alpha+1}},$$

and the conclusion follows as before.

In order to prove a result in the opposite direction we need

LEMMA 5.5. *Let E be a compact set contained in a hypercube Q_0, and such that $h^*(E) < \infty$, for some Hausdorff function h. Let μ be a mass distribution on E and let E_1 be the subset of all points x of E for which*

$$\frac{\mu(Q_n)}{h(\delta_n)} \to 0,$$

as $n \to \infty$, where Q_n is the hypercube of side δ_n containing x, defined as in the proof of Lemma 5.4. Then $\mu(E_1) = 0$.

We divide Q_0 into the $2^{m(n+1)}$ hypercubes Q_n^j of $\mathscr{M}_n$ as in the proof of

Lemma 5.4 and define

$$\phi_n(x) = \frac{\mu(Q_n^j)}{h(\delta_n)}, \qquad x \in Q_n^j.$$

The function $\phi_n(x)$ is clearly Borel measurable and hence so is the set E_1 of all points x of E for which

$$\phi_n(x) \to 0 \quad \text{as} \quad n \to \infty. \tag{5.4.6}$$

Suppose now that $\mu(E_1) = 2\mu_0 > 0$. Then in view of Egorov's theorem† there exists a subset E_2 of E_1, such that $\mu(E_2) > \mu_0$ and (5.4.6) holds uniformly on E_2. Choose now ε so small that

$$2\,.\,3^m \varepsilon h^*(E) < \mu_0.$$

Then we deduce that there exists a positive integer n_0, such that for $x \in E_2$, and $n \geqslant n_0$, we have

$$\phi_n(x) < \varepsilon. \tag{5.4.7}$$

We next note that in view of the definition of $h^*(E)$ we can find a system of open hypercubes R_ν of sides $d_\nu < \delta_{n0}$, covering E and such that

$$\Sigma\, h(d_\nu) < 2h^*(E).$$

Let $R = R_\nu$ be one of these hypercubes of side d, and suppose that $\delta_n \leqslant d < \delta_{n-1}$, where $n > n_0$. Then R can meet at most 3^m of the hypercubes Q_n^j of $\mathscr{M}_n$, so that these cover R and have sides $\delta_n \leqslant d$. If we sum over these Q_n^j we have

$$\Sigma\, h(\delta_n) \leqslant 3^m h(\delta_n) \leqslant 3^m h(d).$$

Thus by replacing each of the R_ν by a system of hypercubes Q_n^j of $\mathscr{M}_n$ for some n depending on R_ν we obtain a covering of E by the Q_n^j for $n \geqslant n_0$, which we relabel $Q^{(\nu)}$. Also if l_ν is the length of the side of $Q^{(\nu)}$ we have

$$\Sigma\, h(l_\nu) \leqslant 2\,.\,3^m h^*(E).$$

We now select those $Q^{(\nu)}$ which meet E_2 and note that these cover E_2. Also in view of (5.4.7) we have for such $Q^{(\nu)}$

$$\mu(Q^{(\nu)}) < \varepsilon h(l_\nu).$$

Thus if Σ' denotes a sum over those ν for which $Q^{(\nu)}$ meets E_2 we have

$$\mu(E_2) \leqslant \Sigma'\, \mu(Q^{(\nu)}) \leqslant \varepsilon\, \Sigma'\, h(l_\nu) \leqslant 2\,.\,3^m \varepsilon h^*(E) < \mu_0.$$

This gives a contradiction, which proves Lemma 5.5.

† Egorov [1911].

We are now in a position to prove a result in the opposite direction to Theorem 5.13. This is

THEOREM† 5.14. *If E is a compact set in R^m, such that $l_\alpha(E) < \infty$, where $\alpha \geqslant 0$, then $C_\alpha(E) = 0$.*

COROLLARY. *If α_0 is the dimension of E, then $C_\alpha(E) = 0$ for $\alpha > \alpha_0$ and $C_\alpha(E) > 0$ for $\alpha < \alpha_0$. If $C_\alpha(E) = 0$ for $\alpha < 1$, then E is totally disconnected.*

Let μ be a unit mass distribution on E and set

$$V_\alpha(x) = \int K_\alpha(x - \xi)\, \mathrm{d}\mu e_\xi.$$

We proceed to show that

$$V_\alpha(x) = -\infty, \tag{5.4.8}$$

at all points of E outside E_1, so that, since $\mu(E_1) = 0$ by Lemma 5.5, we have

$$I_\alpha(\mu) = \int_E V_\alpha(x)\, \mathrm{d}\mu(x) = -\infty.$$

Since this is true for all measures μ, it will follow that $I_\alpha = -\infty$, and $C_\alpha(E) = 0$.

It remains to prove (5.4.8). Suppose that x is a point of E not in E_1. Then there exists $\eta > 0$, and a sequence $Q^{(j)}_{n_p} = R_p$ say of hypercubes of sides $d_p = \delta_{n_p}$ containing x and such that

$$\mu(R_p) > \eta h_\alpha(d_p),$$

where

$$h_\alpha(t) = (\log 1/t)^{-1}, \quad \text{if} \quad \alpha = 0, \qquad h_\alpha(t) = t^\alpha \quad \text{if} \quad \alpha > 0.$$

If the point x has positive mass (5.4.8) is evident. Otherwise, given $p > 0$, we can choose $\varepsilon_p > 0$, such that, if R'_p denotes the part of R_p distant more than ε_p from x, we have

$$\mu(R'_p) > \tfrac{1}{2}\eta h(d_p).$$

Also, by taking a subsequence if necessary, we may assume that $d_{p+1}\sqrt{m} < \varepsilon_p$, so that the sets R'_p for different p are disjoint. Thus

$$V_\alpha(x) \leqslant \sum_p \int_{R'_p} K_\alpha(x - \xi)\, \mathrm{d}\mu e_\xi + O(1).$$

† Erdös and Gillis [1937] first proved this result for the case $\alpha = 0$.

If $\alpha = 0$, we have on R'_p,

$$K_\alpha(x - \xi) \leqslant -\log\left(\frac{1}{d_p\sqrt{m}}\right),$$

so that

$$\int_{R'_p} K_\alpha(x - \xi)\,\mathrm{d}\mu e_\xi \leqslant -\log\left(\frac{1}{d_p\sqrt{m}}\right)\mu(R'_p) \leqslant -\tfrac{1}{2}\eta\,\frac{\log(1/d_p\sqrt{m})}{\log(1/d_p)} < -\frac{\eta}{4},$$

for $p > p_0$. This yields (5.4.8). Similarly if $\alpha > 0$

$$\int_{R'_p} K_\alpha(x - \xi)\,\mathrm{d}\mu e_\xi < -(d_p\sqrt{m})^{-\alpha}\mu(R'_p) < -\tfrac{1}{2}\eta(\sqrt{m})^{-\alpha}.$$

Thus we again deduce (5.4.8). This completes the proof of Theorem 5.14.

The corollary follows at once. In fact if α_0 is the dimension of E, and $\alpha > \alpha_0$, then $l_\alpha(E) = 0$ by Lemma 5.3 so that $C_\alpha(E) = 0$ by Theorem 5.14. Also if $\alpha < \alpha_0$, so that $\alpha_0 > 0$, choose α_1 so that $\alpha < \alpha_1 < \alpha_0$. Then by Lemma 5.3 we have, with $h(t) = t^{\alpha_1}$, $h^*(E) = \infty$. Also

$$\int_0^1 \frac{h(t)\,\mathrm{d}t}{t^{\alpha+1}} = \int_0^1 \frac{\mathrm{d}t}{t^{1+\alpha-\alpha_1}} < \infty.$$

Thus $C_\alpha(E) > 0$, by Theorem 5.13.

Hence if $C_\alpha(E) = 0$ for some $\alpha < 1$, we see that $\alpha_0 < 1$, so that E is totally disconnected by Theorem 5.12.

Theorems 5.13 and 5.14 in fact permit a fairly accurate appraisal of whether or not $C_\alpha(E) = 0$ in terms of Hausdorff measure. If $\alpha = 0$, we see that a necessary condition for $C_\alpha(E) = 0$ is that $h^*(E) = 0$ when

$$h(t) = (\log 1/t)^{-1}(\log\log 1/t)^{-(1+\delta)}, \qquad \delta > 0,$$

and a sufficient condition is that $h^*(E) < \infty$, when $h(t) = (\log 1/t)^{-1}$. If $\alpha > 0$ a necessary condition is that $h^*(E) = 0$, when $h(t) = t^\alpha(\log 1/t)^{-(1+\delta)}$ and a sufficient condition is that $h^*(E) < \infty$, when $h(t) = t^\alpha$. These results are the strongest that are true in general. No complete description of capacity in terms of Hausdorff measure is possible (see e.g. Carleson [1967] Chapter IV).

5.4.2. An application to bounded regular functions.

Suppose that E is a compact set in the open plane and let N be a neighbourhood of E. We say that E is a Painlevé null set (P.N. set) if and only if every function regular and bounded in $N - E$ can be continued as a regular function into the whole of N.

Necessary and sufficient conditions for P.N. sets seem hard to find. However we can prove

THEOREM 5.15. *For E to be a P.N. set it is sufficient*† *that $l_1(E) = 0$ and necessary*‡ *that $C_1(E) = 0$, i.e. a fortiori that $h^*(E) = 0$ for every Hausdorff function $h(t)$, such that*

$$\int_0^1 t^{-2}h(t)\,\mathrm{d}t < \infty.$$

Theorem 5.15 shows in particular that if α_0 is the dimension of E, then E is a P.N. set if $\alpha_0 < 1$, but not if $\alpha_0 > 1$.

Suppose first that $C_1(E) > 0$. Then in view of Theorem 5.8. there exists a unit mass distribution μ in E, such that for all complex z

$$\int \frac{\mathrm{d}\mu(\xi)}{|z-\xi|} \leqslant 2V' = \frac{2}{C_1(E)} = \eta_1, \quad \text{say.}$$

Set

$$f(z) = \int \frac{\mathrm{d}\mu(\xi)}{z-\xi}.$$

Then it follows that $f(z)$ is regular in the closed plane outside E. Also

$$f(z) = \frac{1}{z}\int \mathrm{d}\mu(\xi) + \frac{O(1)}{z^2} = \frac{1}{z} + \frac{O(1)}{z^2}, \quad \text{as} \quad z \to \infty,$$

so that $f(z)$ is not constant. If $f(z)$ could be analytically continued into E, $f(z)$ would be an integral function, and we should get a contradiction from Liouville's Theorem. Thus E is not a P.N. set.

Conversely suppose that $l_1(E) = 0$, and that $f(z)$ is regular and bounded in $N - E$. By hypothesis we can enclose E in a union of squares C_n, of sides d_n, such that $\Sigma\, d_n < \varepsilon$. Also, C_n can be included in an open disk $D_n = D(a_n, d_n)$, so that the union of these disks contains E. By the Heine–Borel Theorem a finite number $D_1, \ldots, D_m$ of these open disks suffices to cover E. We may assume that E is at a distance greater than 2ε from the closed complement of N, so that each disk D_n, which meets E lies entirely in N together with its circumference.

Let $G = \bigcup_{n=1}^{m} D_n$. Then G is an open neighbourhood of E, and $f(z)$ is regular and bounded in $G - E$. Also the boundary of G consists of a finite number of arcs of circles, whose total length is at most 2ε. We first fix such a neighbourhood $G = G_0$, with $\varepsilon = \varepsilon_0$, and then, given $\varepsilon_1 > 0$, construct a new neighbourhood G_1 as above corresponding to ε_1, so that $E \subset G_1 \subset \bar{G}_1 \subset G_0$.

† This result due to Painlevé appears first to have been published by Zoretti [1905]. It was rediscovered by Besicovitch [1932] and others.

‡ See Carleson [1967, p. 73] for a somewhat more general result.

Let D_0 be one of the domains of G_0, let D_1 be the union of all those components of G_1, which lie in D_0 and let C_0, C_1 be the boundaries of D_0, D_1 respectively. Then it follows from the Cauchy integral formula that for $z \in D_0 - D_1$, we have

$$f(z) = \int_{C_0} \frac{f(\zeta)\,d\zeta}{\zeta - z} - \int_{C_1} \frac{f(\zeta)\,d\zeta}{\zeta - z} = F_0(z) - F_1(z), \quad \text{say.}$$

We now vary D_1 and so $F_1(z)$, while keeping C_0, and so $F_0(z)$ fixed. Then, provided D_1 does not contain z, $F_0(z)$ and $f(z)$ and so $F_1(z)$ remain unaltered. Suppose that δ is the distance of z from E and $\varepsilon_1 < \frac{1}{4}\delta$. Then since all the circles of D_1 contain points of E and have radii less that ε_1, C_1 is distant at least $\frac{1}{2}\delta$ from z. Thus

$$|F_1(z)| \leqslant \int_{C_1} \frac{|f(\zeta)|\,|d\zeta|}{|\xi - z|} < \frac{2M}{\delta} \int_{C_1} |d\xi| < \frac{2M}{\delta} 2\pi\varepsilon_1,$$

if M is an upper bound for $f(z)$ in N and ε_1 a bound for the sum of the radii of C_1. Since ε_1 is arbitrary, we deduce that $F_1(z) = 0$, so that

$$f(z) = \frac{1}{2\pi i} \int_{C_0} \frac{f(\zeta)\,d\zeta}{\zeta - z}.$$

This provides the required analytic continuation into the interior of C_0 and similarly into the whole of G_0 and so in particular onto E. Thus E is a P.N. set. This completes the proof of Theorem 5.15.

The most interesting open case concerns compact sets of finite linear measure. Ivanov [1963] has shown† that if E lies on a sufficiently smooth curve then E is a P.N. set if and only if E has zero linear measure. On the other hand Vitushkin [1959] and Garnett [1970] have given examples of plane sets of Cantor type which are P.N. sets but have positive linear measure. We recall that sets of finite linear measure have $C_1(E) = 0$ by Theorem 5.14, so that Theorem 5.15 does not help here. One (rather bold) conjecture would be that a plane compact set of E of finite linear measure is a P.N. set if and only if E is irregular in the sense of Besicovitch [1938], i.e. if E intersects every rectifiable curve in a set of linear measure zero.

We shall see (Theorem 5.18) that the corresponding problems for s.h. and harmonic functions are much simpler. Bounded harmonic functions near a compact set E in R^m can always be continued as harmonic functions onto E, if and only if $C_{m-2}(E) = 0$.

† For linear sets this result is due to Denjoy [1909] and for sets on analytic curves to Ahlfors and Beurling [1950].

5.5. THE EXTENDED MAXIMUM OR PHRAGMÉN–LINDELÖF PRINCIPLE

The $(m-2)$-polar or simply polar sets play a most important role in the study of s.h. functions, since for many classes of such functions we can ignore the behaviour near such sets without significantly weakening our theorems. The function constructed in Theorem 5.11 plays a key role in this connection. We proceed to prove

THEOREM† 5.16. *Suppose that $u(x)$ is s.h. and bounded above in a domain D in R^m. Let F be the frontier of D, let E be a polar subset of F and suppose that*

$$\overline{\lim}\, u(x) \leqslant M \tag{5.5.1}$$

as x approaches any point ξ of F not in E from inside D. If D is unbounded the point at ∞ may be included in E if $m = 2$, but not if $m > 2$. Then if $F = E$, $u(x)$ is constant and otherwise

$$u(x) < M \text{ in } D \quad \text{or} \quad u(x) \equiv M \text{ in } D. \tag{5.5.2}$$

The important thing about Theorem 5.16 is that (5.5.1) need not hold at all points of F, but only at all points with certain exceptions. This is the typical situation in a Phragmén-Lindelöf principle.

Sometimes we need not require that $u(x)$ is bounded above, but only that $u(x)$ does not grow too rapidly near the exceptional points.

To prove Theorem 5.16 we assume first that D is bounded, or $m > 2$. We construct a function $\omega(x)$ which is s.h. and non-positive in D and such that

$$\omega(x) = -\infty \quad \text{in} \quad E.$$

This is possible by Theorem 5.11.

Let x_0 be a point of D. We assume that $\omega(x_0) > -\infty$. Consider now

$$u_\varepsilon(x) = u(x) + \varepsilon\omega(x),$$

where ε is a small positive number. Then if ξ is any point of F we have

$$\overline{\lim}\; u_\varepsilon(x) \leqslant M$$

as x approaches ξ from inside D. For if ξ is not a point of E this follows for (5.5.1) and the fact that $\omega(x) \leqslant 0$. If $\xi \in E$, then

$$\lim u_\varepsilon(x) = -\infty,$$

since $u(x)$ is bounded above in D. Thus we deduce from the ordinary maximum principle for the s.h. function $u_\varepsilon(x)$, that

$$u_\varepsilon(x) \leqslant M \text{ in } D$$

† Ascoli [1928].

and in particular

$$u_\varepsilon(x_0) \leqslant M,$$

i.e.

$$u(x_0) \leqslant M - \varepsilon\omega(x_0).$$

Since $\omega(x_0)$ is finite and ε is arbitrary, we deduce that

$$u(x_0) \leqslant M.$$

This is true for every $x_0 \in D$. It then follows from the ordinary maximum principle that

$$u(x) < M \quad \text{or} \quad u(x) \equiv M \text{ in } D.$$

It remains to consider the more difficult case when D is unbounded and $m = 2$. Since E is a 0-polar set, it follows that E is the countable union of compact polar sets E_n. Each of these has linear measure zero by Theorem 5.13. Thus we can include E_n in the union of squares the sum of whose sides is at most $\varepsilon 2^{-n}$, so that E can be included in the union of squares, the sum of whose sides is at most $\varepsilon \sum_1^\infty 2^{-n} = \varepsilon$. Thus E has linear measure zero. Hence, by Theorem 5.12, the set of r for which $S(0, r)$ does not meet E is everywhere dense on the real axis. We call such r normal. Since D is unbounded $S(0, r)$ meets D for all $r > r_0$. We define for any normal value of r

$$B(r) = \sup_{x \in D,\ |x| = r} u(x).$$

Suppose first that $B(r) \leqslant M$ for some arbitrarily large normal r. Let x_0 be a point of D, choose $r > |x_0|$ and such that $B(r) \leqslant M$, and let D_r be the subdomain of $D \cap D(0, r)$ which contains x_0. Then we can apply the case of Theorem 5.16, which we have already proved, to D_r and deduce that $u(x_0) \leqslant M$. This is true for every x_0 in D so that (5.5.2) holds in this case. Next we assume that

$$B(r) > M, \quad \text{for} \quad r \text{ normal and} \quad r > r_0.$$

We note that in this case there exists x in D, such that $|x| = r$ and $u(x) = B(r)$. In fact we may select a sequence of points x_0 on $S(0, r) \cap D$, such that

$$u(x_n) \to B(r), \text{ as } n \to \infty,$$

and by selecting a subsequence if necessary we may suppose that $x_n \to x$ as $n \to \infty$. The point x cannot be a point of E, since $S(0, r)$ does not meet E. Nor can x be another point of F, since at such points (5.5.1) holds. Thus x must be a point of D and since u is u.s.c. in D, we deduce that $u(x) \geqslant B(r)$ i.e. $u(x) = B(r)$.

We deduce that $B(r)$ is increasing for all sufficiently large r. In fact if $B(r) > M$

we may apply the result we have already proved to $u(x)$ in all the subdomains of $D \cap D(0, r)$ and deduce from (5.5.2) that

$$u(x) \leqslant B(r) \text{ in } D \cap D(0, r).$$

Thus in particular we have for $\rho < r$, $B(\rho) \leqslant B(r)$.

Suppose now that $B(r)$ is finally constant. Then we can find x_1 on $|x| = r_1$, such that $u(x_1) = B(r_1) = B$, and $B(r) = B$ for $r > r_1$. Then $u(x) \equiv B$ in that component D_r of $D \cap D(0, r)$, which contains x_1. Letting $r \to \infty$, we deduce that $D_r \to D$ so that $u(x) \equiv B$ in D. Thus $u(x)$ is constant in this case.

If $B(r)$ is not finally constant then

$$B(r) \to B, \text{ as } r \to \infty,$$

and for any fixed r

$$B(r) < B.$$

We set

$$u_\varepsilon(x) = u(x) - \varepsilon \log (|x|/r),$$

and apply the case of Theorem 5.16 we have already proved to all the subdomains of $(r < |x| < R) \cap D$. We deduce that if D_0 is such a domain, we have in D_0

$$u_\varepsilon(x) \leqslant \max (B(r), B(R) - \varepsilon \log (R/r)) \leqslant \max (B(r), B - \varepsilon \log (R/r)) = B(r),$$

if R is chosen big enough for fixed ε. Since for fixed x, we may choose ε as small as we please and then choose $R > |x|$, we deduce that $u(x) \leqslant B(r)$, $|x| > r$, i.e. $B(R) \leqslant B(r)$, $R > r$, so that $B(R)$ is finally constant contrary to hypothesis. Thus (5.5.2) holds in all cases unless $u(x)$ is constant.

If E contains the whole of F, then (5.5.1) need not hold at any point, so that we can replace M by any smaller number in this argument. In particular we may choose $M < u(x_0)$ for some x_0 in D. Then (5.5.2) is false, so $u(x)$ is necessarily constant. Otherwise (5.5.1) holds for at least one boundary point ξ of F, and so, even if $u(x) = B =$ constant, we must have $B \leqslant M$. This completes the proof of Theorem 5.16.

It is worth noting that the point at ∞ plays an entirely different role in the cases $m = 2$ and $m > 2$. If $m = 2$, it acts as a polar set, and if $u(x)$ is s.h. and bounded above outside a compact polar set in the open plane, it follows from Theorem 5.16 that $u(x)$ is constant. If (5.5.1) holds at all finite boundary points including at least one, and $u(x)$ is bounded above, then we deduce (5.5.2). These results are false if $m > 2$ as is shown by the function

$$u(x) = -|x|^{2-m}$$

which has the upper bound zero in the open plane but is bounded above by a negative constant on any compact subset of the open plane.

5.5.1. Uniqueness of the conductor potential

We can use Theorem 5.16 to prove a significant extension to Theorem 5.8. We have in fact

THEOREM† 5.17. *Suppose that $\alpha = m - 2$ in Theorem 5.8. Then we may replace* (5.2.7) *by*

$$u(x) \geqslant V \text{ in } R^m, \tag{5.5.3}$$

so that

$$u(x) = V \text{ n.e. } \alpha \text{ on } E. \tag{5.5.4}$$

The function $u(x)$ is uniquely determined as a function s.h. in R^m, harmonic outside E, satisfying (5.3.3) *and* (5.3.4) *and*

$$u(x) = \log|x| + o(1), \quad as \quad x \to \infty, \quad if \quad m = 2 \tag{5.5.5}$$

or

$$u(x) \to 0, \quad as \quad x \to \infty, m > 2. \tag{5.5.5'}$$

In particular the conductor potential is unique.

Let G be the unbounded complementary domain of E^*, let F be the frontier of G and let F_0 be those points of F, where

$$u(x) > V.$$

By Theorem 5.8. F_0 is an $(m-2)$-polar set.

We now consider the function

$$v(x) = -u(x),$$

in a component $G(R)$ of $G \cap D(0, R)$, when R is a large positive number. Then since $\alpha = m - 2$, $-K_\alpha(x)$ is harmonic away from the origin and so $u(x)$ is harmonic outside E^*. Also at the boundary points y of E^*, other than those of F_0, $u(x)$ is continuous. For $u(x)$ is u.s.c. in any case on E^* and in R^m, since $K_\alpha(x)$ is u.s.c. Also $u(y) = V$ and $u(x) \geqslant V$ at all other points x of E^*, so that $u(x)$ is continuous at y as a function on E^*. Hence it follows from Theorem 5.1 that $u(x)$ is continuous at y also as a function in R^m, and so the same is true of $v(x)$. Thus

$$v(x) \to -V, \text{ as } x \to y \text{ from } G(R),$$

where y is any boundary point of $G(R)$ not in F_0.

† Frostman [1935].

Also if R is chosen large enough we have on $S(x_0, R)$,

$$v(x) < 0, \text{ if } m = 2$$

$$v(x) < \varepsilon, \text{ if } m > 2,$$

where ε is a small positive number, which may be chosen to be less than $-V$. Thus

$$\overline{\lim}\, v(x) \leqslant -V$$

for any boundary point of $G(R)$ outside a polar set, and since $v(x)$ is s.h. in $G(R)$, and bounded above by (5.2.7), we deduce from Theorem 5.16 that

$$v(x) \leqslant -V, \quad \text{i.e.} \quad u(x) \geqslant V$$

in $G(R)$ and hence outside E^*. In view of (5.2.4) this inequality thus holds in the whole of R^m. Using (5.2.5) we deduce (5.5.4).

We show next that (5.5.5) or respectively (5.5.5′) holds. Suppose first that $m = 2$. Thus

$$\begin{aligned} u(x) &= \int_{E^*} \log|x - y|\, d\mu(y) \\ &= \int_{E^*} \left\{\log|x| + \log\left|1 - \frac{y}{x}\right|\right\} d\mu(y) \\ &= \log|x|\mu(E^*) + O(|x|^{-1}) = \log|x| + O(|x|^{-1}) \end{aligned}$$

as required, since $\mu(E^*) = 1$. This proves (5.5.5). On the other hand if $m > 2$

$$u(x) = \int_{E^*} -|x - y|^{2-m}\, d\mu(y)$$

and now (5.5.5′) is obvious.

We can now prove the uniqueness of the function $u(x)$. In fact let G be a complementary domain of E, and let x_0 be a frontier point of G, such that $u(x_0) = V$. Then since $u(x)$ is u.s.c.

$$\overline{\lim_{x \to x_0}}\, u(x) \leqslant V,$$

and in view of (5.5.3) we have that

$$u(x) \to V, \text{ as } x \to x_0 \text{ from inside } G.$$

Thus this result holds for all finite boundary points x_0 of G apart from a polar set F_0.

Suppose now that $u_1(x)$ is another function s.h. in R^m and harmonic outside

E and satisfying (5.5.3), (5.5.4) and (5.5.5) or (5.5.5'). Set

$$h(x) = u(x) - u_1(x).$$

Then $h(x)$ is harmonic and bounded outside E in view of (5.5.5) and (5.5.5') and

$$h(x) \to 0$$

as x approaches any boundary point of G including ∞, except for a polar set, which is the union of the polar sets corresponding to $u(x)$ and $u_1(x)$. Also $h(x)$ is bounded in G. Thus by Theorem 5.16 applied to $h(x)$ and $-h(x)$ we see that $h(x) \equiv 0$ in G and so outside E. Also by (5.5.4) $h(x) = 0$, n.e. in E, so that $h(x) = 0$ n.e. in R^m.

Finally let x_0 be a point of the polar set F, where $h(x)$ is not zero. If $u(x_0) = u_1(x_0) = -\infty$, there is nothing to prove. Otherwise $h(x_0)$ is well defined and since $u(x)$, $u(x_0)$ are s.h. and F has zero m-dimensional measure it follows from the definition in section 2.1 that

$$h(x_0) = \lim_{r \to 0} \frac{1}{d_m r^m} \int_{D(x_0, r)} h(x)\, dx = 0,$$

where d_m is a constant. Thus for every point in R^m we have

$$u(x) = u_1(x).$$

This completes the proof of Theorem 5.17.

5.5.2. Polar sets as null sets

Two other consequences of Theorem 5.16 can be proved very simply. We have

THEOREM† 5.18. *Suppose that E is a polar set in R^m, that G is open and that $u(x)$ is bounded above and satisfies the conditions* (i) *to* (iii) *for subharmonicity of section* 2.1 *at all points of $G - E$. Then $u(x)$ can be uniquely extended as a* s.h. *function to G.*

If $u(x)$ and $-u(x)$ satisfy the above conditions $u(x)$ can be extended as a harmonic function to G.

We note that since E is polar, E has zero $(m - 1)$-dimensional measure. Thus sphere averages for $u(x)$ are unaffected by the values of $u(x)$ on E and the above hypotheses make sense.

We prove first the first half and define for any $x_0 \in E \cap G$

(5.5.6) $$u(x_0) = \overline{\lim}\, u(x), \text{ as } x \to x_0 \text{ from } G - E.$$

† Brelot [1934].

Since E is polar, E has no interior points so that the definition makes sense. Clearly $u(x)$ is u.s.c. in G. We may assume G to be bounded and construct a s.h. function $\omega(x)$, such that $\omega(x) < 0$ in G, $\omega(x) = -\infty$ in $E \cap G$, $\omega(x_0) > -\infty$, where x_0 is a preassigned point of $G - E$. This is possible by Theorem 5.11.

We have to show that if $C(y, R)$ is any ball in G and $v(x)$ is harmonic in $D(y, R)$ and continuous in $C(y, R)$, and if

$$u(x) \leqslant v(x) \text{ in } S(y, R),$$

then the same inequality holds in $D(y, R)$.

We suppose first that x_0 is a point in $D(y, R)$, but not in E. Consider

$$h_\varepsilon(x) = u(x) + \varepsilon\omega(x) - v(x).$$

This function is certainly s.h. in $D(y, R)$. For, at points not in E, $u(x)$, $\omega(x)$, and $-v(x)$ satisfy the conditions for subharmonicity and at points of E, $h_\varepsilon(x)$ tends to $-\infty$.

Again at all points of $S(y, R)$

$$h_\varepsilon(x) \leqslant u(x) - v(x) \leqslant 0.$$

Thus

$$h_\varepsilon(x) \leqslant 0 \quad \text{in} \quad D(y, R).$$

Also for $x = x_0$ not in E, we may now let ε tend to zero and deduce

$$u(x_0) \leqslant v(x_0).$$

In view of (5.5.6) and the continuity of $v(x)$ this inequality remains true at all points of E. We deduce just as in the proof of Theorem 2.5, that $u(x)$ satisfies the mean value inequality and so $u(x)$ is s.h.

The extension is unique. In fact since our extended function $u(x)$ is s.h. we have for $x_0 \in G$

$$u(x_0) = \lim_{\rho \to 0} \frac{1}{d_m \rho^m} \int_{D(x_0,\, \rho)} u(x)\, dx, \tag{5.5.7}$$

where d_m is the volume of a unit ball in R^m. This follows for instance from the definition in Section 2.1 of s.h. functions. In this equation we can omit from consideration all points of the set E, since E has zero m-dimensional measure. Now the equation (5.5.7) provides an alternative definition for $u(x)$ at all points of E, so that the extension is unique.

Suppose now that $u(x)$ and $-u(x)$ satisfy the above conditions. Then by what we have just proved we may define $u(x)$ by (5.5.7) on E and the resulting function is s.h. in G. The same argument can be applied to $-u(x)$, so that $-u(x)$ as extended by means of (5.5.7) is also s.h. in G. Thus $u(x)$ and $-u(x)$ are s.h. in G and so by Theorem 2.9 $u(x)$ is harmonic in G. Also the extension is again unique since it must be given by (5.5.7). This completes the proof of Theorem 5.18.

For the conclusion of Theorem 5.18 the condition that E is polar is necessary. Suppose in fact that E is a compact set of positive $(m - 2)$-capacity and let

$$V_{m-2}(x) = \int K_{m-2}(x - y)\, d\mu(y)$$

be the corresponding conductor potential. Then $u(x) = -V_{m-2}(x)$ is harmonic outside E and bounded in any fixed ball. For $V_{m-2}(x)$ is s.h. in R^m and so bounded above in any fixed ball and $V_{m-2}(x)$ is bounded below by Theorem 5.8. Also

(5.5.8) $$u(x) > C \text{ in } R^m,\ u(x) \to C, \text{ as } x \to \infty,$$

where

$$C = 0 \text{ if } m > 2,\ C = -\infty, \text{ if } m = 2.$$

Thus $u(x)$ cannot admit any s.h. extension to E. For otherwise $u(x)$ would be s.h. and so u.s.c. in R^m and so have an upper bound M, such that $M > C$. Thus, in view of (5.5.8), $u(x)$ must attain the value M for some x_0 in R^m and, in view of the maximum principle Theorem 2.3, $u(x) \equiv M$ in R^m, which contradicts (5.5.8). A fortiori $u(x)$ cannot have a harmonic extension to E. We note that Theorem 5.18 supplies the analogue of Theorem 5.15 for bounded harmonic functions or s.h. functions which are bounded above. However Theorem 5.18 is much more complete that Theorem 5.15.

5.6. POLAR SETS AND THE PROBLEM OF DIRICHLET

Our next aim is to show that the irregular points for the problem of Dirichlet (see Section 2.6.2) form a polar set. It will follow that for an arbitrary domain in R^m and continuous boundary values the problem of Dirichlet has a unique solution, if we require the desired boundary behaviour at the regular points only.

We need first a result, which shows that we may weaken somewhat the conditions for regularity. This is

† Bouligand [1924].

THEOREM† 5.19. *Suppose that D is a domain in R^m, ζ_0 a boundary point of D, N a neighbourhood of ζ_0, and $N_0 = N \cap D$. Then ζ_0 is a regular boundary point of D if there exists $\psi(x)$ such that*

(i) $\psi(x)$ is s.h. in N_0;
(ii)′ $\psi(x) < 0$ in N_0;
(iii) $\psi(x) \to 0$ as $x \to \zeta_0$ *from inside* N_0.

In particular $\zeta_0 = \infty$ is always regular if D is unbounded and $m > 2$.

The function $\psi(x)$ satisfies all the conditions for the barrier function in Section 2.6 except that (ii) has been replaced by the much simpler condition (ii)′.

Assume first that ζ_0 is finite and that N includes the closed ball $C(\zeta_0, r)$. Let $0 < \rho < r$,

$$S_0 = S(\zeta_0, \rho) \cap D,$$

and let σ_0 be the surface area of S_0. We choose a compact subset e_1 of S_0, whose surface area is σ_1, and whose boundary within S_0 has zero surface area. We may arrange for instance for e_1 to be the union of a finite number of closed spherical caps.

We now define $v_\rho(x)$ to be a bounded harmonic function in $D(\zeta_0, \rho)$, which has boundary values on $S(\zeta_0, \rho)$ given by

$$\begin{aligned} v_\rho(x) &= 1 \text{ for } x \in S_0 - e_1 \\ &= 0 \text{ in } S(\zeta_0, \rho) - \overline{(S_0 - e_1)}. \end{aligned}$$

Such a function is given by

$$v_\rho(x) = \int_{S_0 - e_1} K(x, y)\, d\sigma(y),$$

where $K(x, y)$ denotes the Poisson–kernel on $S(\zeta_0, \rho)$ and $d\sigma(y)$ denotes surface area on $S(\zeta_0, \rho)$. We have

$$v_\rho(\zeta_0) = v_0 > 0,$$

where v_0 is the ratio of the area of $S_0 - e_1$ to that of $S(\zeta_0, \rho)$. We choose the area of e_1 so large that $v_0 < \rho/r$.

We now consider on the boundary Γ_0 of $N_1 = D(\zeta_0, r) \cap D$

$$f(\zeta) = |\zeta - \zeta_0|,$$

and attempt to solve the problem of Dirichlet in the domains of N_1 for these boundary values. In other words just as in Section 2.6, we define the class $U(f)$ of functions s.h. in N_1 and satisfying

$$\varlimsup u(x) \leqslant f(\zeta) \tag{5.6.1}$$

as x approaches any boundary point ζ of N_1 from inside N_1. We set

$$v(x) = \sup_{u \in U(f)} u(x),$$

and recall from Lemma 2.3 that $v(x)$ is harmonic in N_1 and

$$0 \leqslant v(x) \leqslant r, \qquad x \in N_1. \tag{5.6.2}$$

On the other hand the function $u(x) = |x - \zeta_0|$ is itself s.h. in N_1 and satisfies (5.6.1). Thus we can sharpen (5.6.2) to

$$|x - \zeta_0| \leqslant v(x) \leqslant r. \tag{5.6.3}$$

The function

$$\omega(x) = -v(x)$$

will now satisfy all the properties (i), (ii) and (iii) required for the barrier in Section 2.6.2, provided that we can prove that $\omega(x) \to 0$, i.e. that

$$v(x) \to 0, \text{ as } x \to \zeta_0. \tag{5.6.4}$$

This we proceed now to do.

Since $\psi(x)$ is s.h. and $\psi(x) < 0$ in N_0, $\psi(x)$ is u.s.c. on the compact set e_1 and so $\psi(x)$ attains its negative least upper bound, $-m_1$ say, in e_1. Let $u(x) \in U(f)$ and consider

$$h(x) = u(x) - \rho + \frac{r\psi(x)}{m_1} - rv_\rho(x).$$

Then $h(x)$ is s.h. in $N_\rho = D \cap D(\zeta_0, \rho)$. At boundary points ζ of N_ρ, we have

$$\overline{\lim_{x \to \zeta}}\, h(x) \leqslant 0. \tag{5.6.5}$$

In fact if ζ is a frontier point of D, then

$$\overline{\lim_{x \to \zeta}}\, u(x) \leqslant |\zeta - \zeta_0| \leqslant \rho,$$

and so (5.6.5) holds, since $\psi(x) < 0$, and $v_\rho(x) > 0$. Again if $\zeta \in e_1$ we have by (5.6.2)

$$\overline{\lim_{x \to \zeta}}\, u(x) \leqslant v(\zeta) \leqslant r, \frac{r\psi(\zeta)}{m_1} \leqslant -r,$$

so that (5.6.5) holds again. Finally if $\zeta \in S_0 - e_1$, then

$$\lim_{x \to \zeta} h(x) \leqslant r - \lim_{x \to \zeta} rv_\rho(x) = r - r = 0.$$

Thus (5.6.5) holds in all cases and we deduce that

$$h(x) \leqslant 0, \qquad u(x) \leqslant \rho + rv_\rho(x) + \frac{r\psi(x)}{m_1}, \qquad x \in N_\rho$$

and since this is true for every $u(x) \in U(f)$, we obtain

$$v(x) \leqslant \rho + rv_\rho(x) + \frac{r\psi(x)}{m_1}, \; x \in N_\rho.$$

We let $x \to \zeta_0$ in this and deduce that

$$\overline{\lim_{x \to \zeta_0}}\, v(x) \leqslant \rho + rv_\rho(\zeta_0) \leqslant 2\rho.$$

Since ρ can be chosen as small as we please we deduce (5.6.4). Thus $-v(x)$ satisfies all the conditions for a barrier at ζ_0, and ζ_0 is regular for the problem of Dirichlet as required.

It remains to consider the case $\zeta_0 = \infty$. If $m > 2$, the function $\omega(x) = -|x|^{2-m}$ satisfies all the conditions for the barrier function of Section 2.6.2, so that ∞ is always regular in this case. If $m = 2$ we write z for the complex number corresponding to x, and set

$$Z = z^{-1}.$$

Then D is transformed onto a domain D_0 and ∞ is a regular boundary point of D, if and only if 0 is a regular boundary point of D_0, since conformal mappings transform s.h. and harmonic functions into s.h. and harmonic functions, by Theorem 2.8, corollary. Thus we may apply the above theory to D_0 with the point $\zeta_0 = 0$, and Theorem 5.19 is proved in all cases.

We deduce

THEOREM† 5.20. *Let D be a domain in R^m and let E be the intersection of the complement of D with a fixed ball $C(x_0, r)$, chosen so that D is in the unbounded complementary domain of E. Then a frontier point ζ_0 of D in $D(x_0, r)$ is regular for the problem of Dirichlet provided that* $\mathrm{Cap}_{m-2}(E) > 0$ *and*

$$u(\zeta_0) = V,$$

where $u(x)$ is the conductor potential of E and $V = V_{m-2}$ the corresponding equilibrium value.

In particular‡ the set of irregular points for the problem of Dirichlet is a polar set on the frontier of D.

Suppose in fact that $u(\zeta_0) = V$. Then the conductor potential $u(x)$ is har-

† Evans [1933].
‡ This last result is due to Kellogg [1928].

monic outside E and so in particular in

$$N_0 = D(x_0, r) \cap D.$$

Also in view of Theorem 5.17 we have

$$u(x) \geqslant V \text{ in } N_0$$

and since $u(x)$ is harmonic and not constant in the unbounded complementary domain of E, which includes D, we deduce that

$$u(x) > V \text{ in } N_0.$$

Also since $u(x)$ is a potential and so u.s.c. in R^m, we deduce that

$$u(x) \to V, \text{ as } x \to \zeta_0 \text{ from inside } N_0.$$

Thus the function $\psi(x) = V - u(x)$, satisfies the hypotheses of Theorem 5.19 and so ζ_0 is regular for the problem of Dirichlet.

Suppose now that F is the complement of D in R^m. We express F as the union of a countable number of compact sets E_ν whose diameter is smaller than that of D, so that D certainly lies in the unbounded component of the complement of E_ν. Thus all the finite boundary points of D are regular for the problem of Dirichlet except for those points for which the conductor potential in E_ν does not take its equilibrium value for some ν. In view of Theorem 5.17 this exceptional set of irregular points is the subset of a polar set and so is polar. Also, if D is unbounded, ∞ is a regular point if $m > 2$, and a polar set if $m = 2$. This completes the proof of Theorem 5.20.

We can now prove a considerably more general solution of the problem of Dirichlet. This result is due to Wiener [1924].

THEOREM 5.21. *Suppose that D is a domain in R^m, with frontier Γ and that $f(x)$ is bounded on Γ and continuous there outside a polar set E_0. (E_0 may include ∞ if $m = 2$, but not if $m > 2$.)*

Then if Γ is polar, so that $m = 2$ and D is unbounded, all bounded harmonic functions in D are constant.

Otherwise there exists a unique function $v(x)$ harmonic and bounded in D, equal to $f(x)$ on Γ and continuous in $\bar{D}$ outside a polar subset E_1 of Γ. In fact E_1 may be taken to be $E_0 \cup E$, where E is the set of irregular boundary points of D.

We construct the function $v(x)$ as in Section 2.6. It then follows that $v(x)$ is harmonic and bounded in D. Further if we define $v(x) = f(x)$ on Γ it follows from Theorem 2.10 that $v(x)$ is continuous in $\bar{D}$ outside the set $E_1 = E_0 \cup E$. By Theorem 5.20 E is a polar set and so therefore is E_1. Thus $v(x)$ has the required properties as asserted in Theorem 5.21. If D is unbounded we include the point at ∞ in E_1 if $m = 2$, but not if $m > 2$. In view of Theorem 5.19, the point at ∞ is regular if $m > 2$.

Next if $m = 2$ and Γ is polar, it follows from Theorem 5.16 that $v(x)$ is constant in D. Suppose then that Γ is not polar, so that $E_1 = E \cup E_0$ is a proper subset of Γ. It remains to show that $v(x)$ is unique. Suppose that $v_1(x)$ is another function satisfying the conclusions of Theorem 5.21. Set

$$h(x) = v(x) - v_1(x).$$

Then $h(x)$ is harmonic and bounded in D and

$$h(x) \to 0 \text{ as } x \to \zeta \text{ in } \Gamma \text{ from inside } D,$$

except for ζ lying on the polar set E_1. It now follows from the extended maximum principle Theorem 5.16, applied to $h(x)$ and $-h(x)$, that

$$h(x) \leqslant 0 \text{ in } D, \text{ and } h(x) \geqslant 0 \text{ in } D,$$

so that $h(x) \equiv 0$ in D. This concludes the proof of Theorem 5.21.

Suppose now that D and Γ are as in Theorem 5.21. We shall say that a point ζ_0 of Γ is *irregular* (for the problem of Dirichlet) if there exists a function $f(\zeta)$ on Γ, satisfying the conditions of Theorem 5.21 and continuous at ζ_0, such the corresponding function $v(x)$ of Theorem 5.21 is discontinuous at ζ_0. We can describe more closely the nature of the irregular boundary points and also prove a converse to Theorem 5.20. This is

THEOREM 5.22. *Suppose that D is a domain in R^m, and that ζ_0 is a finite point on the frontier Γ of D. Then the following conditions are equivalent*

(i) *ζ_0 is irregular for the problem of Dirichlet.*

(ii) *If E is the intersection of the complement of D with any fixed ball $C(\zeta_0, r)$, such that D lies in the unbounded complementary domain of E, then either $\mathrm{Cap}_{m-2}(E) = 0$, or else, if $u(x)$ is the conductor potential of E and V its equilibrium value, we have*

$$u(\zeta_0) > V. \tag{5.6.6}$$

(iii) *There exists no barrier function at ζ_0 satisfying the conditions of Section 2.6.2.*

If D is unbounded, $\zeta_0 = \infty$ is irregular if and only if $m = 2$ and (iii) holds. Hence the set of irregular points on Γ is a polar F_σ set.

We have proved in Theorem 2.10 that (i) implies (iii) and in Theorem 5.20 that (i) implies (ii). Further it follows from Theorems 5.19, 5.20 that if (ii) is false for some set E, then there exists a barrier function at ζ_0 and so ζ_0 is regular. Thus (iii) implies (ii). Hence it remains to show that (ii) implies (i). This we proceed to do.

Suppose first that the whole frontier Γ including ∞ is polar. Then $m = 2$

and all s.h. functions in D which are bounded above are constant by Theorem 5.16. In this case all the criteria (i), (ii) and (iii) are satisfied and there is nothing to prove. Thus we may assume that Γ contains at least one regular boundary point $\zeta_1 \neq \zeta_0$.

Suppose first that $\mathrm{Cap}_{m-2}E = 0$. Then we define

$$f(\zeta) = \tan^{-1}|\zeta - \zeta_0|$$

on the boundary of D, and proceed to construct the function $v(x)$ with these boundary values as in Theorem 5.21. Then $v(x)$ is harmonic, nonnegative and bounded in D. Also $v(x) > 0$ near a regular boundary point ζ_1 of D, so that, by the maximum principle, $v(x) > 0$ in D. Since $\mathrm{Cap}_{m-2}(E) = 0$, we can extend $v(x)$ as a harmonic function into a neighbourhood of ζ_0 in view of Theorem 5.18. Thus by the maximum principle $v(\zeta_0) > 0$ and

$$v(x) \to v(\zeta_0) \neq f(\zeta_0), \text{ as } x \to \zeta_0 \text{ from inside } D.$$

Thus ζ_0 is an irregular point.

Suppose next that $\mathrm{Cap}_{m-2}(E) > 0$, but that (5.6.6) holds. Suppose first that $m > 2$. Let $u(x)$ be the conductor potential of E, let Γ be the frontier of D and assign boundary values $f(\zeta) = V$ at points ζ in $E \cap \Gamma$ and $f(\zeta) = u(\zeta)$ at all other points of Γ (and in particular $f(\infty) = 0$). Let $v(x)$ be the corresponding solution of the problem of Dirichlet as given by Theorem 5.21.

We note that by Theorem 5.17 $u(x) = V$, and so by Theorem 5.1 $u(x)$ is continuous, at all points of $E \cap \Gamma$ outside a polar set. Thus $f(\zeta)$ is continuous and coincides with $u(\zeta)$ at all points of Γ outside a polar set. Also the function $u(x)$ is harmonic and bounded in D, continuous in $\bar{D}$ outside a polar set and coincides with $f(\zeta)$ on Γ outside a polar set. Thus by Theorem 5.21 we have $v(x) = u(x)$.

To conclude our proof it is enough to show that if (5.6.6) holds $u(x)$ cannot tend to V as $x \to \zeta_0$ from inside D. Suppose in fact contrary to this that

(5.6.7) $$u(x) \to V, \text{ as } x \to \zeta_0 \text{ from inside } D.$$

Then given $\varepsilon > 0$, we have for all sufficiently small ρ

(5.6.8) $$u(x) < V + \varepsilon, \text{ n.e. in } D(\zeta_0, \rho).$$

For this conclusion holds for points of D in view of (5.6.7). Also points of $D(\zeta_0, \rho)$ outside D lie in E and so satisfy $u(x) = V$ outside a polar set, by Theorem 5.17. Since a polar set has m-dimensional measure zero, it follows that

$$\int_{D(\zeta_0, \rho)} \{u(x) - V\}\, dx \leqslant o(\rho^m), \text{ as } \rho \to 0,$$

and since $u(x)$ is s.h. this implies that $u(\zeta_0) \leqslant V$. This contradicts (5.6.6).

If $m = 2$, the argument must be modified slightly. If ζ is a point of E, let

$E(\zeta)$ be the disk $|z - \zeta| < \frac{1}{2}|\zeta - \zeta_0|$. Then clearly a finite set of these disks cover the part of E for which $|\zeta - \zeta_0| \geqslant (1/n)$, and so a countable set of the $E(\zeta)$ cover E apart from the point ζ_0. Since E is not polar it follows that for at least one $\zeta = \zeta_1$ the set

$$E_1 = E \cap \overline{E(\zeta_1)}$$

is not polar and so has positive capacity. Let $u_1(x)$ be the conductor potential of E_1 and $u(x)$ that of E. We define

$$f(\zeta) = V - u_1(\zeta) \text{ on } E,$$

$$f(\zeta) = u(\zeta) - u_1(\zeta) \text{ at all other points of } \Gamma.$$

Then $f(\zeta)$ is bounded on Γ, in view of (5.5.5) and continuous outside a polar set. So we may solve the problem of Dirichlet as in Theorem 5.21 and we see that

$$v(x) = u(x) - u_1(x)$$

provides the required solution. We prove just as before that, in view of (5.6.6), (5.6.7) cannot hold. Since $v_1(x)$ is harmonic and so continuous at ζ_0, we deduce that $v(x)$ cannot tend to $f(\zeta_0)$ as $x \to \zeta_0$ from inside D, so that ζ_0 is irregular. This completes the proof that (ii) implies (i), so that (i), (ii) and (iii) are equivalent.

Finally we cover the finite points of Γ by a countable number of closed balls $C(x_\nu, r_\nu)$ such that $2r_\nu$ is less than the diameter of D. If E_ν is the intersection of the complement of D with $C(x_\nu, r_\nu)$, and F_ν is the set of points in $\Gamma \cap E_\nu$ where (5.6.6) holds, then F_ν is an F_σ set, since potentials are u.s.c. Hence so is the set of irregular points, which is $\bigcup F_\nu$. The set is polar by Theorem 5.20. This completes the proof of Theorem 5.22.

5.7. GENERALIZED HARMONIC EXTENSIONS AND GREEN'S FUNCTION

Wiener's generalized solution of the problem of Dirichlet, which was obtained in Theorem 5.21 has many important applications. In particular it enables us to extend the whole of the theory of Sections 3.6 to 3.8 to arbitrary domains D in R^m. The condition that D is regular can be completely omitted at the cost of having exceptional polar sets on the boundary of D, which include the set of irregular points. Since many proofs are almost unaltered from those of Chapter 3, once the modifications in the definitions are made, we shall confine ourselves in this section to giving the modified definitions and statements of the theorems, while leaving the proofs to the reader unless they are substantially different from those given in Chapter 3.

5.7.1. Harmonic extensions

Let D be a domain in R^m, whose frontier Γ is not polar and suppose that $f(\zeta)$ is continuous on Γ. Let $v(x)$ be the unique function, whose existence is asserted in Theorem 5.21. Then we shall say that

$v(x)$ is the harmonic extension of $f(\zeta)$ from Γ into D.

We note that $v(x)$ is characterized by the following properties.

(i) $v(x)$ is harmonic and bounded in D (and lies between the bounds for $f(\zeta)$).

(ii) If ζ is any regular point of Γ, then

$$v(x) \to f(\zeta), \text{ as } x \to \zeta \text{ from inside } D. \tag{5.7.1}$$

In particular (5.7.1) holds for all points ζ of Γ outside a polar F_σ set Γ_0, which is independent of the particular function f.

We can next extend our definition to semi-continuous functions $f(\zeta)$ just as in Section 2.72, by noting that if $f_n(\zeta)$ is a monotonic sequence of continuous functions on Γ with harmonic extensions $u_n(x)$, the $u_n(x)$ tend to a limit $u(x)$, which is harmonic or identically infinite in D. Also $u(x)$ depends only on the limit $f(\zeta)$ of the sequence $f_n(\zeta)$ and not on the particular sequence. The proof is identical with that given in Theorem 2.17 and enables us to define the harmonic extension of semi-continuous functions $f(\zeta)$ from Γ to D.

With these definitions the statement and proof of Theorem 3.10 remain valid without change for arbitrary domains in R^m, whose complement is not polar. For each such domain D, a harmonic measure $\omega(x, e)$ is defined for Borel sets e on the frontier Γ of D and points x in D, satisfying the conditions of Theorem 3.10, such that the harmonic extension of $f(\zeta)$ from Γ to D is given by

$$u(x) = \int_\Gamma f(\zeta)\, d\omega(x, e_\zeta), \tag{5.7.2}$$

for continuous or semicontinuous functions f. We can then use (5.7.2) to define† the harmonic extension of f to D in all cases where f is a finite Borel function integrable w.r.t. $\omega(x, e)$. The condition for this integrability is again independent of the point x. The following result is worth noting.

THEOREM 5.23. *If D is any domain in R^m, F the frontier of D, and e any polar set on F, then e has harmonic measure zero. In particular the set of irregular points has harmonic measure zero. A single point ζ_0, has zero harmonic measure, except possibly if $m > 2$ and $\zeta_0 = \infty$.*

† Brelot [1939a]. Brelot also showed that even in this general case the solution coincides with that of Perron given by (2.6.1).

Suppose first that e is an arbitrary compact set on the frontier of D. For any point ζ, let

$$\delta(\zeta) = \delta(\zeta, e)$$

be the distance from ζ to e, and define

$$\delta_n(\zeta) = \max(1 - n\delta(\zeta), 0).$$

Then clearly $\delta_n(\zeta)$ is a continuous decreasing sequence of functions on the boundary Γ of D and

$$\delta_n(\zeta) \to \chi_e(\zeta), \text{ as } n \to \infty,$$

where $\chi_e(\zeta)$ is the characteristic function of e. Thus if $v_n(x)$ is the harmonic extension of $\delta_n(\zeta)$, we deduce that

$$v_n(x) \to \omega(x, e)$$

as $n \to \infty$, where $\omega(x, e)$ is the harmonic measure of e.

Suppose now that ζ_0 is a regular point in the complement d of e, with respect to the frontier of D. Then for $n \geqslant n_0$ $\delta_n(\zeta) = 0$ near ζ_0, and so there exists a neighbourhood N_0 of ζ_0, such that

$$v_{n_0}(x) < \varepsilon, \qquad x \in N_0.$$

Thus

$$\omega(x, e) < \varepsilon, \qquad x \in N_0.$$

Since $\omega(x, e) \geqslant 0$ in D, we deduce that

$$\omega(x, e) \to 0 \text{ as } x \to \zeta_0 \text{ from inside } D,$$

where ζ_0 is any regular point of d. If e is a polar set it follows from Theorem 5.21 that there is only one bounded harmonic function with this property, namely the function zero, i.e. $\omega(x, e) = 0$ in this case.

More generally if $e \subset \bigcup_{\nu=1}^{\infty} e_\nu$, where the e_ν are compact polar sets it follows from the properties of measures that

$$\omega(x, e) \leqslant \sum_{\nu=1}^{\infty} \omega(x, e_\nu) = 0.$$

Thus our conclusion holds for any polar set e.

The above argument shows in particular that any finite frontier point has harmonic measure zero. If $m = 2$ we can extend this conclusion to $\zeta_0 = \infty$ by conformal mapping. This completes the proof of Theorem 5.23.

If $m > 2$, $\zeta_0 = \infty$ may have positive harmonic measure. As an example we take for D the domain $|x| > 1$, and for e the point at ∞.

Then

$$\omega(x, e) = 1 - |x|^{2-m}.$$

However for this to happen the complement of D must be very small at ∞. For let $\omega(x)$ be the harmonic measure of $\zeta_0 = \infty$ in D, and suppose that $\omega(x)$ is not identically zero. Then a component, D_0, of $\omega(x) > \varepsilon$ in D cannot be bounded when $\varepsilon > 0$. For otherwise, if ζ is a frontier point of D_0 we should have

$$\overline{\lim_{x \to \zeta}}\, \omega(x) \leqslant \varepsilon, \tag{5.7.3}$$

if ζ is a point of D, since then $\omega(x)$ is continuous at ζ and

$$\overline{\lim_{x \to \zeta}}\, \omega(x) = 0,$$

if ζ is a regular frontier point of D. Thus (5.7.3) holds at all points ζ except for a polar set and so, by Theorem 5.16, we have $\omega(x) \leqslant \varepsilon$ in D_0, giving a contradiction. Thus, since $\omega(x)$ is bounded in D_0, it follows from the proof of Theorem 4.15, that if $\theta(r)$ is the $(m-1)$-dimensional area of the intersection of the complement of D_0 with $S(0, r)$, then

$$\theta(r) = o(r^{m-1}).$$

In particular, if D_1 is a domain of R^m complementary to D_0, then the harmonic measure of $\zeta_0 = \infty$ is zero in D_1.

5.7.2. The generalized Green's function

We can now extend the notion of the Green's function to arbitrary domains D in R^m. Let Γ be the frontier of D and suppose first that $m = 2$ and that Γ is polar. In this case it follows from Theorem 5.16 that the only s.h. functions in D which are bounded above are the constants. We say that D has no Green's function in this case. In all other cases we can define a Green's function $g(x, \xi, D)$. We can proceed just as in Section 3.7, except when $m = 2$ and D has no exterior point. In order to take in this case however it is convenient to proceed differently.

Given D and a point ξ in D, we shall say that $g(x, \xi, D)$ is the (generalized) Green's function of D if $g(x, \xi, D)$ has the following properties.

(i) g is harmonic in D except at the point $x = \xi$;
(ii)′ If ζ is any boundary point of D, apart from a polar set E, then

$$g(x, \xi) \to 0 \text{ as } x \to \zeta \text{ from inside } D,$$

and if ζ is a point of E, $g(x, \zeta)$ remains bounded as $x \to \zeta$ from inside D;

(iii) $g + \log|x - \xi|$ remains harmonic at $x = \xi$ if $m = 2$,
$g - |x - \xi|^{2-m}$ remains harmonic at $x = \xi$ if $m > 2$.

The conditions (i) and (iii) are the same as the corresponding conditions (i) and (iii) for the classical Green's function in Section 1.5.1. However (ii)′ is weaker than (ii), but reduces to (ii) if all the frontier points of D are regular. Thus the classical Green's function is a special case of the generalized Green's function. We proceed to prove†

THEOREM 5.24. *If D is any domain in R^m, whose frontier is not polar, then the (generalized) Green's function of D exists and is unique.*

We prove first the uniqueness. Suppose that $g_1(x)$ and $g_2(x)$ satisfy the conditions (i), (ii)′ and (iii) for $g(x, \xi, D)$. Let

$$h(x) = g_1(x) - g_2(x).$$

Then $h(x)$ is harmonic in D and if ζ is a frontier point of D not belonging to an exceptional set E, we have that

$$h(x) \to 0, \text{ as } x \to \zeta \text{ from inside } D. \tag{5.7.4}$$

Next we show that $h(x)$ is bounded in D. Suppose contrary to this that x_n is a sequence of points in D and that

$$h(x_n) \to \infty, \text{ as } n \to \infty.$$

By choosing a subsequence if necessary we may assume that $x_n \to x$ as $n \to \infty$, where x is a point in R^m or $x = \infty$. Then x cannot be a point of D since h is harmonic and so continuous at points of D. Also x cannot be a frontier point of D, since by (ii)′ $g_1(x_n)$ and $g_2(x_n)$ and so $h(x_n)$ remain bounded as $x_n \to x$. Thus we have our contradiction. Also it follows from our hypothesis that the frontier of D is not polar, while the exceptional set $E_1 \cup E_2$ corresponding to g_1, g_2 is polar. Thus (5.7.4) holds for at least one point ζ (possibly ∞). It now follows from Theorem 5.16 that $h(x) \leqslant 0$ and $h(x) \geqslant 0$, so that $h(x) \equiv 0$ in D. Thus $g(x, \xi, D)$ is unique if it exists.

Next we prove the existence. Suppose first that $m > 2$, and set

$$f(x) = -|x - \xi|^{2-m}$$

on the frontier Γ of D. Then $f(x)$ is bounded and continuous on Γ, including possibly ∞. Thus by Theorem 5.21 the function $f(x)$ possesses a harmonic extension $u(x)$ into D, such that $u(x)$ is bounded in D and

$$u(x) \to f(\zeta)$$

† Bouligand [1924] introduced the generalized Green's function and proved that the exceptional set in (ii)′ is precisely the set of irregular boundary points.

as x tends to any regular boundary point ζ from inside D. Thus

$$g(x, \xi, D) = u(x) + |x - \xi|^{2-m}$$

satisfies the conditions (i), (ii)′ and (iii) for the Green's function.

Suppose finally that $m = 2$. Let Γ_N be the intersection of the complement of D with $|x| \leqslant N$. Then if Γ_N were polar for every N, it would follow that Γ was also polar, since $x = \infty$ is a polar set. This would lead to a contradiction. Thus Γ_N has positive capacity for some $N > 0$. We choose such an N and define the corresponding conductor potential

$$V(x) = \int_{\Gamma_N} \log|x - \zeta| \, d\mu e_\zeta.$$

Let V be the equilibrium potential, and define

$$f(x) = \log|x - \xi| - V, \; x \in \Gamma_N,$$

$$f(x) = \log|x - \xi| - V(x), \; x \in \Gamma - \Gamma_N.$$

Then $f(x)$ is continuous except at those points of $\Gamma_N \cap (|x| = N)$, where $V(x) \neq V$, i.e. apart from a polar set by Theorem 5.17. Thus $f(x)$ can be uniquely harmonically extended to a function $u(x)$, and also, since $f(x)$ is uniformly bounded on Γ, $u(x)$ is uniformly bounded in R^2. We can set

$$g(x, \xi, D) = u(x) + V(x) - \log|x - \xi|.$$

We note that as x approaches any boundary point ζ_0 of D on Γ, $g(x, \xi, D)$ remains bounded. Further

$$g(x, \xi, D) \to 0, \text{ as } x \to \zeta_0,$$

provided that ζ_0 is regular for D, and $V(x)$ is continuous at ζ_0, i.e. for all ζ_0 apart from a polar set. Thus g satisfies (ii)′ and also (i) and (iii). This completes the proof of Theorem 5.24. We next prove a result of Frostman [1935].

THEOREM 5.25. *Suppose that $u(x)$ is* s.h. *in a domain D in R^m, and possesses there a harmonic majorant $v(x)$, so that $u(x) \leqslant v(x)$ and $v(x)$ is harmonic in D. Then for x in D*

$$u(x) = h(x) - \int_D g(x, \xi) \, d\mu e_\xi, \tag{5.7.5}$$

where $g(x, \xi)$ is the generalized Green's function of D $h(x)$ is the least harmonic majorant of $u(x)$ in D, and μ is the Riesz measure of u in D.

In order to prove Theorem 5.25 we need the following

LEMMA 5.6. *The Green's function $g(x, \xi, D)$ is the lower bound of all functions $g(x)$ satisfying* (i) *and* (iii) *of section* (5.7.1) *and*

$$g(x) > 0 \text{ in } D. \tag{5.7.6}$$

Further if D_n is a sequence of domains with compact closures and regular for the problem of Dirichlet, such that $D_n \subset D_{n+1}$, and $\bigcup_{n=1}^{\infty} D_n = D$, then

$$g(x, \xi, D_n) \to g(x, \xi, D) \text{ as } n \to \infty. \tag{5.7.7}$$

We first show that $g(x, \xi, D)$ satisfies (5.7.6). In fact from the definition $-g(x)$ is s.h. in D, and bounded above and

$$\overline{\lim}\, g(x, \xi, D) \leqslant 0$$

as $x \to \zeta$ from inside D, for all ζ on the frontier Γ of D apart from a polar set. Hence, by Theorem 5.16, $-g \leqslant 0$ in D and since g is not constant, (5.7.6) holds.

Next suppose that $u(x)$ is any other function satisfying (i) and (iii) of (5.7.1) for $g(x)$ and in addition $u(x) > 0$ in D. Consider

$$h(x) = g(x) - u(x).$$

Then $h(x)$ is harmonic in D and $h(x)$ remains bounded above as x approaches any boundary point ζ of D. Also in view of (ii)′

$$\overline{\lim_{x \to \zeta}}\, h(x) \leqslant 0$$

for all ζ on Γ apart from a polar set. Thus $h(x) \leqslant 0$ in D by Theorem 5.16. Thus $g(x, \xi, D)$ is a lower bound of all functions $g(x)$ satisfying the conditions (i), (iii) and (5.7.6), and since $g(x, \xi, D)$ itself satisfies these conditions, $g(x, \xi, D)$ is the greatest lower bound, and this classifies the Green's function.

Next let D_n be the sequence of domains of Lemma 5.6 and set $g_n(x) = g(x, \xi, D_n)$, $g_0(x) = g(x, \xi, D)$. Then clearly $g_n(x)$ increases with n. In fact if $m > n, x \in D_n$, then $g_m(x) - g_n(x)$ is harmonic in D_n and if Γ_n is the boundary of D_n, $g_n(x)$ vanishes continuously on Γ_n, while $g_m(x) \geqslant 0$ on Γ_m. Thus $g_m(x) - g_n(x) \geqslant 0$ in D_n. Thus for fixed x and ξ in D, $g_n(x)$ is an increasing sequence of positive harmonic functions near x and so $g_n(x) \to G(x)$, as $n \to \infty$, where $G(x)$ is harmonic and satisfies (i) and (iii) and (5.7.6) in D, or else $G(x) \equiv +\infty$. To see that (iii) holds for instance we note that, if $\xi \in D_n$, $g_m(x) - g_n(x)$ is harmonic near $x = \xi$ and hence so is $G(x) - g_n(x)$.

Hence in view of the first part of the Lemma, it is sufficient to prove that if $g(x)$ satisfies (i), (iii) and (5.7.6) then $g(x) \geqslant G(x)$ in D. For then $G(x)$ will be the greatest lower bound of all such functions $g(x)$ and so $G(x) = g(x, \xi, D)$

by the first part. We now note that by hypothesis

$$h_n(x) = g(x) - g_n(x) \geqslant 0$$

on the boundary Γ_n of D_n, and $h_n(x)$ is harmonic in D_n. Thus $h_n(x) \geqslant 0$, i.e.

$$g_n(x) \leqslant g(x) \text{ in } D_n.$$

Taking x fixed and letting n tend to ∞, we deduce that $G(x) \leqslant g(x)$ as required Since a finite function $g(x) = g(x, \xi, D)$ exists, we deduce that $G(x)$ is not $+\infty$, and in fact $G(x) = g(x, \xi, D)$ as required. This proves Lemma 5.6.

We now embark on the proof of Theorem 5.25. If D is a domain whose boundary is a polar set, then $m = 2$ and D admits no s.h. functions which are bounded above other than constants. With the hypotheses of Theorem 5.26 we deduce that $u(x) - v(x) =$ constant, so that $u(x)$ is harmonic in D and its own least harmonic majorant. Thus Theorem 5.26 holds in this case, and the integral vanishes in (5.7.5).

Next we assume that the boundary of D is not polar, so that the Green's function $g(x, \xi, D)$ exists. Let D_n be a sequence of domains satisfying the conditions of Lemma 5.6. We may take for D_n for instance the union of a finite number of compact hyperballs. Let E be a fixed compact subset of D, suppose that D_1 is so chosen that $E \subset D_1$, and set

$$\phi_n(x) = \int_E g(x, \xi, D_n)\, d\mu e_\xi.$$

Then, as was shown in the proof of Theorem 3.15,

$$\phi_n(x) \to 0, \text{ as } x \to \zeta, \tag{5.7.8}$$

where ζ is any point on the boundary Γ_n of D_n. Suppose that $v(x)$ is any harmonic majorant of $u(x)$, and consider

$$h(x) = u(x) + \phi_n(x) - v(x).$$

It follows from Riesz' representation theorem that $h(x)$ is s.h. in D_n. Also by (5.7.8)

$$\lim h(x) \leqslant 0,$$

as x approaches any point of Γ_n from inside D_n. Thus by the maximum principle $h(x) \leqslant 0$, i.e.

$$v(x) \geqslant u(x) + \phi_n(x).$$

Now let $n \to \infty$. Then the sequence $g(x, \xi, D_n)$ tends monotonically to $g(x, \xi, D)$ for each fixed x, and all the functions concerned are positive. Thus in view of the monotonic convergence condition (3.3.1) for integrals, we deduce that for each fixed x

$$\phi_n(x) \to \phi(x) \quad \text{as} \quad n \to \infty,$$

where

$$\phi(x) = \int_E g(x, \xi, D) \, d\mu e_\xi.$$

Thus we have for any compact subset E of D,

$$v(x) \geqslant u(x) + \int_E g(x, \xi, D) \, d\mu e_\xi.$$

We now take instead of E any of the domains D_n, and define $g_n(x) = g(x, \xi, D_n)$, $x, \xi \in D_n$, $g_n(x) = 0$, otherwise.

Then if x is so chosen that $u(x) > -\infty$, we have

$$\int_D g_n(x) \, d\mu e_\xi \leqslant v(x) - u(x) < +\infty.$$

Also since $g_n(x)$ increases with n, we can apply monotonic convergence again and deduce that

$$\int_D g(x, \xi, D) \, d\mu e_\xi \leqslant v(x) - u(x) < +\infty.$$

Thus

$$v(x) \geqslant u(x) + \int_D g(x, \xi, D) \, d\mu e_\xi = h(x) \tag{5.7.9}$$

say. To prove Theorem 5.25 we have to show that $h(x)$ is the least harmonic majorant of $u(x)$. Evidently $h(x)$ is a majorant for $u(x)$ and in view of (5.7.9) $h(x)$ cannot exceed any harmonic majorant of $u(x)$. Thus it remains to prove only that $h(x)$ is harmonic in D.

To see this we assume that x lies in D_1 and write

$$h(x) = u(x) + \int_{D_1} g(x, \xi, D) \, d\mu e_\xi + \sum_{n=1}^{\infty} h_n(x),$$

where

$$h_n(x) = \int_{D_{n+1} - D_n} g(x, \xi, D) \, d\mu e_\xi.$$

Then the $h_n(x)$ are positive harmonic functions, and hence so is $\Sigma_1^\infty h_n(x)$, since this sum is not identically infinite. Also in view of the Riesz representation theorem

$$h_0(x) = u(x) + \int_{D_1} g(x, \xi)\, d\mu e_\xi = u(x) + \int_{D_1} \{g(x, \xi) + K_m(x - \xi)\}\, d\mu$$
$$- \int_{D_1} K_m(x - \xi)\, d\mu$$

is harmonic in D_1 and so is $h(x) = \sum_{n=0}^{\infty} h_n(x)$. This completes the proof of Theorem 5.25.

5.7.3. The symmetry property of the Green's function

THEOREM 5.26. *Suppose that D is a domain in R^m, whose complement is not polar. Then the Green's function $g(x, \xi, D)$ satisfies*

$$g(x, \xi, D) = g(\xi, x, D).$$

In particular g is a harmonic function of ξ for fixed x in D, $\xi \neq x$.

We assume first that D is a domain with compact closure and that D is regular for the problem of Dirichlet. Then we have

$$g(x, \xi) = -K(x - \xi) + \int_\Gamma K(\eta - \xi)\, d\omega(x, e_\eta), \tag{5.7.10}$$

where $K(x)$ is given by (3.5.1). In fact

$$h(x) = g(x, \xi) + K(x - \xi)$$

is harmonic in D and continuous in $\bar{D}$ and so by Theorem 3.10 we have

$$h(x) = \int_\Gamma h(\eta)\, d\omega(x, e_\eta).$$

Also, on Γ, we have $g(\eta, \xi) = 0$, $h(\eta) = K(\eta - \xi)$, so that (5.7.10) follows.

We now fix x and vary ξ in (5.7.10) and write $\phi(\xi) = g(x, \xi, D)$. Then, since the positive measure $\omega(x, e_\eta)$ is distributed on Γ, it follows that $\phi(\xi)$ is defined and harmonic in D except at the point $\xi = x$, in view of Theorem 3.6. Also $\phi(\xi) + K(x - \xi)$ remains harmonic at $\xi = x$, and $\phi(\xi) \geqslant 0$ in D. Thus by Lemma 5.6

$$\phi(\xi) \geqslant g(\xi, x, D),$$

i.e. $g(x, \xi) \geqslant g(\xi, x)$. Similarly $g(\xi, x) \geqslant g(x, \xi)$, so that $g(x, \xi) = g(\xi, x)$. For general domains D the same conclusion now follows from Lemma 5.6. This proves Theorem 5.26.

5.7.4. The extended Green's function and the Poisson–Jensen formula

We finish the section by proving an extended version of Theorem 3.14. The condition that the frontier of D is regular and has zero m-dimensional measure can be completely omitted.

THEOREM 5.27. *Suppose that D is a bounded domain in R^m. Let Γ_1, Γ_2 be the sets of regular and irregular frontier points of D, so that $\Gamma = \Gamma_1 \cup \Gamma_2$ is the frontier of D. Let $D_0 = R^m - \bar{D}$ be the exterior of D. Then for $\xi \in D$ there exists† a unique function $g(x, \xi)$ which coincides with the Green's function for $x \in D$, vanishes for $x \in D_0$ and is* s.h. *in $R^m - \{\xi\}$. For $x_0 \in \Gamma$*

$$g(x_0, \xi) = \overline{\lim}\, g(x, \xi), \tag{5.7.11}$$

where the upper limit is taken as $x \to x_0$ from inside D.

Further if u is s.h. *in $\bar{D}$ with Riesz mass μ then for $x \in D$*

$$u(x) = \int_{\Gamma_1} u(\zeta)\, d\omega(x, e_\zeta) - \int_{D \cup \Gamma_2} g(\eta, x)\, d\mu e_\eta, \tag{5.7.12}$$

where $\omega(x, e)$ is harmonic measure at x.

We return to the equation (5.7.10) of the previous section. The equation is valid for any bounded D, since the right-hand side clearly satisfies the conditions (i), (ii)′ and (iii) of Section 5.7.2. Since the function $K(x - \xi)$ is continuous on Γ we also deduce that the upper limit in (5.7.11) is zero at all the points of Γ_1. On the other hand at a point x_0 of Γ_2 the limit is positive. For otherwise the function $-g(x, \xi)$ would yield a function satisfying the conditions (i), (ii)′ and (iii) for $\psi(x)$ of Theorem 5.19, so that x_0 would be a regular frontier point of D, contrary to hypothesis. Thus $g(x_0, \xi)$ is positive on Γ_2 and zero on Γ_1.

For general x we do not use (5.7.11) as a definition of g but instead define $g(x, \xi)$ by

$$g(x, \xi) = -K(x - \xi) + \int_\Gamma K(\eta - x)\, d\omega(\xi, e_\eta). \tag{5.7.13}$$

In view of (5.7.10) and Theorem 5.26 the right-hand side is indeed equal to the Green's function for x and ξ in D. Also, for x in D_0, $K(\eta - x)$ is harmonic in $\bar{D}$, so that the right-hand side of (5.7.13) is zero in this case. Finally, in view of Theorem 3.7, the right-hand side of (5.7.13) is a s.h. function of x in $R^m - \{\xi\}$ for fixed ξ. Thus (5.7.13) yields an extended Green's function $g(x, \xi)$ satisfying the conditions of Theorem 5.27. We next show that any such function must be given by (5.7.11) for $x_0 \in \Gamma$.

† For these properties of $g(x, \xi)$ see Brelot [1955].

Suppose first that $x_0 \in \Gamma_1$. Then, in view of Theorem 5.20, Γ_2 is a bounded polar set and so, by Theorem 5.11, there exists a function $h(x)$, s.h. in R^m, finite at x_0 and $-\infty$ on Γ_2. By subtracting a constant we may also assume that $h(x) < 0$ on Γ. For any positive ε set

$$g_\varepsilon(x) = g(x, \xi) + \varepsilon h(x).$$

Then evidently

$$\overline{\lim}\, g_\varepsilon(x) < 0,$$

as x approaches any point x_1 of Γ from D_0 or D. Thus x_1 is the centre of a ball $N(x_1)$, such that

$$g_\varepsilon(x) < 0, \quad x \in N(x_1) - \Gamma.$$

The union of the balls $N(x_1)$ forms an open set N, containing Γ and such that

$$g_\varepsilon(x) < 0 \quad \text{in } N - \Gamma.$$

Since $g_\varepsilon(x)$ is s.h. in N it now follows from the maximum principle that

$$g_\varepsilon(x) < 0 \text{ in } N.$$

For otherwise $g_\varepsilon(x)$ would attain it maximum in N on the compact set Γ. In particular

$$g_\varepsilon(x_0) < 0, \quad \text{i.e.} \quad g(x_0, \xi) < -\varepsilon h(x_0).$$

Since ε is arbitrary we deduce that $g(x_0, \xi) \leqslant 0$, so that (5.7.11) holds on Γ_1.

Thus $g(x_0, \xi)$ is determined outside the polar set Γ_2. It now follows from Theorem 5.18 and (5.5.6) that on Γ_2 g must be given by (5.7.11). This proves the uniqueness of $g(x, \xi)$ in Theorem 5.27 and incidentally that the functions given by (5.7.11) and (5.7.13) are identical on Γ.

We can write (5.7.13) as

$$\int_\Gamma K(\xi - \eta)\, d\omega(x, e_\xi) = K(x - \eta) + g(\eta, x), x \in D, \eta \in R^m - \{x\}.$$

This is the required extension of Lemma 3.9. The proof of Theorem 5.27 is now completed just as in Section 3.7.3. Since $g(\eta, x) = 0$ outside $D \cup \Gamma_2$, we can confine the second integral in (5.7.12) to $D \cup \Gamma_2$. Also, by Theorem 5.23, Γ_2 has harmonic measure zero, so that the first integral in (5.7.12) need be taken only over Γ_1. Thus from the point of view of the Poisson–Jensen formula the irregular frontier points behave like interior points of D.

We cannot extend Theorem 5.27 to unbounded domains without making additional assumptions about the behaviour of $u(x)$ near ∞. For instance the function $u(x) = \alpha x_1$ is harmonic in R^m and zero on the subspace $x_1 = 0$. Thus $u(x) = 0$ on the frontier of the half space D_1 given by $x_1 > 0$, but $u(x)$

is not identically zero as the Poisson–Jensen formula would imply. However if $m = 2$, Theorem 5.27 extends to unbounded domains D which have exterior points, provided that $u(x)$ remains s.h. at ∞. For such domains can be mapped onto bounded domains by a conformal map of the closed plane.

We also note that Theorems 3.15 to 3.17 can be extended, though the irregular boundary points require some modification. Thus, with the hypotheses of Theorem 5.27 if follows from Theorem 5.25 that the least harmonic majorant of $u(x)$ in D is

$$\int_{\Gamma_1} u(\xi)\, d\omega(x, e_\xi) - \int_{\Gamma_2} g(\eta, x)\, d\mu e_\eta,$$

which is equal to the harmonic extension of $u(x)$ from Γ into D if and only if $\mu(\Gamma_2) = 0$. This condition is satisfied for instance if u is finite at every point of Γ_2. For if $\mu(\Gamma_2) > 0$ in this case we could find a compact subset E of Γ_2 on which $u(x)$ is bounded below and such that $\mu(E) > 0$. Thus E would have positive capacity contrary to the fact that Γ_2 is polar.

Similar remarks apply to Theorems 3.16 and 3.17. Thus for $\alpha > 0$ the functions $\alpha \log |z|$ are all harmonic in the punctured disk D, given by $0 < |z| < 1$, s.h. in the open plane and equal on the boundary of D, without being equal to each other.

5.8. CAPACITABILITY AND STRONG SUBADDITIVITY

We now develop the important results of Choquet [1955] regarding the capacitability of analytic and in particular Borel sets. Accordingly we consider a set function $\phi(F)$ defined on the closed (and so compact) subsets of a compact metric space S. We may think of S for instance as a closed ball in R^m. Our set function $\phi(F)$ will be required to satisfy the following conditions.

(i) $0 \leqslant \phi(F) < \infty$. ($\phi$ is finite and positive.)
(ii) If $F_1 \subset F_2$ then $\phi(F_1) \leqslant \phi(F_2)$ ($\phi(F)$ increases with F)
(iii) Given F and $\varepsilon > 0$, there exists F_1 containing F in its interior such that

$$\phi(F_1) < \phi(F) + \varepsilon,$$

($\phi(F)$ is upper semi-continuous).
(iv) We have for any pair of compact sets F_1, F_2

$$\phi(F_1 \cup F_2) + \phi(F_1 \cap F_2) \leqslant \phi(F_1) + \phi(F_2).$$

($\phi(F)$ is strongly subadditive).

Evidently (i) and (ii) imply that $\phi(F) \leqslant \phi(S) = M_0 < \infty$ say, so that ϕ is uniformly bounded. It is clear that if $g(x)$ is a continuous strictly increasing nonnegative function in $[0, M_0]$, such that $g(0) = 0$, then $g\{\phi(F)\}$ satisfies

(i) to (iii) if and only if $\phi(F)$ does. It will appear that we can choose g in such a way that if $\phi(F)$ is capacity, then $g(\phi(F))$ also satisfies (iv). In the case of capacity in the plane slight modifications have to be made which however do not affect the final conclusions.

We note that α-capacity in R^m as defined in Section 5.1 evidently satisfies (i) and (ii) if we confine ourselves to the subsets of a fixed closed ball. It follows from Theorem 5.5 that these capacities also satisfy (iii). It is much less obvious that α-capacity in R^m or any suitable function of it satisfies (iv) and we proceed to obtain such a result only in the case $\alpha = m - 2$. We note also that any measure which is finite in the compact space S necessarily satisfies (i) to (iv) with equality in (iv), so that Choquet's theory includes as a special case the result that Borel sets are measurable, with respect to any such measure. For we shall see that if inner and outer ϕ-capacity are defined analogously to the way indicated in section 5.2. then analytic sets and in particular Borel sets are ϕ-capacitable.

5.8.1. Strong subadditivity

We proceed to show that certain set-functions are strongly subadditive, so that the theory we are about to develop will apply to them. It is not easy to proceed directly to $(m - 2)$-capacity and we proceed to the analogue of capacity for finite balls. We first need

LEMMA 5.7. *Suppose that E is a compact set, N a neighbourhood of E, and that $D = N - E$ is connected. Let Γ be the set of frontier points of D on E and suppose that $u(x)$ is harmonic and bounded in D, and*

$$u(x) \to a, \quad as \quad x \to \xi \text{ from inside } D \tag{5.8.1}$$

for all points ξ of Γ apart from possibly a polar set. Suppose further that $u(x) > a$ in D. Then $u(x)$ possesses an extension as a s.h. *function satisfying*

$$u(x) \geq a \tag{5.8.2}$$

to the whole of N.

We define $u(x) = a$ at all points of E apart from the irregular boundary points of D. Since these form a polar set our result follows from Theorem 5.18.

Let $D_0 = D(0, R_0)$. We shall call a function $V(x)$ a potential on D_0, if $V(x)$ is s.h. in D_0 and harmonic outside a compact subset of D_0 and

$$V(x) \to 0, \quad \text{uniformly as} \quad |x| \to R_0. \tag{5.8.3}$$

Then clearly the least harmonic majorant of $V(x)$ in D_0 is zero and so Theorem

5.25 gives

$$V(x) = -\int_E g_{R_0}(x, \xi)\, d\mu e_\xi, \tag{5.8.4}$$

where E is the compact subset of $D(0, R)$ on which V is not harmonic. We denote by $M(V)$ the total Riesz mass of $V(x)$. In the notation of Section 3.9. we have for $R_0 - \delta < r < R_0$, when δ is small

$$M = n(r) = \frac{1}{d_m} r^{m-1} \frac{d}{dr} N(r) = \frac{r^{m-1}}{d_m c_m} \frac{d}{dr} \frac{1}{r^{m-1}} \int_{S(0,r)} V(x)\, d\sigma(x) = r^{m-1} \frac{d}{dr} I(r)$$

say. Since by (5.8.4) $V(x)$ remains harmonic on $S(0, R_0)$ we can set $r = R_0$ in this result. Also by (5.8.3) $I(R_0) = 0$. Thus we obtain

$$M(V) = \lim_{r \to R_0} -R_0^{m-1} \frac{I(r)}{(R_0 - r)}. \tag{5.8.5}$$

We deduce that $M(V)$ increases with decreasing V, i.e. if

$$V_1 \leqslant V_2, \quad \text{then} \quad M(V_1) \geqslant M(V_2). \tag{5.8.6}$$

Also $M(V)$ only depends on the behaviour of V in the immediate neighbourhood of $S(0, R_0)$.

Let E be a compact subset of D_0 and let Δ_0 be that component of $D_0 - E$ which has $S(0, R_0)$ as part of its boundary. We define the conductor potential $V_E(x)$ of E to be the harmonic function in Δ_0 which has boundary values 0 on $S(0, R_0)$ and -1 n.e. on the part of the frontier of Δ_0 which lies in D_0. By Lemma 5.7. $V_E(x)$ has a s.h. extension onto D_0 and so the quantity $M(V_E(x))$ can be defined, which we denote now simply by $M(E)$. We proceed to prove that $M(E)$ is strongly subadditive.

Lemma 5.8. *If E_1, E_2 are compact subsets of D_0, we have*

$$M(E_1 \cup E_2) + M(E_1 \cap E_2) \leqslant M(E_1) + M(E_2),$$

where $M(E)$ is defined as above.

We note first that if $E_1 \subset E_2$, then

$$V_{E_1}(x) \geqslant V_{E_2}(x) \text{ outside } E_2, \text{ and so } M(E_1) \leqslant M(E_2). \tag{5.8.7}$$

In fact $V_{E_1}(x) \geqslant -1$, everywhere and

$$V_{E_2}(x) \to -1, \quad \text{as} \quad x \to \xi$$

from the complementary domain D_2 of E_2, where ξ is any boundary point of D_2 apart from a polar set. Thus

$$\overline{\lim}\ V_{E_2}(x) - V_{E_1}(x) \leqslant 0$$

as x approaches any boundary point of D_2 apart from a polar set. Since $V_{E_2} - V_{E_1} \leqslant 1$ in D_2, we deduce the first inequality of (5.8.7) and so the second by (5.8.6).

Consider now

$$\phi(x) = V_{E_1 \cup E_2} + V_{E_1 \cap E_2} - V_{E_1} - V_{E_2}$$

in the domain Δ complementary to $E_1 \cup E_2$ having $S(0, R_0)$ on its frontier. Then $\phi(x)$ is harmonic in Δ. Also since the functions $V_E(x)$ are potentials, so that (5.8.3) holds, $\phi(x)$ vanishes on $S(0, R_0)$. Next $\phi(x)$ is clearly uniformly bounded in Δ.

Finally let ξ be a boundary point of Δ on $E_1 \cup E_2$, which is not on any of the polar exceptional sets for E_1, E_2, $E_1 \cup E_2$ or $E_1 \cap E_2$. Then we shall see that

$$\text{(5.8.8)} \qquad \underline{\lim}\, \phi(x) \geqslant 0, \quad \text{as} \quad x \to \xi \quad \text{from inside } \Delta.$$

We consider various special cases. Suppose first that $\xi \in E_1 \cap E_2$. Then V_{E_1}, V_{E_2}, $V_{E_1 \cap E_2}$ and $V_{E_1 \cup E_2}$ all tend to -1, as $x \to \xi$ from inside Δ, so that (5.8.8) holds.

Next suppose that ξ belongs to E_1 but not to E_2. Then

$$V_{E_1 \cup E_2}(x) \to -1, V_{E_1}(x) \to -1, \quad \text{as} \quad x \to \xi \text{ from inside } D.$$

On the other hand since $E_1 \cap E_2 \subset E_2$, we deduce from (5.8.7) that

$$V_{E_1 \cap E_2}(x) - V_{E_2}(x) \geqslant 0$$

in Δ. Thus again (5.8.8) holds. The same conclusion applies similarly if ξ belongs to E_2 but not to E_1. Thus (5.8.8) holds in all cases for all boundary points ξ of D outside a polar set, and we deduce from the extended maximum principle, Theorem 5.16, that

$$\phi(x) \geqslant 0 \quad \text{in} \quad \Delta.$$

In view of (5.8.5) we deduce that

$$M(E_1 \cup E_2) + M(E_1 \cap E_2) - M(E_1) - M(E_2) = \lim_{r \to R_0-} \; - \frac{I(r, \phi)}{R_0 - r} \leqslant 0,$$

which gives Lemma 5.8.

We deduce at once

THEOREM 5.28. *If $C(F)$ denotes $(m-2)$-capacity in R^m, where $m > 2$, then $\phi(F) = C(F)^{m-2}$, satisfies the axioms (i) to (iv) of Section 5.8.*

We have already noted that $\phi(F)$ satisfies axioms (i) to (iii), provided that we confine ourselves to subsets of a given compact set. Now define $V_E(x) = V_E(x, R)$ as before for a given ball $D(0, R)$ and let $R \to \infty$. The resulting functions $V_E(x, R)$ decrease with increasing R, in the unbounded complementary domain D of E, and so tend to a harmonic limit $V_E(x, \infty)$, such that

$$V_E(x, \infty) \geqslant -1.$$

Since $V_E(x, R)$ decreases with increasing R, we deduce that

$$V_E(x, \infty) \to -1, \quad \text{as} \quad x \to \xi \text{ from inside } D$$

for all boundary points ξ of D apart possibly from a polar set.

Next, since $V_E(x, R)$ decreases and remains bounded as $R \to \infty$ through integral values we have for any fixed r, such that $S(0, r)$ lies in D

$$V_E(x, R) \to V_E(x, \infty) \quad \text{and} \quad \frac{\partial}{\partial r} V_E(x, R) \to \frac{\partial}{\partial r} V_E(x, \infty).$$

Hence if $M(E, R)$ denotes the Riesz mass of $V_E(x, R)$, we deduce that

$$M(E, R) = \frac{r^{m-1}}{d_m c_m} \frac{\mathrm{d}}{\mathrm{d}r} \frac{1}{r^{m-1}} \int_{S(0,r)} V_E(x, R)\, \mathrm{d}\sigma(x) \to M(E, \infty).$$

Thus it follows from Lemma 5.8. that $M(E, \infty)$ is strongly subadditive.

Next we note that for $m > 2$

$$g_R(x, \xi) = |x - \xi|^{2-m} - \{|\xi|\, |x - \xi'|/R\}^{2-m},$$

where ξ' is the inverse point of ξ with respect to $S(0, R)$. Thus

$$g_R(x, \xi) \to |x - \xi|^{2-m}, \quad \text{as} \quad R \to \infty,$$

uniformly for bounded x and ξ. In particular

$$V_E(x, R) = -\int_E g_R(x, \xi)\, \mathrm{d}\mu_R(e_\xi) \to -\int_E |x - \xi|^{2-m}\, \mathrm{d}\mu(e_\xi) = V_E(x, \infty),$$

where μ is a weak limit of the measures μ_R corresponding to $D(0, R)$. Thus $V_E(x, \infty)/M(E, \infty)$ is the potential due to a unit mass, which is $-1/M(E, \infty)$ n.e. on the finite boundary of D and zero at ∞. Thus, by Theorem 5.17, $V_E(x, \infty)/M(E, \infty)$ is the equilibrium potential of E and so by the definitions of Section 5.1, we deduce that

$$C_{m-2}(E) = M(E, \infty)^{1/(m-2)}.$$

Since $M(E, \infty)$ is strongly subadditive Theorem 5.28 follows.

The situation in the case $m = 2$ is more complicated. In this case

$$g_R(x, \xi) = -\log\left|\frac{R(x - \xi)}{R^2 - \bar{x}\xi}\right| = -\log|x - \xi| + \log R + \frac{O(1)}{R},$$

uniformly as $R \to \infty$ for bounded x and ξ. In this case it is convenient to consider

$$u_E(x, R) = \frac{V_E(x, R)}{M(E, R)} + \log R = \frac{1}{M(E, R)} \int_E (-g_R(x, \xi) + \log R)\, d\mu_R(e_\xi)$$

$$\to \int_E \log |x - \xi|\, d\mu\,(e_\xi) = u_E(x, \infty),$$

where μ is a weak limit of the measures $\mu_R(e)/M(E, R)$. Thus μ is a unit measure. Also n.e. on the finite boundary of D we have

$$u_E(x, R) = -\frac{1}{M(E, R)} + \log R.$$

It follows that $u_E(x, \infty)$ is the conductor potential of E, so that

$$\log C_0(E) = \log R - \frac{1}{M(E, R)} + o(1).$$

Thus in this case

(5.8.9) $$\frac{1}{M(E, R)} = \log\left(\frac{R}{C_0(E)}\right) + o(1), \quad \text{as} \quad R \to \infty.$$

This gives

$$M(E, R) = \frac{1}{\log R - \log (C_0(E)) + o(1)} = \frac{1}{\log R} + \frac{\log C_0(E) + o(1)}{(\log R)^2}.$$

If we substitute this in Lemma 5.8, we deduce that $\log C_0(E)$ is subadditive However this result is much less useful, since if $E_1 \cap E_2 = 0$, the left hand side in the inequality (iv) is $-\infty$ for this set-function. Fortunately the subadditivity of $M(E, R)$ together with (5.8.9) will still enable us to obtain our conclusions.

5.8.2. Outer capacities

The conditions (i) to (iv) of Section 5.8 are used by Choquet only to extend $\phi(E)$ as a set-function satisfying certain conditions to arbitrary subsets of S. The resulting set function is called an outer capacity. More precisely an outer capacity $\phi^*(E)$ is a set function defined on arbitrary subsets E of a compact space S and satisfying the following conditions

(i) $0 \leqslant \phi^*(E) < \infty$.
(ii) If $E_1 \subset E_2$, then $\phi^*(E_1) \leqslant \phi^*(E_2)$.
(iii) Given E and $\varepsilon > 0$, there exists an open set G containing E, such that $\phi^*(G) < \phi^*(E) + \varepsilon$.

(iv)′ Given any sequence of sets E_n, such that $E_n \subset E_{n+1}$, we have $\phi^*(\bigcup E_n) = \lim_{n\to\infty} \phi^*(E_n)$.

The conditions (i) and (ii) are the same as the corresponding conditions for ϕ extended to general sets, and (iii)′ is a slight modification of (iii).

However (iv)′ is quite different from (iv) and (iv) is in fact only used to lead to (iv)′. The relation between the two sets of conditions is brought out by

LEMMA 5.9. *Suppose that $\phi(F)$ is any set function on compact subsets of a compact metric space S, satisfying the conditions* (i) *to* (iii) *of Section* 5.8. *Then $\phi(F)$ can be uniquely extended to a set function $\phi^*(E)$ on arbitrary subsets of S, satisfying* (i), (ii), (iii)′ *and* (iv)′ *in the special case when the E_n are compact and E_n lies in the interior of E_{n+1}. We shall call this case the restricted form of* (iv)′.

Let G be an open set and let $\phi(F)$ be the set function of Lemma 5.9. Denote compact sets by F. Then we must have

$$\alpha = \sup_{F\subset G} \phi(F) = \phi^*(G). \tag{5.8.10}$$

In fact we must have $\phi^*(G) \geqslant \phi^*(F) = \phi(F)$ for $F \subset G$, so that $\phi^*(G) \geqslant \alpha$. On the other hand let F_n be a sequence such that F_n is in the interior of F_{n+1} and $\bigcup F_n = G$. We may take for F_n for instance the set of all points of G distant at least $1/n$ from the complement of G. Then if (iv)′ is to hold for compact sets F_n, we must have

$$\phi^*(G) = \lim \phi(F_n) \leqslant \alpha.$$

Thus $\phi^*(G) = \alpha$.

It is easy to see that in fact if F_n is any such sequence

$$\phi(F_n) \to \alpha,$$

so that with our definition the restricted form of (iv)′ holds. For let F be any compact subset of G. Then the interiors G_n of the F_n cover F, since the F_n cover F and $F_n \subset G_{n+1}$. Hence by the Heine–Borel Theorem a finite number of the G_n, G_1 to G_N say, cover F. Thus

$$\phi(F_{N+1}) \geqslant \phi(F)$$

and

$$\lim \phi(F_{N+1}) \geqslant \phi(F).$$

Since F is an arbitrary compact subset of G, we deduce that

$$\lim \phi(F_n) \geqslant \alpha,$$

and since the opposite inequality is trivial we see that

$$\phi(F_n) \to \alpha.$$

Thus this definition of $\phi^*(G)$ for open sets, is implied by and implies (iv)′ for compact sets E_n, such that E_n lies in the interior of E_{n+1}. We now define for arbitrary sets E

$$\phi^*(E) = \inf \phi^*(G) \tag{5.8.11}$$

where the inf is taken over all open sets containing E. This definition is forced by (ii) and (iii)′. Also in view of (iii), we see that $\phi^*(E) = \phi(E)$ if E is compact. Thus (5.8.10) and (5.8.11) lead to the unique extension $\phi^*(E)$ of $\phi(F)$ to arbitrary sets. It is clear that our new definition satisfies (i), (ii) and (iii)′.

We have seen that set functions ϕ satisfying (i), (ii) and (iii) lead uniquely to set functions ϕ^* satisfying (i), (ii), (iii)′ and the restricted form of (iv)′. The converse is also true. Any set function ϕ^* satisfying (i), (ii) and (iii)′ clearly leads, when restricted to compact sets, to (i), (ii) and (iii). Only (iii) is not quite obvious. To see this let F be a compact set and let G be an open set such that

$$\phi^*(G) < \phi^*(F) + \varepsilon.$$

Then we can find a compact set F_1 containing F in its interior and contained in G. This yields

$$\phi^*(F_1) \leqslant \phi^*(G) < \phi^*(F) + \varepsilon$$

as required.

In view of the correspondence between ϕ and ϕ^* we henceforth just write ϕ instead of ϕ^* for arbitrary sets. If ϕ is only given on compact sets, but satisfies (i), (ii) and (iii) we assume ϕ extended to arbitrary sets as in Lemma 5.9. We can now prove

LEMMA 5.10. *Suppose that ϕ is a set-function satisfying* (i) *to* (iv) *of Selection 5.8 and extended to arbitrary sets as in Lemma 5.9. Let E_n, G_n be sets such that G_n is open, $E_n \subset G_n$ and $\phi(G_n) < \phi(E_n) + \varepsilon_n$. Then*

$$\phi\left(\bigcup_{n=1}^{N} G_n\right) < \phi\left(\bigcup_{n=1}^{N} E_n\right) + \sum_{n=1}^{N} \varepsilon_n.$$

We suppose first that the E_n are compact and in this case prove the result by induction on N. Let H be a compact set in $G_1 \bigcup G_2$. Then each point of H can be enclosed in a closed ball C which lies either in G_1 or in G_2. A finite number of these balls covers H, of which we may suppose that $C_1, C_2, \ldots, C_p$ lie in G_1 and $C_{p+1}, \ldots, C_q$ lie in G_2. Then if we set

$$H_1 = H \bigcap \bigcup_{n=1}^{p} C_n, \qquad H_2 = H \bigcap \bigcup_{n=p+1}^{q} C_n,$$

H_1, H_2 are compact, $H_1 \subset G_1$, $H_2 \subset G_2$, and $H_1 \cup H_2 = H$.
We wish to prove that

$$\phi(H_1 \cup H_2) \leqslant \phi(E_1 \cup E_2) + \varepsilon_1 + \varepsilon_2.$$

To do this we may replace H_j by the bigger sets $H_j \cup E_j$ which are still compact and contained in G_j, and so we suppose that

$$E_1 \subset H_1 \subset G_1, \qquad E_2 \subset H_2 \subset G_2.$$

We now note that in view of strong subadditivity

$$\phi(H_1 \cup H_2) + \phi(E_1) \leqslant \phi(E_1 \cup H_2) + \phi(H_1).$$

For $E_1 \subset (E_1 \cup H_2) \cap H_1$, and $E_1 \cup H_2 \cup H_1 = H_1 \cup H_2$. Similarly

$$\phi(E_1 \cup H_2) + \phi(E_2) \leqslant \phi(E_1 \cup E_2) + \phi(H_2).$$

Adding we obtain

$$\phi(H_1 \cup H_2) + \phi(E_1) + \phi(E_2) \leqslant \phi(E_1 \cup E_2) + \phi(H_1) + \phi(H_2),$$

whence

$$\phi(H) = \phi(H_1 \cup H_2) \leqslant \phi(E_1 \cup E_2) + \varepsilon_1 + \varepsilon_2.$$

Taking the upper bound over all compact subsets, H of $G_1 \cup G_2$, we obtain Lemma 5.10 for $N = 2$, and compact sets E_1, E_2.

Still assuming the E_n to be compact we now prove the general case by induction on N. We assume the result holds for N and write

$$G = \bigcup_{n=1}^{N} G_n, \; E = \bigcup_{n=1}^{N} E_n, \; \varepsilon = \sum_{n=1}^{N} \varepsilon_n.$$

Then $E \subset G$ and

$$\phi(G) < \phi(E) + \varepsilon.$$

Also $E_{N+1} \subset G_{N+1}$ and by hypothesis

$$\phi(G_{N+1}) < \phi(E_{N+1}) + \varepsilon_{N+1}.$$

Thus since Lemma 5.10 holds for $N = 2$, we deduce that

$$\phi(G \cup G_{N+1}) < \phi(E \cup E_{N+1}) + \varepsilon + \varepsilon_{N+1},$$

so that the lemma holds for $N + 1$ when the E_n are compact.

Suppose next that the E_n are open. Then since

$$\phi(G_n) < \phi(E_n) + \varepsilon_n,$$

it follows from the definition of ϕ for open sets, that we can find a compact

subset F_n of E_n, such that
$$\phi(G_n) < \phi(F_n) + \varepsilon_n.$$
Hence by the case we have just considered
$$\phi\left(\bigcup_{n=1}^{N} G_n\right) < \phi\left(\bigcup_{n=1}^{N} F_n\right) + \sum_{n=1}^{N} \varepsilon_n \leqslant \phi\left(\bigcup_{n=1}^{N} E_n\right) + \sum_{n=1}^{N} \varepsilon_n.$$
This proves Lemma 5.10 when the E_n are open.

Finally suppose that the E_n are arbitrary. Let G be an open set containing $\bigcup_{n=1}^{N} E_n$, and set $F_n = G \cap G_n$. Then $E_n \subset F_n \subset G_n$, and
$$\phi(G_n) < \phi(E_n) + \varepsilon_n \leqslant \phi(F_n) + \varepsilon_n.$$
Also the F_n are open. Hence by what we have just proved
$$\phi\left(\bigcup_{n=1}^{N} G_n\right) \leqslant \phi\left(\bigcup_{n=1}^{N} F_n\right) + \sum_{n=1}^{N} \varepsilon_n \leqslant \phi(G) + \sum_{n=1}^{N} \varepsilon_n.$$
This is true for any open set G containing $\bigcup_{n=1}^{N} E_n$ and hence by taking lower bounds over all such G, we deduce finally
$$\phi\left(\bigcup_{n=1}^{N} G_n\right) \leqslant \phi\left(\bigcup_{n=1}^{N} E_n\right) + \sum_{n=1}^{N} \varepsilon_n,$$
as required.

We deduce

LEMMA 5.11. *Suppose that the set-function ϕ satisfies the conditions* (i) *to* (iv) *of Section* 5.8. *Then ϕ extended as in Lemma* 5.9 *is an outer capacity.*

We have to show that $\phi(E)$ satisfies condition (iv)′ for outer capacities, since we have already seen that the other conditions are satisfied. Suppose then that E_n is an expanding sequence of sets and given E_n, choose open sets G_n, such that $E_n \subset G_n$ and
$$\phi(G_n) < \phi(E_n) + \varepsilon 2^{-n}, n = 1, 2, \ldots$$
This is possible in view of (iii)′. Let F be a compact subset of $G = \bigcup_{n=1}^{\infty} G_n$. Then in view of the Heine–Borel theorem F is already a subset of $\bigcup_{n=1}^{N} G_n$ for some finite N. Hence, using Lemma 5.10 we deduce
$$\phi(F) \leqslant \phi\left(\bigcup_{n=1}^{N} G_n\right) \leqslant \phi\left(\bigcup_{n=1}^{N} E_n\right) + \sum_{n=1}^{N} \varepsilon 2^{-n} \leqslant \phi(E_N) + \varepsilon.$$

Write $\alpha = \lim_{n \to \infty} \phi(E_n)$, $E = \bigcup_{n=1}^{\infty} E_n$. Then we deduce that

$$\phi(F) \leqslant \alpha + \varepsilon.$$

This is true for any compact subset F of G, and hence

$$\phi(G) \leqslant \alpha + \varepsilon.$$

Since E is contained in G, this yields

$$\phi(E) \leqslant \alpha + \varepsilon,$$

and since ε is arbitrary $\phi(E) \leqslant \alpha$. The opposite inequality is obvious since $E_n \subset E$ and so $\phi(E) \geqslant \phi(E_n)$ for every n. This proves Lemma 5.11.

We finish this sub-section by proving

THEOREM 5.29. *The capacities $C_{m-2}(E)$ for subsets of a fixed compact set in R^m are outer capacities for $m \geqslant 2$.*

In view of Theorem 5.28 we know that for $m > 2$, the set functions $C_{m-2}(E)^{m-2}$ satisfy conditions (i) to (iv) of Section 5.8 and so are outer capacities in view of Lemma 5.11. Hence it is evident that the $C_{m-2}(E)$ then also satisfy the conditions (i), (ii), (iii)′ and (iv)′ and so these quantities are also outer capacities.

This argument breaks down in the case $m = 2$, and in this case we argue as follows. We saw earlier that $C_0(E)$ even for subsets of R^m satisfies the conditions (i), (ii) and (iii) of Section 5.8 and so the corresponding extended function satisfies (i), (ii) and (iii)′ for outer capacity. It remains to see that (iv)′ is satisfied. Suppose then that E_n is an expanding sequence of plane sets, such that $E = \bigcup_{n=1}^{\infty} E_n$ is bounded. The argument of Lemma 5.11 applies to the set function $M(E) = M(E, R)$ discussed in Lemma 5.8, and hence, in view of Lemma 5.11, $M(E, R)$ suitably extended to arbitrary sets, is an outer capacity. (We did not prove upper semi-continuity, but this can be dealt with just as in Theorem 5.5. The properties (i) and (ii) are obvious.) Hence the same conclusion holds for the set function

$$\phi(R, E) = R\, e^{-1/M(E, R)}.$$

Also in view of (5.8.9) we have, uniformly for sets in a fixed ball,

$$\phi(R, E) \to C_0(E), \text{ as } R \to \infty.$$

Since $\phi(R, E)$ is an outer capacity, we know that for fixed R

$$\phi(R, E_n) \to \phi(R, E).$$

Given $\varepsilon > 0$, we choose R so large that for E and all the sets E_n we have

$$|\phi(R, E_n) - C_0(E_n)| < \varepsilon, \qquad |\phi(R, E) - C_0(E)| < \varepsilon.$$

Having fixed R we choose n so large that

$$\phi(R, E) < \phi(R, E_n) + \varepsilon.$$

Thus

$$C_0(E) < C_0(E_n) + 3\varepsilon,$$

and so

$$\alpha = \lim_{n\to\infty} C_0(E_n) \geqslant C_0(E) - 3\varepsilon.$$

Since ε is arbitrary, we deduce that $\alpha \geqslant C_0(E)$, i.e. $\alpha = C_0(E)$, since we obviously have $\alpha \leqslant C_0(E)$. This completes the proof of Theorem 5.29.

The result† extends to the capacities $C_\alpha(E)$ for $0 \leqslant \alpha < m$.

5.8.3. Capacitability

The capacity of the previous section is called an outer capacity, since the capacity of a general set E is by (iii)′ the lower bound of the capacities of open sets G containing E.

We shall say that a set E is capacitable w.r.t. the set function ϕ if $\phi(E)$ is the upper bound of $\phi(F)$ for compact sets F contained in E. Thus a set E is capacitable if and only if given $\varepsilon > 0$, there exists a compact set F and an open set G, such that

(5.8.12) $$F \subset E \subset G, \qquad \phi(G) < \phi(F) + \varepsilon.$$

We shall proceed to prove Choquet's remarkable theorem that for an outer capacity ϕ satisfying (i), (ii), (iii)′ and (iv)′, and in particular for $C_{m-2}(E)$ in R^m, all Borel sets are capacitable.

It follows from (iv)′ that open sets are capacitable and from (iii)′ that compact sets are capacitable. It is natural to try to prove that capacitable sets form a σ-ring, by proving that countable unions and intersections of capacitable sets are capacitable. It would then follow that all Borel sets are capacitable. This is the way that the classical proofs proceed for Lebesgue measure. However this method breaks down immediately as the intersection of two capacitable sets is not in general capacitable.

† See Kishi [1957] for the case $\alpha > m - 2$. It follows from a result of Fuglede [1965], that Kishi's solution is also valid for $0 \leqslant \alpha < m - 2$. For unbounded sets and $\alpha = 0$, it is necessary to use a result of Choquet [1958]. For general capacities see also Aronszaijn and Smith [1955–1956].

Example

Let E be the class of subsets of the closed square $|x| \leqslant 1$, $|y| \leqslant 1$, on the (x, y) plane, and let $\phi(E)$ be the Lebesgue measure of the projection of E on the x-axis. Clearly $\phi(E)$ defined as above on compact sets satisfies the axioms (i) to (iv) of Section 5.8. In fact if E_1, E_2 are compact sets and F_1, F_2 their projections on the x axis then $F_1 \cup F_2$ is the projection of $E_1 \cup E_2$ but the projection of $E_1 \cap E_2$ is contained in $F_1 \cap F_2$. Thus if m denotes Lebesgue measure

$$\phi(E_1 \cup E_2) + \phi(E_1 \cap E_2) \leqslant m(F_1 \cup F_2) + m(F_1 \cap F_2)$$
$$= m(F_1) + m(F_2) = \phi(E_1) + \phi(E_2).$$

Thus (iv) holds for ϕ. The other axioms are trivial. Now let E_1 be any Lebesgue non-measurable set in the interval $[0, 1]$ of the x-axis, so that E_1 is not capacitable with respect to ϕ. Let E_2 be the whole segment $[0, 1]$, and let E_3 be the segment $y = 1, 0 \leqslant x \leqslant 1$.

Then $E_1 \cup E_3$ and E_2 are both capacitable with capacity one, but $E_1 = (E_1 \cup E_3) \cap E_2$ is not capacitable.

The above example forces us to choose a different way of dealing with capacitability. Following Choquet we prove that analytic sets are capacitable and that all Borel sets are analytic. The following approach is due to Carleson [1967, Chapter I].

Suppose that to every finite set of non-negative integers $(n_1, n_2, \ldots, n_p)$ there is associated a compact set $A_{n_1, n_2, \ldots, n_p}$ in R^m.

By means of the "Souslin operation" these sets generate the set

$$A = \bigcup_{n_1, \ldots, n_p, \ldots} A_{n_1} \cap A_{n_1, n_2} \cap \ldots \cap A_{n_1, n_2, \ldots, n_p} \cap \ldots \tag{5.8.13}$$

The sets A arising in this way for different choices of the $A_{n_1, \ldots, n_p}$ are called *analytic*. We prove

LEMMA 5.12. *Any Borel set in R^m is analytic.*

It is evident that compact sets are analytic, since if E is compact we may choose $A_{n_1, n_2, \ldots, n_p} = E$ for every combination of positive integers $(n_1, n_2, \ldots, n_p)$ and then $A = E$.

Next if $A^{(k)}$ is a sequence of analytic sets, generated by $A^{(k)}_{n_1, n_2, \ldots, n_p}$, then $\bigcup A^{(k)}$ is obtained by performing the Souslin operation on the sets $A^{(n_1)}_{n_2, n_3, \ldots, n_p}$.

Finally we show that $\bigcap A^{(k)}$ is also analytic. To see this we proceed as follows. We order the pairs of integers (k, p) by means of the diagonal process (1, 1), (2, 1), (1, 2), (3, 1), (2, 2), (1, 3), ... Let t be the integer associated with

(k, p) in this way, so that $t = 5$ is associated with $(2, 2)$ for instance. Then given any infinite sequence of positive integers $(n_1, n_2, \ldots, n_t, \ldots)$, we find for each pair (k, p) the associated integer t and define $n_{k,p} = n_t$. Also we now set

$$B_{n_1, n_2, \ldots, n_t} = A^{(k)}_{n_{k,1}, n_{k,2}, \ldots, n_{k,p}}.$$

In fact the sequence $n_1, n_2, \ldots$ is replaced by the double sequence

$$\begin{array}{l} n_1, n_3, n_6, \ldots \\ n_2, n_5, n_9, \ldots \\ n_4, n_8, n_{13}, \ldots \end{array}$$

For instance

$$B_{n_1, n_2, n_3, n_4, n_5} = A^{(2)}_{n_2, n_5}.$$

With this definition it is clear that the Souslin operation carried out on the sets $B_{n_1, n_2, \ldots}$ yields $\bigcap A^{(k)}$.

Thus open and compact sets are analytic. Let now $\mathscr{F}$ be the smallest class of sets containing all compact sets and closed under the operations of countable unions and intersections. To show that $\mathscr{F}$ contains the Borel sets it is in view of Section 3.1 enough to show that $\mathscr{F}$ is a σ-ring. To see this it is enough to show that if $A \subset \mathscr{F}$, then the complement A' of A in R^m also belongs to $\mathscr{F}$. This can be shown as follows. Suppose that A_n is a sequence of sets which are known with their complements to belong to $\mathscr{F}$. Then $A = \bigcap A_n \in \mathscr{F}$ and

$$A' = \bigcup_{n=1}^{\infty} A'_n \in \mathscr{F}.$$

Also compact sets belong to $\mathscr{F}$ together with their open complements. Hence under the generating operations for $\mathscr{F}$ we only obtain sets which belong with their complements to $\mathscr{F}$ and so $\mathscr{F}$ is a σ-ring and so contains the Borel sets. The above arguments now show that all the sets in $\mathscr{F}$ are analytic, so that the Borel sets are analytic. This proves Lemma 5.12.

We have next

LEMMA 5.13. *Assume that A is defined as above by* (5.8.13). *Let $\{h_i\}$ be an arbitrary sequence of positive integers and define*

$$F_p = \bigcup_{n_i \leqslant h_i} A_{n_1} \cap A_{n_1, n_2} \cap \cdots \cap A_{n_1, \ldots, n_p} \tag{5.8.14}$$

Then

$$F = \bigcap_p F_p \tag{5.8.15}$$

is a compact subset of A.

Clearly F_p is the union of a finite number of compact sets and so F_p is compact for every p. Thus F is also compact. It remains to show that $F \subset A$. To see this suppose that $x \in F$. Then to every p there correspond integers $n_{ip} \leqslant h_i$, for $i = 1, 2, \ldots, p$ such that

$$x \in A_{n_{1p}} \bigcap A_{n_{1p}, n_{2p}} \bigcap \cdots \bigcap A_{n_{1p}, \ldots, n_{pp}}.$$

We choose a sequence p_ν so that the limit

$$\lim_{\nu \to \infty} n_{ip_\nu} = m_i$$

exists for each i. This is possible since $n_{ip} \leqslant h_i$ for all p. Thus we see that

$$x \in A_{m_1} \bigcap A_{m_1, m_2} \bigcap \cdots \bigcap A_{m_1, m_2, \ldots m_p},$$

so that $x \in A$. Thus $F \subset A$ as required.

We can now prove

THEOREM 5.30. *If $\phi(E)$ is an outer capacity, then bounded analytic sets are capacitable for $\phi(E)$.*

Let A be an analytic set defined by (5.8.13) and let $A^{(h)}$ be the set

$$A^{(h)} = \bigcup_{\substack{n_1, n_2, \ldots \\ n_1 \leqslant h}} A_{n_1} \cap A_{n_1, n_2} \cap \cdots \cap A_{n_1, \ldots, n_j} \cap \cdots$$

Then $A^{(h)}$ expands with increasing h and $A^{(h)} \to A$ as $h \to \infty$. Thus by (iv)′ of Section 5.8.2 we can, for any given $\varepsilon > 0$, choose $h = h_1$ so that $A_1 = A^{(h_1)}$ satisfies

$$\phi(A_1) \geqslant \phi(A) - \frac{\varepsilon}{2}.$$

Suppose that $A_1, A_2, \ldots, A_{p-1}$ have already been defined so that

$$\phi(A_j) \geqslant \phi(A_{j-1}) - \varepsilon . 2^{-j}, j = 1 \text{ to } p - 1.$$

Define

$$A_p^{(h)} = \bigcup_{\substack{n_1 \leqslant h_1, \ldots, n_{p-1} \leqslant h_{p-1}, \\ n_p \leqslant h}} A_{n_1} \cap A_{n_1, n_2} \cap \cdots \cap A_{n_1, n_2, \ldots, n_j} \cap \cdots$$

Then $\phi(A_p^{(h)})$ increases to $\phi(A_{p-1})$ as $h \to \infty$ and so we can choose $h = h_p$, so that

$$A_p = A_p^{(h_p)} \text{ satisfies } \phi(A_p) \geqslant \phi(A_{p-1}) - \varepsilon . 2^{-p}.$$

It follows that

$$\phi(A_p) \geqslant \phi(A) - \varepsilon, \qquad p = 1, 2, \ldots$$

The sequence h_p being defined we form F_p by means of (5.8.14). Then $A_p \subset F_p$, so that

$$\phi(F_p) \geqslant \phi(A) - \varepsilon, \, p = 1, 2, \ldots$$

Further F defined by (5.8.15) is a compact subset of A. Also if G is an arbitrary open set containing F, then G contains F_p for $p \geqslant p_0$ in view of (5.8.15) and so

$$\phi(G) \geqslant \phi(F_{p_0}) \geqslant \phi(A) - \varepsilon.$$

In view of (iii)′ of Section (5.8.2) we deduce that

$$\phi(F) \geqslant \phi(A) - \varepsilon.$$

Thus, since ε is arbitrary and (5.8.11) holds for $\phi = \phi^*$, the condition (5.8.12) for capacitability is satisfied for A, and Theorem 5.30 is proved.

From the foregoing analysis it follows that in order to prove capacitability for a set function ϕ it is only necessary to establish that the conditions (i), (ii), (iii)′ and (iv)′ are satisfied. The main difficulty in practical cases consists in establishing (iv)′. We established this property for bounded sets E in R^m and the set function $C_{m-2}(E)$ by using Choquet's approach via strong subadditivity. As we noted at the end of Section 5.8.2. a corresponding result for more general capacities can be proved by different methods.

Our results yield at any rate

THEOREM 5.31. *If $\alpha = m - 2$ or $\alpha = m - 1$, then bounded analytic sets and so in particular bounded Borel sets are capacitable for $C_\alpha(E)$ in R^m.*

In fact if $\alpha = m - 2$, this follows from Theorems 5.29 and 5.30. If $\alpha = m - 1$, we embed R^m in R^{m+1} and apply the result we have just proved to our sets considered as subsets of R^{m+1}.

It is sometimes convenient to consider the capacities of unbounded sets E in R^m. These are defined formally to be the upper bounds (possibly $+\infty$) of capacities of bounded sets E_0 contained in E.

5.9. SETS WHERE S.H. FUNCTIONS BECOME INFINITE

We close the chapter by a refinement of Theorem 5.11. A set E which is a countable intersection of open sets is called a G_δ set. We proceed to prove

THEOREM† 5.32. *Let $u(x)$ be* s.h. *and not identically* $-\infty$ *in a domain D in* R^m. *Then the subset E of D where* $u(x) = -\infty$ *is a* G_δ *set of capacity zero. Conversely, given any set E of capacity zero in* R^m, *there exists* $u(x)$ s.h. *and not identically* $-\infty$ *in* R^m, *such that* $u(x) = -\infty$ *in E.*

Deny [1947] has proved a refinement of this result according to which, if E is any G_δ set of capacity zero, there exists $u(x)$, such that $u(x) = -\infty$ on E and nowhere else. However this result lies deeper and we omit it here.

The positive part of Theorem 5.32 is now very easy. Let G_n be the subset of G, where $u(x) < -n$. Then $u(x)$ is u.s.c. so that G_n is open. Also the set E where $u(x) = -\infty$ is precisely $\bigcap_{n=1}^{\infty} G_n$, so that E is a G_δ set. Suppose next that E has positive capacity. Since E is a Borel set, it follows from the results of the last section that E is capacitable. Thus there exists a compact subset E_0 of E, having positive capacity. This is however impossible by Theorem 5.10. This proves the first half of Theorem 5.32.

To prove the second half we need

LEMMA 5.14. *Let G be a bounded open set in* R^m *of capacity C. Then there exists a unit mass distribution* μ *on the closure* $\bar{G}$ *of G, such that*

$$V(x) = \int_{\bar{G}} K_{m-2}(x-y)\,d\mu(y) = \begin{cases} \log C, & \text{if } m = 2 \\ -C^{2-m}, & \text{if } m > 2 \end{cases} \tag{5.9.1}$$

for all x in G.

Let E_n be compact subsets of G, such that E_n lies in the interior of E_{n+1}, E_n is regular for the problem of Dirichlet and $\bigcup E_n = G$. Let $V_n(x)$ be the conductor potential of E_n, so that

$$V_n(x) = \int K_{m-2}(x-y)\,d\mu_n(y),$$

where μ_n is a unit mass distribution on E_n. Also since E_n is regular for the problem of Dirichlet it follows that we may choose $V_n(x)$ to be continuous in R^m, constant on E_n, harmonic outside E_n and satisfying (5.5.5) or (5.5.5)′ at ∞. By Theorem 5.17 these conditions determine $V_n(x)$ uniquely. Also on E_n we have

$$V_n(x) = \left\{\begin{array}{ll} \log C_n, & \text{if } m = 2 \\ -(C_n)^{2-m}, & \text{if } m > 2 \end{array}\right\},$$

† Cartan [1945].

where C_n is the capacity of E_n. In view of Theorem 5.29, we see that

$$C_n \to C \text{ as } n \to \infty. \tag{5.9.2}$$

It now follows from Theorem 5.3. that we can find a subsequence of the measures μ_{n_p} weakly converging to a limit unit measure μ.

The measure μ must be distributed in $\bar{G}$.

The corresponding functions $V_n(x)$ converge, at least for x outside $\bar{G}$, to a limit function

$$V(x) = \int_{\bar{G}} K_{m-2}(x - y)\, d\mu(y). \tag{5.9.3}$$

We consider the functions

$$u_n(x) = \begin{cases} V_n(x) - \log C_n, & m = 2 \\ V_n(x)(C_n)^{m-2}, & m > 2. \end{cases}$$

Then $u_n(x)$ is harmonic outside E_n. On the boundary of E_{n+1}, we see that

$$u_{n+1}(x) - u_n(x) \leqslant 0,$$

and the same inequality is true at ∞. It also remains true inside E_{n+1}. Thus $u_n(x)$ is a decreasing sequence of s.h. functions and so tends everywhere to a s.h. limit $u(x)$. Thus also $V_n(x)$ tends everywhere to a s.h. limit $V(x)$ which is given by (5.9.3) at least outside $\bar{G}$. However since $V(x)$ is s.h. in R^m and satisfies the limiting behaviour (5.5.5) or (5.5.5′) it is easy to see that $V(x)$ has a representation of the form (5.9.1) in terms of its own Riesz mass. Since the total mass can be calculated from the behaviour of $V(x)$ on $S(0, R)$ for large R, the total mass must be unity in any such representation.

Also if $x \in G$, then $x \in E_n$ for large n and so

$$V(x) = \lim V_n(x)$$

assumes the values stated in Lemma 5.14, in view of (5.9.2). This proves the lemma.

Now let E be an arbitrary bounded set of capacity zero in R^m, let G_n be a sequence of bounded open sets containing E, such that $G_{n+1} \subset G_n$ and

$$\log C(G_n) < -2^n, \qquad \text{if } m = 2$$
$$C(G_n)^{2-m} > 2^n, \qquad \text{if } m > 2.$$

This is possible by the property (iii)′ of outer capacities, since we can choose the G_n so that $C(G_n) \to 0$ as $n \to \infty$.

It then follows from Lemma 5.14 that there exists a unit mass distribution

μ_n on $\bar{G}_n$, such that, for $x \in G_n$,

$$V_n(x) = \int_{\bar{G}_n} K_{m-2}(x-y)\,d\mu_n(y) \leqslant -2^n.$$

Set

$$\mu = \sum_1^\infty 2^{-n}\mu_n.$$

Then μ is a unit mass distribution and if we set

$$V(x) = \int K_{m-2}(x-y)\,d\mu(y) = \sum_1^\infty 2^{-n} V_n(x),$$

we deduce that $V(x)$ is s.h. in R^m, and $V(x) = -\infty$ on E. This proves Theorem 5.32, when E is bounded.

If E is unbounded, let E_n be the part of E in $|x| \leqslant n$. Then $C(E_n) = 0$, and hence we can construct a mass distribution μ_n of total mass 2^{-n}, such that the corresponding potential is identically $-\infty$ on E_n. If we set $\mu = \sum_{n=1}^\infty \mu_n$, then μ is a unit mass distribution in R^m. Then if we set

$$V(x) = \int K_{m-2}(x-y)\,d\mu(y),\ m > 2,$$

$$V(x) = \int_{|y| \leqslant 1} \log|x-y|\,d\mu(y) + \int_{|y|>1} \log\left|\frac{x-y}{y}\right| d\mu(y),\ m = 2,$$

then the corresponding potential is $-\infty$ on the whole of E_n for each n and so on the whole of E. Also $V(x)$ converges as $y \to \infty$, and so $V(x)$ is s.h. and not identically $-\infty$ in R^m. This completes the proof of Theorem 5.32.

References

Books and papers are referred to in the text by the authors and year of publication. This distinguishes between different works except in one case where the year is followed by a and b. The figures in brackets refer to the pages in the text where the work in question is referred to.

Ahlfors, L. V., Untersuchungen zur Theorie der konformen Abbildungen und der ganzen Funktionen. *Acta Soc. sci. Fenn. Nova* Ser. 1, No. 9 (1930). [170]

Ahlfors, L. V. and Beurling, A., Conformal invariants and function-theoretic null sets. *Acta Math.* **83** (1950), 101–129. [231]

Ahlfors, L. V. and Sario, L., "Riemann Surfaces", Princeton University Press, 1960. [74]

Aronszaijn, N. and Smith, K. T., Functional spaces and functional completion. *Ann. Inst. Fourier* **6** (1955–1956), 125–185. [269]

Ascoli, G., Sulla unicità della soluzione nel problema di Dirichlet. *Rendi Accad. d. Lincei, Roma* (6), **8** (1928), 348–351. [232]

Baernstein, A. II. Integral means, univalent functions and circular symmetrization. *Acta Math.* **133** (1975), 139–169. [xii, 79]

Bauer, H., Harmonische Räume und ihre Potentialtheorie (Lecture Notes in Mathematics 22, Springer, Berlin, 1966). [v]

Besicovitch, A. S., On the fundamental geometric properties of linearly measurable plane sets of points II. *Math. Ann.* **115** (1938), 296–329. [231]

Besicovitch, A. S., On sufficient conditions for a function to be analytic, and on the behaviour of analytic functions in the neighbourhood of non-isolated singular points. *Proc. London Math. Soc.* (2) **32** (1932), 1–9. [230]

Bouligand, G., Domaines infinis et cas d'exception du problème de Dirichlet. *C.R. Acad. Sci. Paris* **178** (1924), 1054–1057. [239, 250]

Brannan, D. A., Fuchs, W. H. J., Hayman, W. K. and Kuran, Ü. A characterisation of harmonic polynomials in the plane. *Proc. London Math. Soc.* (3) **32** (1976) 213–229 [200]

Brelot, M., Étude des fonctions sousharmoniques au voisinage d'un point (Actualités scientifiques et industrielles, No. 134, Hermann 1934). [xi, 237]

Brelot, M., Familles de Perron et problème de Dirichlet. *Acta Szeged*, IX (1939a) 133–153. [xi, 114, 247]

Brelot, M., Sur la théorie moderne du potential. *C.R. de l'Acad. sci., Paris* 209 (1939b), 828. [210]

Brelot, M., Sur la théorie autonome des fonctions sousharmoniques. *Bull. Sci. Math.* **65** (1941), 72–98. [xi, 212]

Brelot, M., A new proof of the fundamental theorem of Kellog-Evans on the set of irregular points in the Dirichlet problem. *Rendi. Circ. Mat. di Palermo* (2) **11** (1955), 112–122. [256]

Brelot, M., Lectures on potential theory (Tata Institute No. 19, 1960). [v]

Brelot, M., Les étapes et les aspects multiples de la théorie du potentiel. *L'Enseignement mathématique* XVIII (1972), 1–36. [v]

Carleson, L., "Selected Problems on Exceptional Sets", Van Nostrand Mathematical studies, No. 13, Van Nostrand, 1967. [v, xii, 229, 230].

Carleson, L., Asymptotic paths for subharmonic functions in R^n. Report No. 1, Institute Mittag-Leffler 1974. [187]

Cartan, H., Sur les systèmes de fonctions holomorphes á variétés linéaires lacunaires et leurs applications. *Ann. sci. école norm. sup.* (3) **45** (1928), 255–346. [131]

Cartan, H., Théorie du potentiel newtonien, énergie, capacité, suites de potentiels. *Bull. Soc. Math.* **73** (1945), 74. [274]

Choquet, G., Theory of capacities. *Ann. Inst. Fourier* **5** (1955), 131–295. [xi, 258]

Choquet, G., Capacitabilité en potentiel logarithmique. *Bull. classe sci. Bruxelles,* **44** (1958), 321–326. [xi, 269]

Dahlberg, B., Mean values of subharmonic functions. *Arkiv för Math.* **10** (1972), 293–309. [ix, 147]

Denjoy, A., Sur les fonctions analytiques uniformes à singularités discontinues. *C.R. Acad. sci. Paris* **149** (1909), 258–260. [231]

Deny, J., Sur les infinis d'un potentiel, *C.R. Acad. sci. Paris* **224** (1947), 524–525 [274]

Deny, J., Un théorème sur les ensembles effilés. *Annales Univ. Grenoble, Sect. sci. Math. Phys.* **23** (1948), 139–142 [131]

Du Plessis, N., "An Introduction to Potential Theory", Oliver and Boyd, Edinburgh, 1970. [v]

Edrei, A., Locally tauberian theorems for meromorphic functions of lower order less than one. *Trans. Amer. Math. Soc.* **140** (1969). [152]

Edrei, A. and Fuchs, W. H. J., The deficiencies of meromorphic functions of order less than one. *Duke Math. J.* **27** (1960), 233–249. [161]

Egorov, D. T., Sur les suites des confuctions mesurables. *C.R. de l'Acad. sci. Paris* **152** (1911), 244–246. [227]

Erdös, P. and Gillis, J., Note on the transfinite diameter. *J. London Math. Soc.* **12** (1937), 185–192. [228]

Evans, G. C., Applications of Poincaré's sweeping out process. *Proc. Nat. Acad. sci.* **19** (1933), 457. [xi, 202, 242]

Evans, G. C., Potentials and positive infinite singularities of harmonic functions. *Monatsh. für Math. u. Phys.* **43** (1936), 419–424. [217]

Friedland, S. and Hayman, W. K., Eigenvalue inequalities for the Dirichlet problem and the growth of subharmonic functions. *Comment. Math. Helv.* (1976) [184, 197]

Frostman, O., Potentiel d'équilibre et capacité des ensembles avec quelques applications à la théorie des fonctions. *Meddel. Lunds. Univ. Mat. Sem.* **3** (1935), 1–118. [205, 208, 213, 223, 225, 235]

Fubini, G., Sugli integrali multipli. *Rendi. accad. d. Lincei, Roma* (5) **16** (1907), 608–614. [17]

Fuglede, B., Le théorème du minimax et la théorie fine du potentiel. *Ann. Inst. Fourier* **15** (1965), 65–88. [269]

Fuglede, B., Finely harmonic functions (Lectures Notes in Mathematics 289, Springer, 1972). [v].

Fuglede, B., Asymptotic paths for subharmonic functions, *Math. Ann.* **213** (1975), 261–274. [x, 187]

Garnett, J., Positive length but zero analytic capacity. *Proc. Amer. Math. Soc.* **24** (1970), 696–699. [231]

Govorov, N. V., On Paley's problem, (Russian), *Funkz. Anal.* **3** (1969), 35–40. [147]

Green, G., Essay on the application of Mathematical Analysis to the theory of Electricity and Magnetism. (Nottingham, 1828). [22]

Hadamard, J., Étude sur les propriétés des fonctions entières et en particulier d'une fonction considérée par Riemann. *J. de Math.* (4) **9** (1893), 171–215. [viii, 146]

Hardy, G. H., On the mean value of the modulus of an analytic function. *Proc. London Math. Soc.* (2) **14** (1915), 269–277. [69]

Harnack, A., Existenzbeweise zur Theorie des Potentials in der Ebene und im Raume. *Leipziger Ber.* 1886, 144–169. [35, 37]

Hausdorff, F., Dimension und äusseres Maasz. *Math. Ann.* **79** (1918), 157–179. [xi, 220]

Hayman, W. K., The minimum modulus of large integral functions. *Proc. London Math. Soc.* (3) **2** (1952), 469–512. [120]

Hayman, W. K., "Multivalent Functions", Cambridge Tracts in Mathematics and Mathematical Physics, No. 48, Cambridge University Press, Cambridge, 1958. [78]

Hayman, W. K., On the growth of integral functions on asymptotic paths. *J. Indian Math. Soc.* **24** (1960), 251–264. [192]

Hayman, W. K., "Meromorphic Functions", Oxford Mathematical Monograph, Clarendon Press, Oxford, 1964. [143]

Hayman, W. K., Power series expansions for harmonic functions. *Bull. London Math. Soc.* **2** (1970), 152–158. [31]

Heins, M., Entire functions with bounded minimum modulus; subharmonic function analogues. *Ann. of Math.* (2) **49** (1948), 200–213. [146]

Heins, M., On a notion of convexity connected with a method of Carleman. *J. Analyse Math.* **7** (1959), 53–77. [170]

Helms, L. L., "Introduction to Potential Theory", Wiley Interscience Pure and Applied Mathematics, 22, New York, 1969. [v]

Ivanov, L. D., On Denjoy's conjecture (Russian). *Uspehi Mat. Nauk* **18** (1963), No. 4 (112), 147–149. [231]

Iversen, F., Sur quelques propriétés des fonctions monogènes au voisinage d'un point singulier. *Öfv. af Finska Vet. Soc. Forh.* **58A**, No. 25 (1915–1916). [x, 185]

Jensen, J. L. W. V., Om konvexe Funktioner og Uligheder mellem Middelvaerdier. *Ngt. Tids. for Mat.* **16B** (1905), 49–68. [42]

Kellogg, O. D., Unicité des fonctions harmoniques, *C.R. Acad. sci. Paris* **187** (1928), 526–527. [242]

Kellogg, O. D., "Foundations of Potential Theory". Grundlehren der Math. Wiss. 31, Springer, Berlin, 1929. [v]

Kennedy, P. B., A class of integral functions bounded on certain curves. *Proc. London Math. Soc.* (3) **6** (1956), 518–547. [146]

Kerékjártó, B.v., "Vorlesungen über Topologie I", Berlin, 1923. [191]

Kiselman, C. O., Prolongement des solutions d'une équation aux dérivées partielles à coefficients constants. *Bull. Soc. Math. France* **97** (1969), 329–356. [31]

Kishi, M., Capacities of Borelian sets and the continuity of potentials. *Nagoya Math. J.* **12** (1957), 195–219. [269]

Landkof, N. S., "Foundations of Modern Potential Theory", *Grundlehren der Math. Wiss.* **180**, Springer, Berlin, 1972. [v]

Lebesgue, H., Conditions de regularité, conditions d'irregularité, conditions d'impossibilité dans le problème de Dirichlet. *C.R. Acad. sci. Paris* **178** (1924), 349–354. [58]

Littlewood, J. E., On inequalities in the theory of functions. *Proc. London Math. Soc.* (2) **23** (1924), 481–519. [74]

Martensen, E., "Potentialtheorie", Teubner, Stuttgart, 1968. [v]

Miles, J. and Shea, D. F., An extremal problem in value distribution theory. *Quart. J. of Math., Oxford* (2) **24** (1973), 377–383. [152]

Nevanlinna, F. and Nevanlinna, R. Über die Eigenschaften einer analytischen Funktion in der Umgebung einer singulären Stelle oder Linie. *Acta Soc. Sci. Fenn.* **50**, No. 5 (1922). [116]

Nevanlinna, R., "Le théorème de Picard-Borel et la théorie des fonctions méromorphes", Paris, 1929. [viii, 120, 125]

Newman, M. H. A., "Elements of the Topology of Plane Sets of Points". 2nd Ed., Cambridge, 1951. [173]

Nguyen-Xuan-Loc and Watenabe, T., Characterization of fine domains for a certain class of Markov processes. *Z. f. Wahrscheinlichkeitstheorie u. verw. Geb.* **21** (1972), 167–178. [187]

Painlevé, P. see Zoretti, L. [230]

Perron, O., Eine neue Behandlung der ersten Randwertaufgabe für $\Delta u = 0$. *Math. Zeit.* **18** (1923), 42–54. [vi, 55]

Petrenko, P., The growth of meromorphic functions of finite lower order (Russian). *Izv. Akad. Nauk. S.S.S.R.* **33** (1969), 414–454. [147]

Poincaré, H., Sur les équations aux dérivées partielles de la physique mathématique. *Amer. J. Math.* **12** (1890), 211–294. [vi]

Poisson, S. D., Mémoire sur le calcul numérique des integrales définies. *Mémoires de l'acad. Royale des sci. de l'institut de France*, vi (1823, published 1827), 571–602, particularly p. 575. [25]

Rado, T., "Subharmonic Functions". Berlin, 1937. [v]

Radon, J., Über lineare Funktionaltransformationen und Funktionalgleichungen. *Sitzungsber. Akad. Wien* (2) **128** (1919), 1083–1121. [83]

Rao, N. V. and Shea, D. F., Growth problems for subharmonic functions of finite order in space. *Proc. London Math. Soc.* (1976). [152]

Riemann, B., Theorie der Abelschen Funktionen. *Crelle's J.* **54** (1857). See e.g. collected works, Dover, 1953, 88–144, particularly p. 97. [30]

Riesz, F., Sur les opérations fonctionelles linéaires, *C.R. Acad. sci. Paris* 149 (1909), 1303–1305. [vii, 82, 94]

Riesz, F., Sur les fonctions subharmoniques et leur rapport à la théorie du potentiel. I *Acta Math.* **48** (1926), 329–343; II ibid. **54** (1930), 321–360. [vii, 104]

Rogosinski, W. W., On the coefficients of subordinate functions. *Proc. London Math. Soc.* (2) **48**)1943), 48–82. [77]

Rudin, W., "Principles of Mathematical Analysis", 2nd. Edn. McGraw-Hill, New York, 1964. [84]

Schwartz, L., "Théorie des distributions", 2 Vols., Paris, 1950–1951. [104]

Selberg, H. L., Über die ebenen Punktmengen von der Kapazität Null. *Ark. Norske Videnskap Akad. Oslo* (*Mat. Naturvid. Kl.*) (1937), No. 10. [217]

Talpur, M. N. M., A subharmonic analogue of Iversen's theorem. *Proc. London Math. Soc.* (3) **31** (1975), 129–148. [x, 170, 180, 187]

Talpur, M. N. M., On the existence of asymptotic paths for subharmonic functions in R^n. *Proc. London Math. Soc.* **32** (1976), 181–192. [x, 170, 185]

Titchmarsh, E. C., "The Theory of Functions", 2nd Edn., Oxford, 1939. [14, 170]

Tsuji, M., "Potential Theory in Modern Function Theory", Maruzen, Tokyo, 1959. [v, 213]

Valiron, G., Sur le minimum du module des fonctions entières d'ordre inférieur à un. *Mathematica* **11** (1935) (Cluj), 264–269. [157]

Vitushkin, A. G., Example of a set of positive length but zero analytic capacity. (Russian). *Dokl. akad. Nauk. S.S.S.R.* **127** (1959), 246–249. [231]

Weierstrass, K., Zur Theorie der eindeutigen analytischen Funktionen. *Math. Abh. der Akad. der Wiss. zu Berlin* 1876, 11–60. [viii, 140]

Wiener, N., Certain notions in potential theory. *J. Math. Mass.* **3** (1924), 24–51. [xi, 243]

Zaremba, S., Sur le principe du minimum, *Krakau Anz.* 1909 (2), 197–264. [vi]

Zoretti, L., Sur les fonctions analytiques uniformes qui possèdent un ensemble parfait discontinu de points singuliers. *J. Math. Pures Appl.* (6) **1** (1905), 1–51. [230]

Index

The index lists some of the main concepts and theorems used in the text together with the page or pages where they are introduced.

R

S

T

U

W